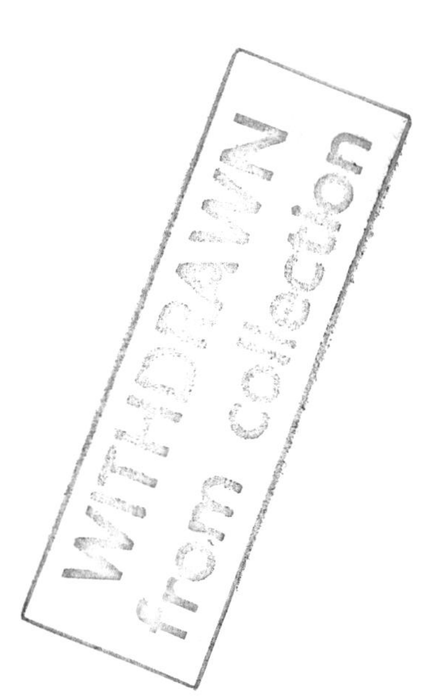
WITHDRAWN
from collection

D1429075
Moulton College
T 6 1 3 1 9

Tilapia Culture

Tilapia Culture

Abdel-Fattah M. El-Sayed

*Oceanography Department, Faculty of Science,
Alexandria University, Alexandria, Egypt*

CABI Publishing

CABI Publishing is a division of CAB International

CABI Publishing
CAB International
Wallingford
Oxfordshire OX10 8DE
UK

Tel: +44 (0)1491 832111
Fax: +44 (0)1491 833508
E-mail: cabi@cabi.org
Website: www.cabi-publishing.org

CABI Publishing
875 Massachusetts Avenue
7th Floor
Cambridge, MA 02139
USA

Tel: +1 617 395 4056
Fax: +1 617 354 6875
E-mail: cabi-nao@cabi.org

A catalogue record for this book is available from the British Library, London, UK.

Library of Congress Cataloging-in-Publication Data

El-Sayed, Abdel-Fattah M., 1950-
Tilapia culture / Abdel-Fattah M. El-Sayed.
p. cm.
Includes bibliographical references (p.) and index.
ISBN-13: 978-0-85199-014-9 (alk. paper)
ISBN-10: 0-85199-014-2 (alk. paper)
1. Tilapia. I. Title

SH167.T54E42 2006
639.3′774--dc22

2005012005

Typeset in Baskerville by AMA DataSet Ltd, UK.
Printed and bound in the UK by Biddles Ltd, King's Lynn.

Contents

Foreword

It seems highly fitting that a truly comprehensive book on tilapia culture should be authored by a scientist at the Alexandria University in Egypt, a country where, arguably, the farming of tilapia has its roots. I do not use the word 'comprehensive' glibly, for Dr El-Sayed brings together information on everything from the history of tilapia culture to the latest production practices being used around the world, covering everything from soup to nuts, or, in this case, morphology to socio-economics. The author has extensively mined the literature as well as obtaining additional information and photographs through his contacts with a number of aquaculturists from virtually all the tilapia-producing regions around the globe. The result is an excellent volume that will serve several purposes. It can be used as an introduction to this fascinating group of fishes for the interested layman, but primarily will serve as a valuable resource for those who are engaged in tilapia research or production, as well as those who teach about aquaculture.

Tilapia aquaculture is somewhat unique in that it occurs in virtually every type of water system in utilization today. Tilapia are produced in everything from ponds fertilized with manure to closed recirculating systems. Production occurs in freshwater and in brackish to ocean salinities. The types of production systems in use today are described in considerable detail and span the gamut from subsistence culture to high-tech aquaponics. Subsistence culture has improved the nutritional plane of large numbers of people in developing nations, while more intensive culture systems produce fish that are now commonly seen on the menus of highly rated white-tablecloth restaurants.

Aquaculture, in general, has been on the receiving end of numerous criticisms over the past several years, and tilapia culture has not escaped the eyes of the critics. Dr El-Sayed devotes an entire chapter to environmental impacts and discusses how those impacts can be reduced. In these days when sustainability and responsible aquaculture are on the minds of so many, discussion of the issues he addresses is of vital importance.

Finally, what with the explosion of literature on tilapia that has appeared in recent years, the publication of a book that brings everything together is extremely timely. Examination of the table of contents should convince the prospective reader that *Tilapia Culture* does a very good job of covering all aspects of the subject and should quickly become a standard reference work on the topic.

Robert R. Stickney
Texas Sea Grant College Program
Texas A&M University
USA

Preface

Tilapia are currently known as 'aquatic chicken' due to their high growth rates, adaptability to a wide range of environmental conditions and ability to grow and reproduce in captivity and feed on low trophic levels. It is thus no surprise that these fish have become an excellent candidate for aquaculture, especially in tropical and subtropical environments. Tilapia culture is believed to have originated more than 4000 years ago, but very little information is available on their culture during those ancient times. The first trials of tilapia culture were recorded in Kenya in the 1920s. Since then, tilapia culture has been established in many tropical and subtropical regions, and even in areas beyond their native ranges, where they have been introduced for various purposes. As a result, considerable attention has been paid to tilapia culture during the past three decades. Consequently, tilapia culture is currently practised in more than a 100 countries all around the globe. A plethora of information has also been generated and published in specialized periodicals on the biology and culture of tilapia during the second half of the last century. Key books have also been published on these fish during the last two decades, in addition to the excellent proceedings of the International Symposium on Tilapia in Aquaculture (ISTA), where six proceedings have already been produced. The World Aquaculture Society (WAS) has also contributed significantly to tilapia culture with an excellent series of Tilapia Culture in the Americas and also through the annual 'tilapia update' series, prepared by R. Stickney and published in *World Aquaculture* magazine, which summarizes published papers on tilapia culture.

However, most tilapia farmers, farm owners/managers, researchers and graduate students in developing countries, where over 90% of farmed tilapia are produced, have little or no access to the accumulating information on tilapia culture. The present book is an attempt to pull together, as far as possible, the scientific publications on tilapia culture in a single volume for the benefit of this target audience. It is hoped that this book will provide these target groups with access to published information, especially in areas where the access is lacking. This task was not easy, because the information published on tilapia was too much and too wide to be collected in a single book. Therefore, I must admit here that I was somewhat selective in my endeavour. Selection was made for those publications that were directly related to the practices and enhancement of tilapia culture. Considerable attention was paid to the most recent research published during the past decade. In fact, over 550 references, representing over 55% of the total references cited in the book, were published during 1995–2004. Moreover, about 30% of those references appeared during 2000–2005.

The book includes 12 chapters covering almost all aspects of tilapia culture. Each chapter ends with a number of 'closing remarks' summarizing the main points covered in the chapter. The book begins with a historical review of global and continental tilapia culture, current state, production and future potential, with emphasis on the major cultured species and major producers. In Chapter 2, the basic biology of tilapia, including taxonomy, body shapes, geographical distribution, introductions and transfers, gut morphology and feeding habits, has been briefly reviewed. Chapter 3 describes the environmental requirements of tilapia. Optimum and critical ranges/levels of temperature, salinity, pH, dissolved oxygen, ammonia and nitrite, photoperiod and turbidity are reviewed. Chapter 4 deals with the culture of tilapia in semi-intensive systems. Full details are presented on pond fertilization, supplemental feeding, polyculture and integrated farming systems. Chapter 5 covers the intensive culture of tilapia in earthen ponds, tanks, raceways, cages, recirculating systems and aquaponics. Chapter 6 discusses tilapia nutrition, including protein, lipid, carbohydrate, vitamin and mineral requirements. Feed sources, digestibility, inclusion levels and feeding methods and frequencies are also described. Reproduction and seed production of tilapia are covered in Chapter 7, with special emphasis on broodstock management, production of monosex tilapia, seed production and larval rearing under different culture systems. Stress and diseases that infect tilapia, both in the wild and in aquaculture environments, including bacterial, parasitic, fungal, viral and non-infectious diseases, in addition to disorders caused by pollution, are discussed in Chapter 8. Chapter 9 deals with the harvesting, processing, marketing and economics of farmed tilapia. The role of tilapia culture in rural development in developing countries is discussed in Chapter 10. Chapter 11 looks at the recent technological innovations in tilapia culture. Modern technologies in reproduction and genetics, namely transgenesis, gynogenesis, androgenesis, cloning, production of genetically male tilapia (GMT) and genetically improved farmed tilapia (GIFT), are all reviewed. The global expansion of tilapia farming at an exceptionally high rate is very likely to pose environmental and socio-economic threats. Chapter 12 considers the environmental impacts of tilapia culture and the best management methods to reduce those impacts.

I hope the reader who is interested in aquaculture in general, or involved in tilapia culture in particular, will find this book useful. I hope also that the objectives for which the book was written have been achieved. Finally, I would very much welcome any feedback from my colleagues and fellow readers.

Professor Abdel-Fattah M. El-Sayed
Alexandria, Egypt, April 2005

Acknowledgements

I would like to thank the following, who have provided me with photographs, reprints, data, advice and unpublished information by personal communication. However, the responsibility for all of the content of the book resides with me as author.

Ahmed H. Al-Harbi: Fish Culture Project, Natural Resources and Environment Research Institute, King Abdulaziz City for Science and Technology, Riyadh, Saudi Arabia.
Donald S. Bailey: University of Virgin Islands, Agricultural Experimental Station, Kingshill, Virgin Islands, USA.
Ibrahim S.H. Bilal: Department of Aridland Agriculture, College of Food Systems, United Arab Emirates University, Al-Ain, United Arab Emirates.
Randall E. Brummett: World Fish Centre (ICLARM), Humid Forest Ecoregional Centre, Yaoundé, Cameroon.
James Diana: School of Natural Resources and Environment, University of Michigan, Ann Arbor, Michigan, USA.
Peter Edwards: Aquaculture and Aquatic Resources Management, School of Environment, Resources and Development, Asian Institute of Technology, Klong Luang, Pathumthani, Thailand.
Gamal O. El-Naggar: World Fish Centre, Regional Research Centre for Africa and West Asia, Abbasa, Abu Hammad, Sharkia, Egypt.
Kevin Fitzsimmons: Department of Soil, Water and Environmental Science, Environmental Research Laboratory, University of Arizona, Tucson, Arizona, USA.
General Authority for Fisheries Resources Development (GAFRD): Madeenet Nasr, Cairo, Egypt.
Dao H. Giap: Aquaculture and Aquatic Resources Management, School of Environment, Resources and Development, Asian Institute of Technology, Klong Luang, Pathumthani, Thailand.
Lake Harvest: Lake Kariba, Zimbabwe.
C. Kwei Lin: Aquaculture and Aquatic Resources Management, School of Environment, Resources and Development, Asian Institute of Technology, Klong Luang, Pathumthani, Thailand.
David C. Little: Institute of Aquaculture, University of Stirling, UK.
Ismail Radwan: Kafr El-Shaikh Fish Farming Cooperative Society, Al-Hamool, Kafr El-Shaikh, Egypt.
James E. Rakocy: University of Virgin Islands, Agricultural Experimental Station, Kingshill, Virgin Islands, USA.
Mahmoud Rizk: World Fish Centre, Regional Research Centre for Africa and West Asia, Abbasa, Abu Hammad, Sharkia, Egypt.
Thesthong Samrit: Aquaculture and Aquatic Resources Management, School of Environment, Resources and Development, Asian Institute of Technology, Klong Luang, Pathumthani, Thailand.

Magdy Soliman: Faculty of Veterinary Medicine, University of Alexandria, Alexandria, Egypt.
Robert R. Stickney: Texas Sea Grant College Program, Texas A&M University, College Station, Texas, USA.
Yang Yi: Aquaculture and Aquatic Resources Management, School of Environment, Resources and Development, Asian Institute of Technology, Klong Luang, Pathumthani, Thailand.

I would also like to thank **Ram C. Bhujel:** Aquaculture and Aquatic Resources Management, School of Environment, Resources and Development, Asian Institute of Technology, Klong Luang, Pathumthani, Thailand; **Gina Conroy:** Maracay, Estado Aragua, Venezuela; **Ahmed M. Darwish:** Harry K. Dupree Stuttgart National Aquaculture Research Center, United States Department of Agriculture, Agriculture Research Service, Stuttgart, Arkansas, USA; and **László Szathmári:** University of West Hungary, Faculty of Agriculture and Food Sciences, Hungary, for providing me with photographs, even though I could not include those photographs in the book. Thanks are also due to **Elham Wassef:** National Institute of Oceanography and Fisheries, Alexandria, Egypt, for reading the draft text and making some valuable editing corrections. The effort made by **Essam Abdel-Mawla:** National Institute of Oceanography and Fisheries, Alexandria, Egypt, in editing many of the photographs included in the book is highly appreciated.

Finally, I could not have finished this task without the patience and support of my wife Azza and my daughters Israa and Ayat.

1

Current State and Future Potential*

1.1. Historical Review

Aquaculture is the farming of aquatic organisms, including fish, molluscs, crustaceans and aquatic plants, in freshwater, brackish-water and seawater environments. Capture fisheries, on the other hand, are the exploitation of aquatic organisms by the public as a common property resource, with or without appropriate licences. Aquaculture's contribution to total global fisheries landings was very low during 1950–1970, ranging from only 638,577 Mt (3.2%) in 1950 to 5.2% in 1970. Global aquaculture production continued to grow to 9.6% in 1980 and 16.3% in 1990. In the 1990s and early 2000s, this production grew at an outstanding rate to reach an annual rate of 32.1% in 2000, 34% in 2001 and 35.2% in 2002 (Fig. 1.1). The average annual compounded growth rate of aquaculture production was 9% per year during 1970–2000, compared with only 1.3% for capture fisheries (Tacon, 2003). Half of total global aquaculture production in 2002 was finfish (25,728,611 Mt).

Tilapia are freshwater fish belonging to the family Cichlidae. They are native to Africa, but were introduced into many tropical, subtropical and temperate regions of the world during the second half of the 20th century (Pillay, 1990). The introduction of tilapia into those areas was for: (i) farming as food fish; (ii) recreational fishing; (iii) aquatic weed control; and (iv) research purposes. Tilapia have many attributes that make them an ideal candidate for aquaculture, especially in developing countries. These include:

1. Fast growth.
2. Tolerance to a wide range of environmental conditions (such as temperature, salinity, low dissolved oxygen, etc.).
3. Resistance to stress and disease.
4. Ability to reproduce in captivity and short generation time.
5. Feeding on low trophic levels and acceptance of artificial feeds immediately after yolk-sac absorption.

Tilapia culture is believed to have originated some 4000 years ago, about 1000 years before carp culture was introduced into China (Balarin and Hatton, 1979). However, other than biblical references and illustrations from ancient Egyptian tombs, very little information is available on their culture during those early times. Current Food and Agriculture Organization (FAO) aquaculture production statistics indicate that about 100 countries practise tilapia culture, since these countries reported tilapia production from aquaculture in 2002 (FAO, 2004). The global development of tilapia culture has passed through three distinctive phases (Fig. 1.2): before 1970; from 1970 to 1990; and from 1990 to now.

BEFORE 1970. The contribution of tilapia production to total global aquaculture production before 1970 was very minor, representing less than 1% of total production. For example, tilapia production in 1969 was only 24,633 Mt, representing 0.76% of total aquaculture production (3,238,079 Mt). Very few countries practised

*Unless otherwise indicated, all data presented in this chapter are derived from FAO *Fishstat Plus* (2004).

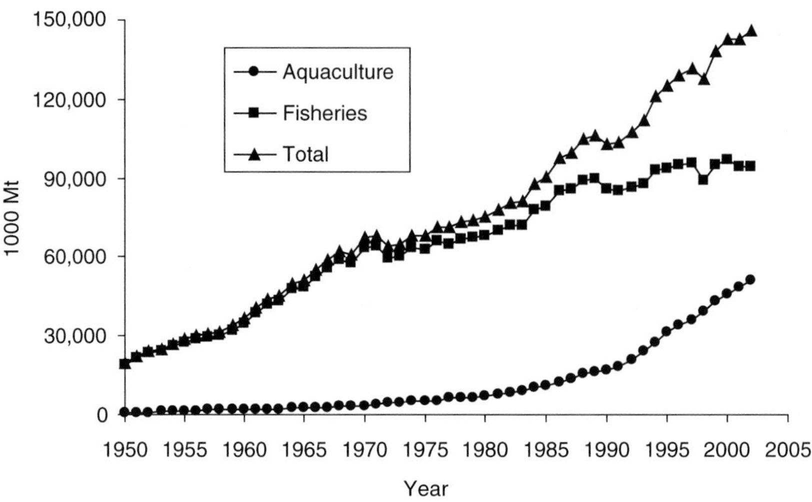

Fig. 1.1. Global fisheries and aquaculture production (1000 Mt) during 1950–2002.

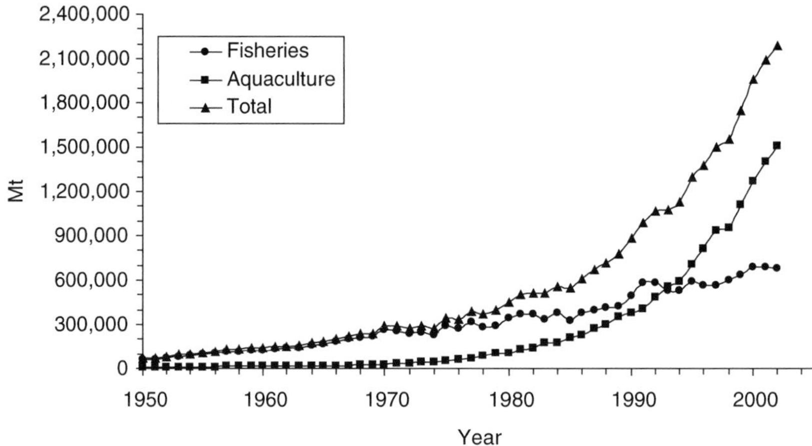

Fig. 1.2. Global production (Mt) of tilapia from aquaculture and capture fisheries during 1950–2002.

tilapia culture during that period. Only seven countries reported tilapia production in 1950, increasing to 12 countries in 1969. Taiwan, China, Egypt, Nigeria, Israel and Thailand were the major tilapia producers. The insignificant contribution of tilapia production during that period was mainly because aquaculture in general, and tilapia culture in particular, was not known as a food production system in most countries.

FROM 1970 TO 1990. Tilapia culture was gradually expanding worldwide during the period from 1970 to 1990. The number of countries practising tilapia culture increased significantly to reach 78 countries in 1990, compared to only 12 countries in 1969. However, tilapia production in many of those countries was very limited. According to FAO aquaculture production statistics, out of those 78 countries, 40 countries produced less than 100 Mt/year each. The production of farmed tilapia gradually increased to reach 383,654 Mt by 1990, representing 2.28% of total aquaculture production in 1990. During that period, the annual growth of tilapia production fluctuated between $<6\%$ and $>28\%$, with an average of 14.2%.

FROM 1990 TO NOW. Tilapia culture has witnessed a huge expansion during the past decade. As a result, the number of countries practising tilapia culture has reached over 100, as mentioned earlier. The production of farmed tilapia has also increased more than 390% to jump from 383,654 Mt in 1990, representing 2.28% of total aquaculture production, to 1,505,804 Mt in 2002, representing 2.93% of total production. The average annual growth of tilapia production during that period approached 12.2%.

1.2. Global Tilapia Production

1.2.1. Capture fisheries

Global landing of tilapia from capture fisheries increased progressively during the 1950s to the 1980s. During the 1990s and early 2000s, the landings were almost stable, fluctuating around 585,000–680,000 Mt/year (Fig. 1.2). Africa is by far the most important tilapia producer from capture fisheries, where it contributed about 70% of global landing in 2002, followed by Asia (18%), North America (9%) and South America (3%) (Figs 1.3 and 1.4). Therefore, it is no surprise that, among the world's top ten tilapia producers from capture fisheries, six are African countries. In addition, Egypt and Uganda, the first and second largest world tilapia producers, landed over 138,000 and 98,000 Mt in 2002, representing 20% and 14% of global landings (Fig. 1.5). The top ten producers included three Asian countries (Thailand, the Philippines and Sri Lanka) and one North American country (Mexico).

Among all tilapia species, Nile tilapia (*Oreochromis niloticus*) is the most important identified species in capture fisheries. In 2002, the production of that species approached 253,871 Mt, representing 37% of total production. Other identified species include Mozambique tilapia (*Oreochromis mossambicus*), blue tilapia (*Oreochromis aureus*), jaguar guapote (*Parachromis managuensis*) and mango (Galilee) tilapia (*Sarotherodon galilaeus*). However, most tilapia catches are not identified. For example, 59% of the catch in 2002 was reported under 'unidentified' cichlids, 'mouthbrooding' cichlids and 'unidentified' tilapias.

1.2.2. Aquaculture

As pointed out earlier, the production of farmed tilapia increased from 28,260 Mt in 1970 to 1,505,804 Mt in 2002. However, these values may

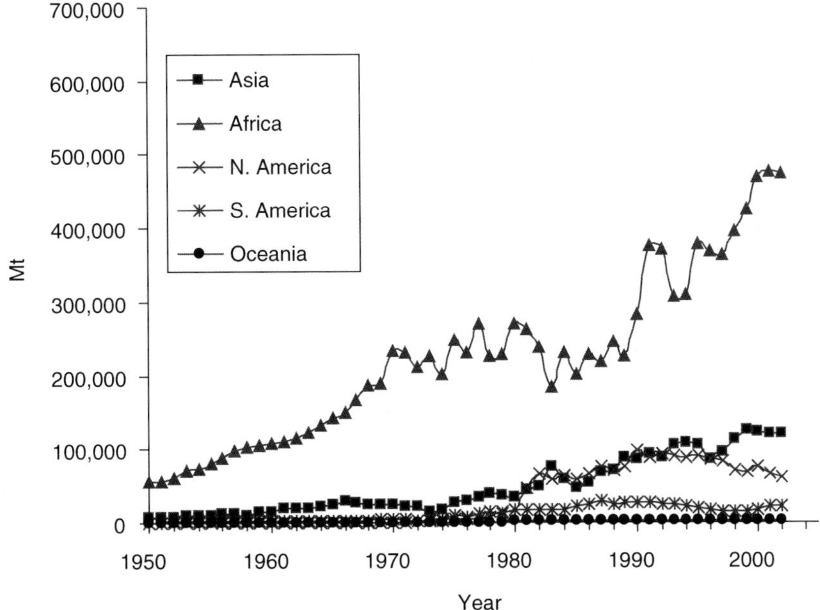

Fig. 1.3. Tilapia capture (Mt) by region during 1950–2002.

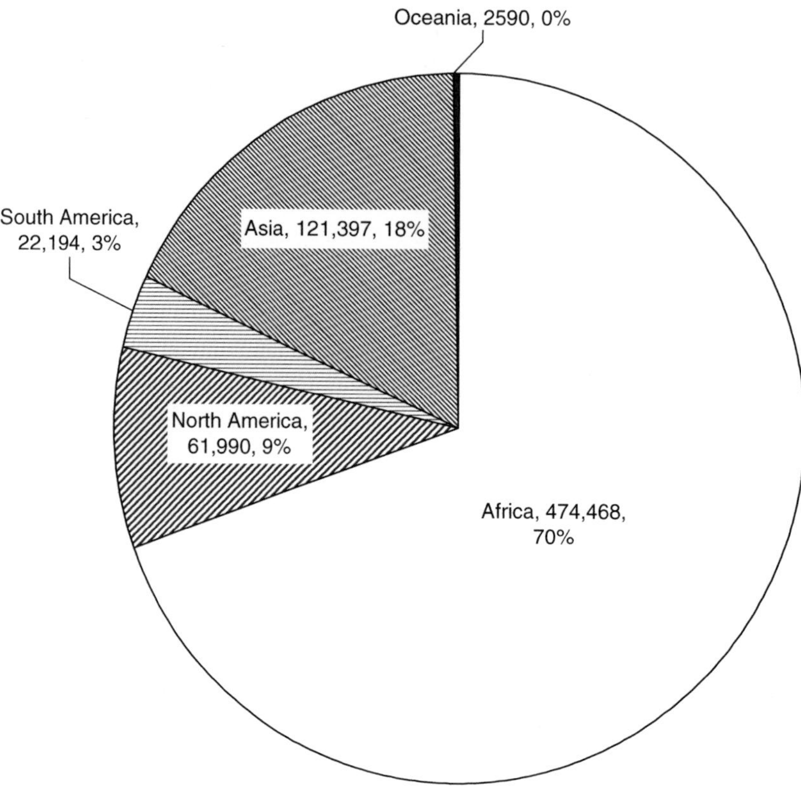

Fig. 1.4. Tilapia production from capture fisheries (Mt, %) by region in 2002.

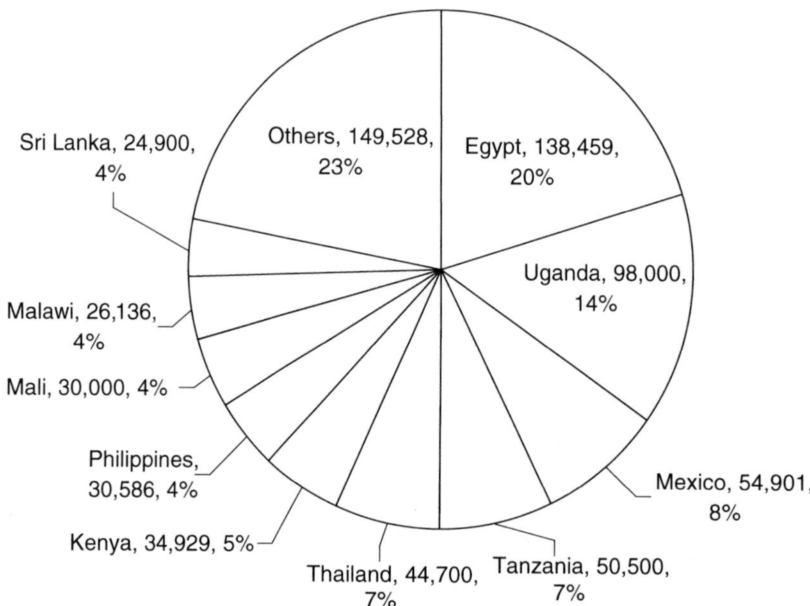

Fig. 1.5. Top ten producers of tilapia from capture fisheries (Mt, %) in 2002.

be much less than the actual amounts produced. The under-reporting of tilapia production can be attributed to the following:

1. The reproductive characteristics of tilapia. The simple, frequent and asynchronous reproduction of tilapia makes the estimate of total production of these fish almost impossible.

2. The poor management of evaluating and utilizing tilapia. The large amounts consumed by tilapia farmers' families and the amounts that are locally marketed informally may make production statistics incorrect and unreliable. For example, it has been reported that about 20% of aquaculture production in rural China is consumed by farmers' families. The global production of farmed tilapia may, therefore, be higher than the officially reported quantities.

The global production trends of farmed tilapia can be divided into two distinctive phases:

- During the 1950s to 1970s tilapia culture grew at a relatively slow rate, where farmed tilapia production was much lower than that of capture fisheries (Fig. 1.2).
- In the 1980s to 1990s, tilapia culture expanded at a much wider and more rapid rate, where the gap between tilapia landings from capture fisheries and aquaculture continued to narrow, until the production of farmed tilapia exceeded the landings from capture fisheries in 1993 (Fig. 1.2). Since then, tilapia culture has been growing at a very high rate, while tilapia landings from capture fisheries are about stable.

Since tilapia can tolerate a wide range of water salinity, they are currently farmed in freshwater, brackish-water and even seawater environments, but freshwater tilapia aquaculture dominates. The production of tilapia from freshwater systems reached 1,312,776 Mt in 2002, representing 87.2% of total farmed tilapia production.

The value of farmed tilapia has also witnessed a great increase during the past two decades. The value increased from about US$154 million in 1984 to US$1800.7 million in 2002. As expected, the value of Nile tilapia represented between 60 and > 70% of the total market value of farmed tilapia during the past decade (Fig. 1.6).

1.2.3. Major producers

Despite the fact that more than 100 countries practised tilapia farming in 2002, only five countries (China, Egypt, the Philippines, Indonesia and Thailand) dominated world production.

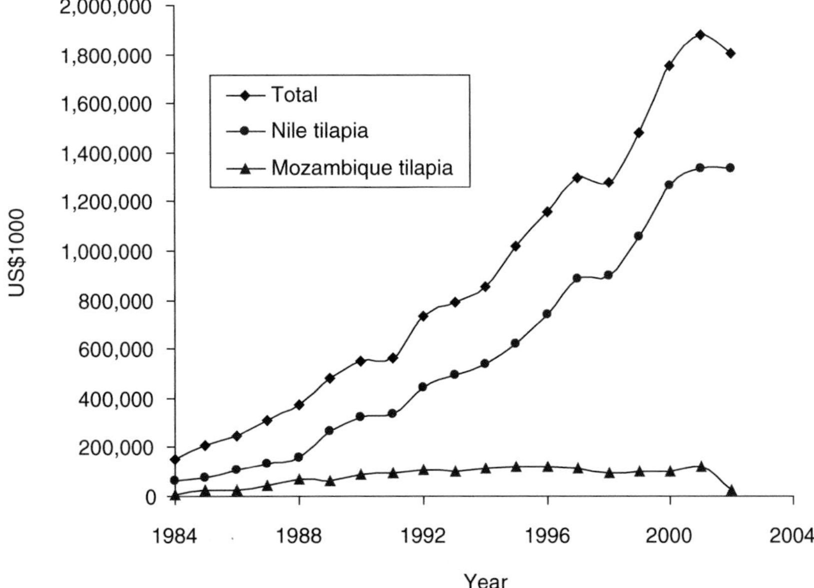

Fig. 1.6. The value of important farmed tilapia (US$1000) during 1984–2002.

Those countries yielded 87% of global tilapia production in 2002. China alone produced 706,585 Mt in 2002, representing 50% of total production, followed by Egypt (12%), the Philippines (9%), Indonesia (8%) and Thailand (7%) (Fig. 1.7). However, the value of farmed tilapia in China in 2002 was only US$706.6 million, representing only 2.15% of the total value of aquaculture production in China (US$32,827 million). This is mainly because tilapia is produced in China as a low-value food fish and used typically for local consumption in rural areas. The values of farmed tilapia in selected major producer nations are given in Fig. 1.8.

1.2.4. Major cultured species

Among cultured fishes of the world, tilapia rank third in terms of production, only after carps and salmonids. According to FAO statistics, 16 tilapia/cichlid groups, in addition to unidentified cichlids, have been used for aquaculture production. However, commercial tilapia culture is currently restricted to about ten species (Table 1.1). Nile tilapia is, by far, the most important farmed tilapia species in the world. It represented more than 80% of total tilapia production during 1970–2002. Nile tilapia also ranked sixth in terms of global farmed fish production in 2002, after silver carp, grass carp, common carp, crucian carp and big-head carp (Fig. 1.9). Mozambique tilapia comes second, with a production of 54,146 Mt in 2002, representing 3.6% of the production of total farmed tilapia. Three-spotted tilapia (*Oreochromis andersonii*), blue tilapia, redbreast tilapia (*Tilapia rendalli*) and longfin tilapia (*Oreochromis macrochir*) are also gaining some popularity in certain parts of the world. The contribution of other tilapia species to global tilapia production is insignificant, while unidentified tilapias represent a significant proportion of the production. In 2002, that category amounted to 227,741 Mt, representing 18.7% of total tilapia production.

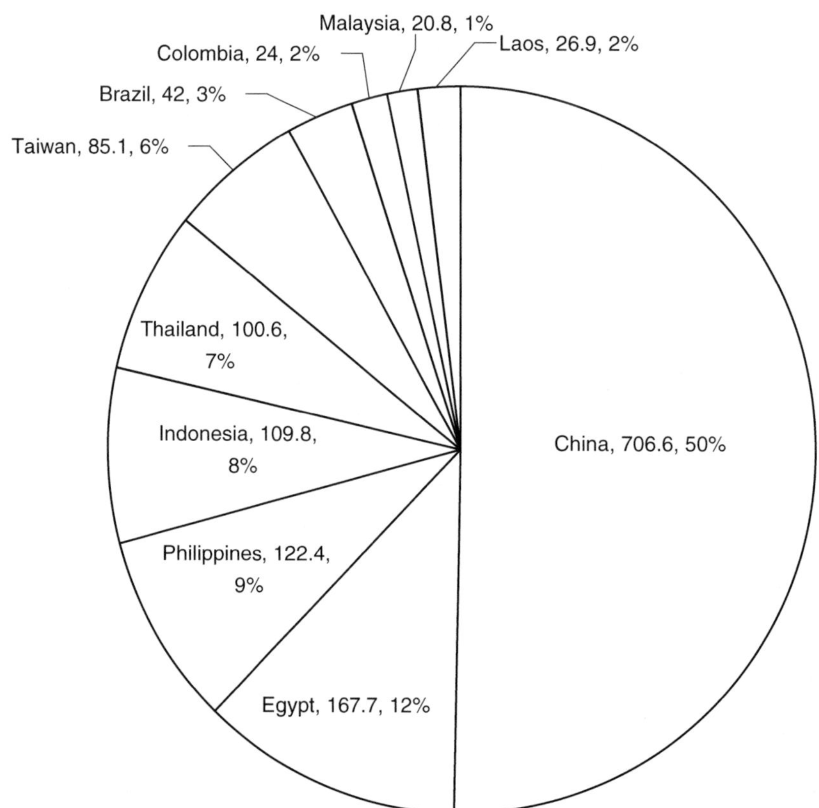

Fig. 1.7. Top ten producers of farmed tilapia (1000 Mt, %) in 2002.

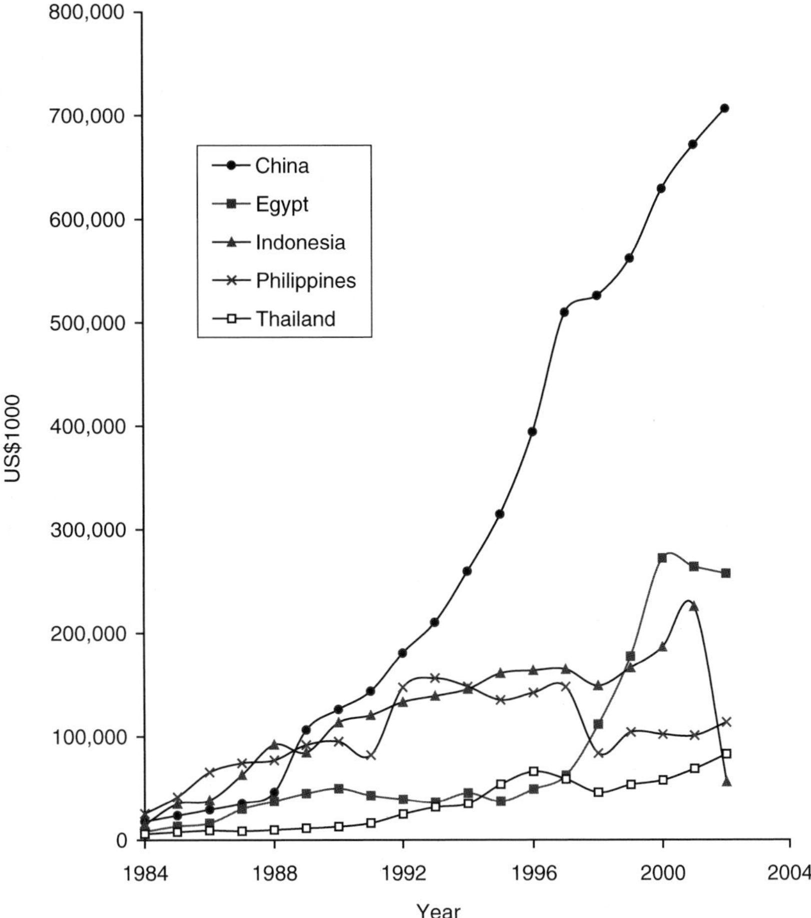

Fig. 1.8. The value of farmed tilapia (US$1000) for selected major producers during 1984–2002.

1.3. Tilapia Production in Asia

Asia is the largest tilapia producer in the world, accounting for 79% of the production of global farmed tilapia in 2002. Tilapia is farmed mainly in freshwater environment in Asia. In 2002, about 95.5% of total farmed tilapia production came from freshwater environments. Twenty-one Asian countries, mainly in South Asia and the Far East, reported tilapia production in 2002. Tilapia culture in Asia has witnessed three developmental phases:

1. **1950–1980**: During this early period, tilapia culture was practised on a very small scale and grew at slow rates. The production gradually increased from only 4,810 Mt in 1950 to reach 88,011 Mt in 1980.
2. **1981–1991**: The production of tilapia witnessed a sharp increase from 109,301 Mt in 1981 to reach 353,686 Mt in 1991, with an over 300% increase.
3. **1992–now**: This period was characterized by an outstanding expansion and development of tilapia culture in Asia. As a result, tilapia production jumped from 421,649 Mt in 1992 to 1,191,611 Mt in 2002 (Table 1.2).

The growth rate in the production of farmed tilapia in Asia during 1950–2002 was among the fastest in world, with an overall average of 20.5% annually. It is noteworthy that the recorded production of tilapia in Asia is lower than the real production, because the production of some other Asian countries, such as Vietnam, Bangladesh,

Table 1.1. Major cultured tilapia species and species production (Mt) in the world during 1950–2002 (from FAO, 2004).

Species	1950	1960	1970	1980	1990	1992	1994	1996	1998	2000	2001	2002
Nile tilapia	1,590	7,736	12,058	41,357	233,601	320,092	425,500	623,652	772,706	1,047,885	1,126,927	1,217,055
Mozambique tilapia	. . .	1	1,186	12,640	42,912	49,327	51,872	56,311	45,822	49,418	59,264	54,146
Three-spotted tilapia	. . .	. . .	. . .	27	1,000	1,800	2,200	2,661	2,689	2,750	2,700	2,700
Blue tilapia	. . .	. . .	. . .	1,012	3,748	3,455	2,368	2,425	844	1,277	1,135	1,350
Redbreast tilapia	. . .	. . .	. . .	. . .	105	524	803	1,043	839	853	877	860
Longfin tilapia	. . .	. . .	. . .	. . .	60	230	350	404	207	210	210	210
Sabaki tilapia	. . .	. . .	. . .	. . .	1	< 0.5	20	20	103	83	63	165
Redbelly tilapia	. . .	. . .	. . .	. . .	8	10	18	20	100	201	201	161
Jaguar guapote	. . .	. . .	. . .	. . .	6	9	31	152	37	40	42	42
Unidentified tilapias	4,128	9,594	15,016	53,500	102,204	104,703	112,203	126,018	130,270	171,617	212,376	227,741
Total	5,718	17,331	28,260	108,536	383,654	488,527	595,535	812,850	953,659	1,274,389	1,404,904	1,505,804

. . . denotes that data are not available, unobtainable or not separately available but included in another category.

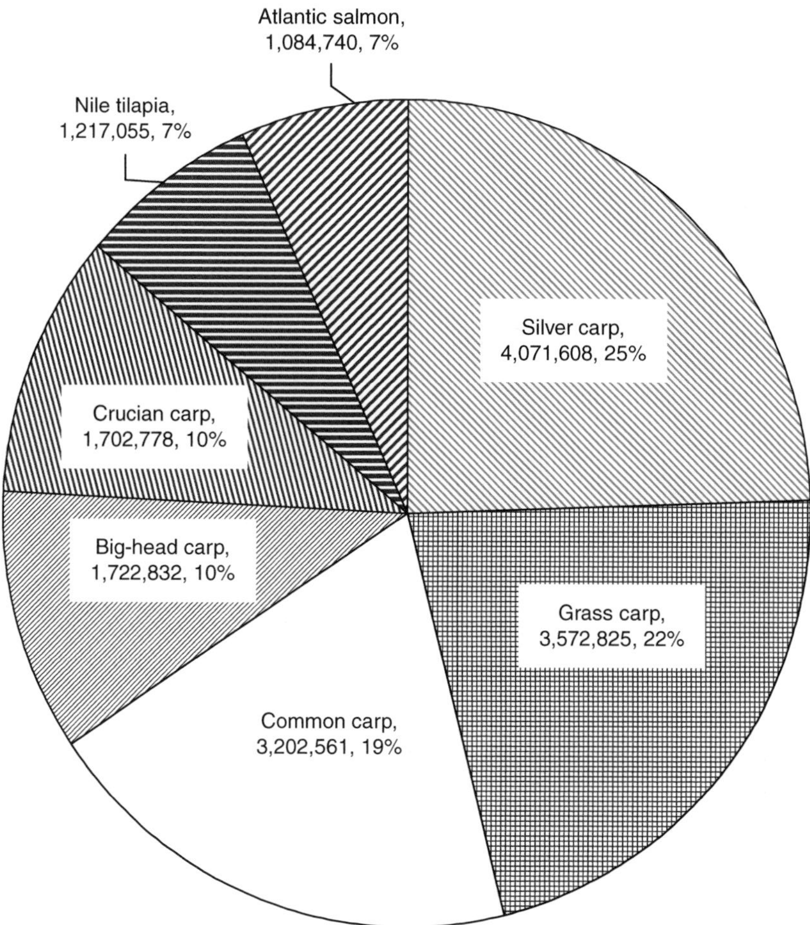

Fig. 1.9. Global production of major farmed fish species (Mt, %) in 2002.

India and Pakistan, was not included in production statistics. In addition, a considerable proportion of produced tilapia is consumed by fish farmers and their families, leading to underestimation of tilapia production in Asia.

1.3.1. Major producers

Despite the fact that Asia is the most important tilapia producer in the world, only 21 Asian countries practise tilapia culture (Table 1.3), with China being the largest producer. In 2002, China contributed about 60% of total Asian tilapia production. When the contribution of China is discounted, the contribution of Asia to global tilapia production declines from 79% to only 32.2% in 2002. All the production of tilapia in China comes from freshwater culture, mainly from semi-intensive culture systems. The production trend in China can be divided into two phases (Fig. 1.10):

- **1950–1988**: During this period, the production of tilapia increased slowly from 660 Mt in 1950 to 39,000 Mt in 1988. Chinese production was exceeded by other Asian countries such as Taiwan, Indonesia and the Philippines.
- **1989–now**: From 1989 onward, the production of farmed tilapia in China increased at an exceptionally high rate, with an average annual growth rate of over 20%. As a result, China currently dominates tilapia production in Asia.

Table 1.2. Continental production (Mt) of farmed tilapia during 1950–2002 (from FAO, 2004).

Year	1950	1960	1970	1980	1990	1992	1994	1996	1998	2000	2001	2002
Africa	908	3,399	4,723	12,456	33,094	35,715	38,164	40,077	67,421	177,202	174,985	193,240
Asia	4,810	13,932	23,337	88,011	333,016	421,649	521,914	715,602	814,407	997,046	1,124,602	1,191,611
N. America	0	0	200	7,963	14,891	18,020	22,308	24,730	25,522	33,418	37,753	45,089
S. America	0	0	0	103	2,307	12,732	12,728	31,849	45,716	66,087	66,966	75,328
Oceania	0	0	0	0	105	211	221	272	393	456	398	346

Table 1.3. Production (Mt) of farmed tilapia by country in Asia during 1950–2002 (from FAO, 2004).

Country	1950	1960	1970	1980	1990	1992	1994	1996	1998	2000	2001	2002
Brunei	...	...	...	...	1	3	3	10	20	14	...	52
Cambodia	...	...	...	...	170	230	200	230	330	370	359	376
China	660	5,003	5,828	9,000	106,071	157,233	235,940	394,303	525,926	629,182	671,666	706,585
Hong Kong	...	...	450	2,120	1,195	980	161	442	1,058	613	641	411
Indonesia	100	100	1,191	14,901	53,768	59,945	64,431	75,473	65,894	85,179	105,106	109,768
Israel	20	95	1,400	2,512	4,795	3,368	5,631	6,399	6,696	7,059	8,217	7,819
Japan	...	...	...	2,392	5,825	4,697	2,125	1,479	885	434	434	400
Jordan	...	...	...	...	40	16	67	135	263	563	540	515
Republic of Korea	...	...	...	...	650	437	448	998	796	787	609	588
Kuwait	...	...	...	...	...	...	...	...	70	30	16	16
Laos	...	...	20	176	1,250	1,345	1,400	2,000	9,549	18,928	22,499	26,872
Lebanon	...	...	...	...	...	...	...	...	...	...	...	25
Malaysia	...	< 0.5	12	366	1,145	4,632	8,507	11,177	12,625	18,471	16,253	20,757
Myanmar	...	...	...	...	...	...	...	...	...	...	...	1,000
The Philippines	...	70	1,417	13,214	76,142	91,173	90,341	79,415	72,023	92,579	106,746	122,390
Saudi Arabia	...	...	...	...	1,926	2,191	2,220	3,614	3,315	3,968	3,981	2,019
Singapore	...	...	...	...	...	...	...	45	150	37	52	142
Sri Lanka	...	...	...	1,097	4,500	3,500	2,500	2,500	3,500	4,390	3,130	3,670
Syria	...	...	...	102	596	1,126	991	1,588	1,372	2,626	3,195	2,571
Taiwan	3,900	8,200	11,287	33,712	52,047	47,226	47,435	44,756	36,126	49,235	82,781	85,059
Thailand	130	464	1,732	8,419	22,895	43,547	59,514	91,038	73,809	82,581	98,377	100,576
Total	4,810	13,932	23,337	88,011	333,016	421,649	521,914	715,602	814,407	997,046	1,124,602	1,191,611

. . . denotes that data are not available, unobtainable or not separately available but included in another category.

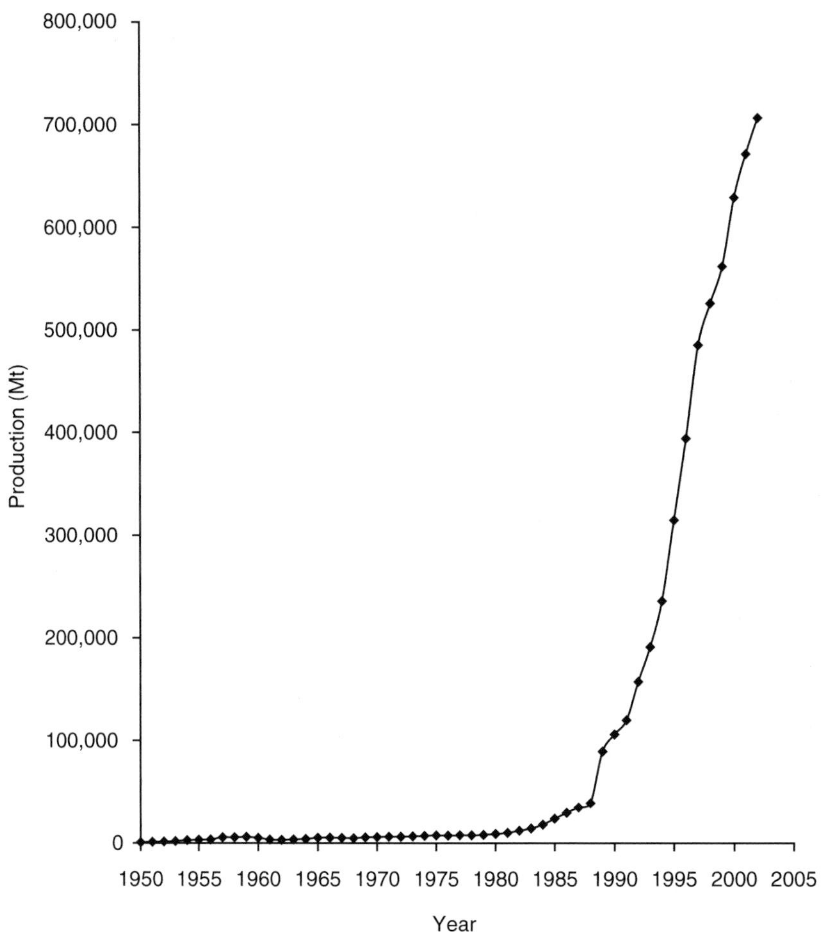

Fig. 1.10. Production (Mt) of farmed tilapia in China during 1950–2002.

Tilapia culture is also growing at a very high rate in some other Asian countries, including the Philippines, Indonesia, Thailand and Taiwan. These four countries, together with China, yielded 94% of Asian tilapia production in 2002. Tilapia farming in other Asian countries, such as Saudi Arabia, Israel, Jordan, Syria, India, Bangladesh and Vietnam, has started to gain considerable attention in recent years.

1.3.2. Major cultured species

It has been reported that Mozambique tilapia was the first tilapia species to be introduced to Asia, into the Indonesian island of Java, in 1939 (Guerrero, 2001). Subsequently, the species was introduced into other Asian countries, during the 1940s to the 1960s, and was considered a prime candidate for aquaculture in Asia. Thus, the average annual production of this species increased at a rate of 26.7% during 1970–1980, compared with 13.1% for Nile tilapia. However, Mozambique tilapia may have suffered from inbreeding problems due to the small number of the original population of founder stocks. The acceptability of Mozambique tilapia to tilapia farmers was also limited because of the problem of overcrowding and poor growth. The dark colour of this fish also reduces its marketability and consumers' acceptance. Subsequently, Nile tilapia attracted attention in the 1960s and early 1970s as an ideal aquaculture substitute for Mozambique tilapia. Currently, Mozambique tilapia is a major culture species only in Indonesia. The Indonesian supply of that species in 2002 was 49,331 Mt, representing 93% of total production of farmed Mozambique tilapia in Asia (and 91.1% of global production). The dominance of the species in Indonesia continued until the late 1990s, when Nile tilapia started to take over. For example, Mozambique tilapia accounted for 57.8% of tilapia production in 1999. Its contribution declined in the following years to reach about 52% of total tilapia production in 2001. In 2002, the production of Nile tilapia in Indonesia exceeded that of Mozambique tilapia and reached 60,437 Mt, representing 55% of total tilapia production. This means that commercial culture of Mozambique tilapia in Asia will probably stop within the next few years.

Nile tilapia currently dominates tilapia culture in Asia, with a production of 1,001,302 Mt in 2002, representing 84% of total tilapia production in the continent (Table 1.4). In terms of countries, 11 Asian countries reported Nile tilapia production in 2002, compared to six countries in 1980. On the other hand, only five countries reported Mozambique tilapia culture in Asia in 2002, with a production of 53,000 Mt, representing 4.4% of total tilapia production. The contribution of other tilapias, including blue tilapia, Galilean tilapia (mango tilapia) (*S. galilaeus*), Wami tilapia (*Oreochromis hornorum*), Zill's tilapia (redbelly tilapia) (*Tilapia zillii*) and redbreast tilapia (Congo tilapia) (*T. rendalli*), is not significant. These species have been introduced to Asia mainly for research and experimental purposes.

In addition to the species mentioned, introgressive hybridization between Nile tilapia and Mozambique tilapia has been reported in several Asian countries. All-male hybrids of Nile tilapia and blue tilapia are also widely cultured in China and Taiwan. Red tilapia, which is generally a hybrid of Mozambique tilapia and Nile tilapia and/or blue tilapia, is also commonly cultured in many Asian countries, including China, Taiwan, Thailand, Indonesia and the Philippines. The hybrids are characterized by high growth rates, preferred colour and high consumer acceptance. Tilapia hybrids and other 'unidentified' tilapia groups accounted for 11.5% of the total production of farmed tilapia in Asia in 2002.

1.4. Tilapia Production in Africa

Despite the fact that tilapia are African fish, tilapia culture in Africa is relatively new, with a low contribution to world tilapia production, being 12.8% in 2002. Farmed tilapia production in Africa, however, has sharply increased during the past few years (Table 1.2). Generally, the trend in tilapia production in Africa can be divided into three phases:

1. **1950–1984**: During this 35-year period, the production of tilapia slowly grew from 908 Mt in 1950 to 15,747 Mt in 1984.
2. **1985–1997**: Production jumped from 15,747 Mt in 1984 to 28,724 Mt in 1985, and continued to grow at a relatively slow rate to reach 43,946 Mt in 1997.
3. **1998–now**: Tilapia production increased from 43,946 Mt in 1997 to 67,421 Mt in 1998, with an increase in growth rate of 53.4%. Another huge increase in tilapia production occurred during 1999–2002, with 119,416, 177,202, 174,985 and 193,240 Mt produced in 1999, 2000, 2001 and 2002, respectively. The average annual growth of tilapia production during the period 1998–2002 was 47.3%, excluding the year 2001, which showed a slight decrease compared to 2000, or 37.6% if the production of the year 2001 is included.

1.4.1. Major producers

In 2001, 32 African countries reported production of farmed tilapia; however, in 2002, some of these countries reported no production. Unlike Asia, most of farmed tilapia production in Africa comes from brackish-water environments. In 2002, tilapia production from brackish water amounted to 138,923 Mt, representing 71.9% of total African tilapia production. Farmed tilapia output in Africa is dominated by a single country: Egypt (Table 1.5). The production of farmed tilapia in Egypt accounted for 86.8% (167,735 Mt) of total production of tilapia in Africa in 2002. If the contribution of Egypt is discounted, Africa's contribution to global tilapia output would decline from 12.8% to only 1.7%. Tilapia culture in Egypt is practised mainly in brackish-water environments in the northern lakes areas along the Mediterranean coast. In 2002, the production of tilapia from brackish-water systems in Egypt reached 138,456 Mt, representing

Table 1.4. Major cultured tilapia species and species production (Mt) reported in Asia during 1950–2002 (from FAO, 2004).

Species	1950	1960	1970	1980	1990	1992	1994	1996	1998	2000	2001	2002
Blue tilapia	...	...	...	...	< 0.5	< 0.5	< 0.5	< 0.5	...	...	...	...
Mozambique tilapia	...	1	1,186	12,637	42,664	48,806	51,489	55,733	44,879	48,532	58,095	53,000
Nile tilapia	890	5,636	9,464	31,781	200,814	296,748	391,204	584,640	700,964	853,630	937,524	1,001,302
Sabaki tilapia	...	...	...	...	...	...	20	20	103	83	63	165
Unidentified tilapias	3,920	8,295	12,687	43,593	89,538	76,095	79,201	75,209	68,461	94,801	128,920	137,144
Total	4,810	13,932	23,337	88,011	333,016	421,649	521,914	715,602	814,407	997,046	1,124,602	1,191,611

... denotes that data are not available, unobtainable or not separately available but included in another category.

Table 1.5. Production (Mt) of farmed tilapia by country in Africa during 1950–2002 (from FAO, 2004).

Country	1950	1960	1970	1980	1990	1992	1994	1996	1998	2000	2001	2002
Burkina Faso	...	...	...	...	3	2	< 0.5	30	40	5	5	5
Burundi	...	...	...	...	30	50	55	50	55	100	100	100
Cameroon	...	...	...	25	80	50	45	50	60	40	100	210
Republic of Central Africa	...	...	...	76	100	337	250	140	80	120	125	...
Democratic Republic of Congo	...	...	...	...	700	730	650	600	1,833	2,073	2,738	2,959
Republic of Congo	...	...	...	...	240	191	121	106	140	200	200	...
Côte d'Ivoire	...	...	...	...	32	44	109	933	495	967	870	725
Egypt	700	2,100	2,500	9,000	24,916	21,505	25,214	27,854	52,755	157,425	152,515	167,735
Ethiopia	...	...	...	...	30	20	30	35	10	< 0.5	< 0.5	...
Gabon	...	...	...	...	2	5	23	59	150	533	102	83
Ghana	...	...	94	251	280	300	330	350	1,350	3,712	4,400	4,400
Guinea	...	...	...	...	< 0.5	< 0.5	< 0.5	< 0.5	< 0.5	< 0.5	< 0.5	...
Kenya	...	...	...	70	405	467	502	500	87	222	412	421
Liberia	...	...	...	...	< 0.5	< 0.5	< 0.5	< 0.5	...	19	12	12
Madagascar	...	...	...	...	40	< 0.5	< 0.5	< 0.5	< 0.5	< 0.5	< 0.5	< 0.5
Malawi	...	...	...	30	50	30	28	20	22	500	532	620
Mali	...	...	...	...	10	20	58	35	35	19	350	708
Mauritius	...	...	...	...	8	12	32	71	40	44	30	20
Mayotte	...	...	...	...	...	...	...	1	2	3	3	...
Mozambique	...	...	...	...	15	12	116	4	< 0.5	< 0.5	< 0.5	77
Niger	...	...	...	...	36	8	17	11	12	15	21	40

Continued

Table 1.5. *Continued.*

Country	1950	1960	1970	1980	1990	1992	1994	1996	1998	2000	2001	2002
Nigeria	208	1,299	2,129	2,952	3,795	7,525	5,500	3,259	4,471	2,705	2,626	4,496
Rwanda	...	...	...	23	154	45	50	90	120	252	381	542
Réunion	...	...	...	...	...	...	...	...	75	88	68	60
Senegal	...	...	...	...	5	5	31	53	3	9	10	22
Sierra Leone	...	...	...	2	20	20	25	30	30	30	30	...
South Africa	...	...	...	...	30	55	60	15	70	110	200	200
Sudan	...	...	...	...	234	200	200	1,000	1,000	1,000	1,000	1,000
Swaziland	...	...	...	...	...	...	...	47	49	38	40	...
Tanzania	...	...	...	...	375	350	150	200	200	210	300	630
Togo	...	...	...	...	22	150	150	21	25	102	120	25
Uganda	...	...	...	...	32	42	108	40	200	600	1,550	1,957
Zambia	...	...	...	27	1,400	3,500	4,280	4,403	3,942	4,020	3,980	3,980
Zimbabwe	...	...	...	...	50	40	30	70	70	2,041	2,165	2,213
Total	908	3,399	4,723	12,456	33,094	35,715	38,164	40,077	67,421	177,202	174,985	193,240

. . . denotes that data are not available, unobtainable or not separately available but included in another category.

82.5% of total tilapia production in the continent. Tilapia culture is also practised in some other countries, including Nigeria, Ghana, Zambia, the Democratic Republic of Congo and Zimbabwe (Table 1.5). The contribution by the rest of African countries is insignificant.

1.4.2. Major cultured species

Seven tilapia species or species groups are used for aquaculture in Africa (Table 1.6). Nile tilapia is by far the most widely cultured species. Nile tilapia was reportedly cultured in 23 African countries out of 32 countries that practised tilapia culture in Africa in 2001. It also accounted for 92.5% of total tilapia production in Africa in 2002. The culture of other tilapia species, namely three-spotted tilapia (Zambia), redbreast tilapia (Malawi), Mozambique tilapia (Malawi), redbelly tilapia (Liberia) and mango tilapia (Liberia), was reported in 2002, though production was very limited. In addition to the species, a considerable proportion of tilapia production is reported under 'unidentified' tilapia and tilapia hybrids. This category comes second after Nile tilapia in terms of production. The production of unidentified tilapias in 2002 amounted to 10,405 Mt, representing 5.4% of total tilapia production.

1.5. Tilapia Production in South America

Tilapia culture is relatively new in South America. It started on a small scale, mainly for subsistence farming, in the early 1970s. The first FAO record of tilapia production in the region was reported in Colombia in 1971, with only 1 Mt of Nile tilapia. The progress of tilapia culture in South America can be divided into three phases:

1. **1971–1982**: During this period, tilapia culture was practised in only two countries, Colombia and Peru, with very little production, ranging from 1 Mt in 1971 to 182 Mt in 1982.
2. **1983–1991**: During this period, the number of countries practising tilapia culture increased gradually to six countries in 1991. Nevertheless, tilapia production was still very low and was dominated by one country: Colombia (87.5%).
3. **1992–now**: The production of cultured tilapia in South America increased from 3475 Mt in 1991 to 12,732 Mt in 1992, with a 266% increase. Another jump occurred in 1995, where the production approached 30,032 Mt, with a 136% increase. The production continued to increase at a high rate, to reach 75,328 Mt in 2002, which represented 5% of global production of farmed tilapia. Consequently, the number of countries practising tilapia culture increased to ten (Table 1.7). The average annual growth rate of tilapia production during 1996–2002 was 15.7%. More than 99% of tilapia production comes from freshwater environments.

1.5.1. Major producers

As mentioned earlier, Colombia was the only country practising tilapia culture in South America in the 1970s, while Peru reported tilapia production starting 1979. In 1980, total production of tilapia was only 103 Mt but increased to 2307 Mt in 1990. In 2002, ten countries reported tilapia production. Interestingly, the first record of tilapia production in Brazil appeared in 1995. During 1995–2002, tilapia production in that country increased from 12,014 Mt to 42,003 Mt. This means that Brazil has one of the fastest growth rates of tilapia production in South America, with an average annual growth rate of 20%. As a result, Brazil contributed 55.8% (42,003 Mt) to tilapia production in the continent in 2002.

Departing from the FAO tilapia production statistics, Kubitza (2004) reported that farm-raised tilapia in Brazil reached 57,000 Mt in 2002. Moreover, he stated that, at present, production is estimated at about 70,000 Mt. Colombia and Ecuador are the second and third major tilapia producers, with a production amounting to 24,000 Mt (31.9%) and 8181 Mt (10.9%) in 2002. Brazil, Colombia and Ecuador accounted for 98.5% of total tilapia production in 2002. This means that the current production of tilapia in the other South American countries is not significant (Table 1.7).

1.5.2. Major cultured species

One of the major problems associated with tilapia culture in South America is that most cultured tilapias are not identified. Many tilapia strains and hybrids are currently used for aquaculture, with minimal control and regulation. As a result, most

Table 1.6. Major cultured tilapia species and species production (Mt) in Africa during 1950–2002 (from FAO, 2004).

Species	1950	1960	1970	1980	1990	1992	1994	1996	1998	2000	2001	2002
Blue tilapia	. . .	. . .	. . .	. . .	12	24	. . .	. . .	. . .	. . .	. . .	. . .
Longfin tilapia	. . .	. . .	. . .	. . .	60	230	350	404	207	210	210	210
Mozambique tilapia	. . .	. . .	. . .	. . .	30	68	75	57	96	55	145	130
Nile tilapia	700	2,100	2,594	9,447	26,996	23,827	27,641	31,667	56,981	165,665	162,711	178,762
Redbelly tilapia	. . .	. . .	. . .	. . .	8	10	18	20	100	201	201	161
Redbreast tilapia	. . .	. . .	. . .	. . .	100	503	803	1,028	839	853	877	860
Three-spotted tilapia	. . .	. . .	. . .	27	1,000	1,800	2,200	2,661	2,689	2,750	2,700	2,700
Unidentified tilapias	208	1,299	2,129	2,982	4,888	9,253	7,077	4,240	6,508	7,454	8,130	10,405
Total	908	3,399	4,723	12,456	33,094	35,715	38,164	40,077	67,421	177,202	174,985	193,240

. . . denotes that data are not available, unobtainable or not separately available but included in another category.

Table 1.7. Production (Mt) of farmed tilapia by country in South America during 1980–2002 (from FAO, 2004).

Country	1980	1990	1991	1992	1993	1994	1995	1996	1997	1998	1999	2000	2001	2002
Brazil	. . .	. . .	. . .	. . .	. . .	. . .	12,014	15,700	16,845	24,062	27,104	32,459	35,830	42,003
Colombia	93	2,040	3,040	11,050	11,046	11,084	16,057	14,026	16,112	17,665	19,842	22,870	24,000	24,000
Ecuador	. . .	21	33	876	912	68	< 0.5	< 0.5	1,730	1,730	4,400	9,201	5,169	8,181
Venezuela	. . .	4	127	400	700	1,103	1,650	1,700	1,936	2,010	2,320	970	1,250	560
Guyana	. . .	30	50	75	100	159	160	160	170	180	369	369	370	370
Peru	10	186	200	250	181	205	< 0.5	47	< 0.5	. . .	. . .	8	223	122
Suriname	. . .	. . .	. . .	. . .	. . .	1	1	1	1	1	50	130	54	54
Bolivia	. . .	. . .	. . .	51	79	68	70	55	40	30	30	30	30	35
Paraguay	. . .	26	25	30	30	40	80	150	210	38	38	40	40	. . .
Argentina	. . .	. . .	. . .	. . .	. . .	. . .	. . .	10	10	. . .	. . .	10	. . .	3
Total	103	2,307	3,475	12,732	13,048	12,728	30,032	31,849	37,054	45,716	54,153	66,087	66,966	75,328

. . . denotes that data are not available, unobtainable or not separately available but included in another category.

tilapia production in South America is reported in the 'unidentified' category. Out of the 75,328 Mt of tilapia produced in 2002, 62,723 Mt were 'unidentified' tilapias (83.3%). Nile tilapia is the second most important cultured category, where it accounted for 16.5% (12,422 Mt) of total tilapia production in 2002. Other tilapia/cichlids, including *Cichlasoma*, green terror (*Aequidens rivulatus*), Mozambique tilapia, redbreast tilapia and velvety cichlids are also cultured in South America, but on a very small scale, with negligible production (Table 1.8). The culture of *Cichlasoma* stopped completely after 1993, while that of green terror, redbreast tilapia and velvety tilapia stopped in the late 1990s.

1.6. Tilapia Production in North America and the Caribbean

Mozambique tilapia were introduced into the Caribbean in 1947 and first came to the USA in 1954 (Fitzsimmons, 2001a). Tilapia culture in North America and the Caribbean is therefore relatively new, beginning in the 1960s and 1970s, on a small scale, mainly for subsistence objectives. The first FAO record of tilapia production appeared in 1970, with 200 Mt from Mexico. During 1984–2002, the production increased progressively at an annual growth rate of 12.75%, to reach 45,089 Mt in 2002 (Table 1.9), which represented about 3% of global tilapia production. More than 99% of tilapia production in North America and the Caribbean comes from freshwater environments.

1.6.1. Major producers

Twenty countries from North America and the Caribbean reported tilapia production in 2002. Unlike Asia, Africa and South America, where a single country in each continent dominates tilapia production, no single country dominates the production in North America and the Caribbean (Table 1.9). Instead, Costa Rica, the USA, Mexico and Jamaica together accounted for 78.7% of tilapia production in 2002 (29.3%, 20%, 16.1% and 13.3%, respectively). Countries such as the Dominican Republic, Guatemala, Honduras, Panama, Cuba and El Salvador are also paying considerable attention to tilapia culture. The production of the rest of North American countries is not significant.

1.6.2. Major cultured species

Nile tilapia is the main tilapia species cultured in North America and the Caribbean. It represented about 54% of tilapia production in the region in 2002. Blue tilapia and the cichlid *Cichlasoma* are also important aquaculture species. On the other hand, the production of Mozambique tilapia has declined from 1747 Mt in 1999 to 733 Mt in 2002. As in the case of South America, several tilapia hybrids and strains are also produced in the region. Therefore, a considerable proportion of produced tilapia is reported in the 'unidentified' category. In 2002, 'unidentified' tilapia represented 38.3% of total tilapia production in the region (Table 1.10).

1.7. Future Potential

1. The ever-increasing global growth rate of tilapia culture, accompanied by the continuous introductions of these fish into new geographical areas, reflects a positive future for tilapia culture. Tilapia are expected to play a substantial role as a food fish to meet the needs of the poor for animal protein in developing countries and will probably become an important cash crop in those countries. More value added for tilapia products will also come from developing countries. This will encourage foreign companies to invest in joint ventures in the main producing countries. It is also expected that the increase in tilapia imports will continue.

2. The genetically improved Nile tilapia developed by the International Center for Living Aquatic Resources Management (ICLARM) and the YY males developed by the University of Wales, Swansea, to produce genetically male tilapia (Mair *et al.*, 1997) will probably become a breakthrough in tilapia culture in the near future. Distribution of these tilapia strains has already begun in many countries around the world.

3. China is very likely to continue dominating global tilapia production. In addition, a significant expansion in tilapia farming in other Asian countries, such as Cambodia, Vietnam, Laos and Thailand, is likely to occur (Dey, 2001). It is also

Table 1.8. Major cultured tilapia species and species production (Mt) in South America during 1993–2002 (from FAO, 2004).

Species	1993	1994	1995	1996	1997	1998	1999	2000	2001	2002
Cichlasoma	355	. . .	. . .	. . .	. . .	. . .	. . .	. . .	. . .	. . .
Green terror	< 0.5	< 0.5	< 0.5	< 0.5	. . .	. . .	. . .	. . .	. . .	. . .
Mozambique tilapia	100	159	160	160	170	180	184	184	183	183
Nile tilapia	3,863	2,925	3,817	3,378	3,328	4,185	7,844	13,143	9,506	12,422
Redbreast tilapia	< 0.5	< 0.5	< 0.5	< 0.5	< 0.5	< 0.5	. . .	. . .	. . .	. . .
Unidentified tilapias	8,730	9,644	26,055	28,311	33,556	41,351	46,125	52,760	57,277	62,723
Velvety cichlids	< 0.5	< 0.5	< 0.5	< 0.5	< 0.5	. . .	. . .	. . .	. . .	. . .
Total	13,048	12,728	30,032	31,849	37,054	45,716	54,153	66,087	66,966	75,328

. . . denotes that data are not available, unobtainable or not separately available but included in another category.

Table 1.9. Production (Mt) of farmed tilapia by country in North America and the Caribbean during 1970–2002 (from FAO, 2004).

Country	1970	1980	1990	1991	1992	1993	1994	1995	1996	1997	1998	1999	2000	2001	2002
Costa Rica	. . .	18	232	1,200	1,350	2,360	2,790	3,800	4,100	4,100	5,398	6,588	8,100	8,500	13,190
Cuba	. . .	1,012	3,733	3,313	3,425	2,337	2,363	2,096	2,418	1,888	540	1,060	730	480	500
Dominican Republic	. . .	. . .	100	374	400	218	1,311	985	177	344	446	445	994	1,666	2,084
El Salvador	. . .	. . .	4	17	36	105	151	196	93	138	278	141	56	29	405
Guatemala	. . .	. . .	161	201	540	486	638	774	1,751	2,074	1,647	3,352	2,361	2,501	2,501
Honduras	. . .	6	120	119	173	184	102	172	190	232	102	135	927	1,900	2,874
Jamaica	. . .	20	3,364	3,100	3,200	3,300	3,400	3,500	3,450	3,400	3,360	4,100	4,500	4,500	6,000
Martinique	. . .	. . .	36	35	50	62	65	30	13	10	10	10	8	8	8
Mexico	200	6,907	5,000	4,500	4,600	4,800	5,439	1,482	4,800	8,318	5,398	7,023	6,726	8,845	7,271
Nicaragua	. . .	. . .	3	4	4	4	4	5	52	45	15	16	24	64	64
Panama	. . .	. . .	49	83	65	77	109	186	115	102	55	634	900	1,181	1,181
Puerto Rico	. . .	. . .	3	164	85	48	43	48	24	6	6	3	15	18	3
Trinidad and Tobago	. . .	. . .	2	3	3	4	4	15	15	13	13	12	21	6	6
USA	. . .	. . .	2,041	2,041	4,082	5,670	5,888	6,838	7,242	7,648	8,251	8,051	8,051	8,051	9,000
Total	200	7,963	14,891	15,158	18,020	19,657	22,308	20,143	24,730	28,458	25,522	31,573	33,418	37,753	45,089

. . . denotes that data are not available, unobtainable or not separately available but included in another category.

Table 1.10. Major cultured tilapia species and species production (Mt) in North America and the Caribbean during 1970–2002 (from FAO, 2004).

Species	1970	1980	1990	1991	1992	1993	1994	1995	1996	1997	1998	1999	2000	2001	2002
Black-belt cichlid	. . .	. . .	6	9	16	120	135	140	78	89	30	32	30	32	32
Blue tilapia	. . .	1,012	3,736	3,317	3,431	2,343	2,368	2,103	2,425	1,890	844	1,324	1,276	1,133	1,350
Cichlasoma nei[a]	. . .	. . .	. . .	. . .	. . .	. . .	35	21	66	71	11	14	11	1,066	1,330
Jaguar guapote	. . .	. . .	6	6	9	36	31	40	152	168	37	48	40	42	42
Mozambique tilapia	. . .	. . .	60	225	240	131	. . .	< 0.5	91	168	517	1,747	497	691	733
Nile tilapia	. . .	26	3,530	3,381	3,551	3,674	3,658	3,868	3,845	3,972	10,333	12,771	15,143	16,941	24,324
Redbreast tilapia	. . .	. . .	5	19	21	11	. . .	. . .	15	15	. . .	. . .	. . .	. . .	. . .
Unidentified tilapias	200	6,925	7,548	8,201	10,752	13,342	16,081	13,971	18,058	22,085	13,750	15,637	16,421	17,848	17,278
Total	200	7,963	14,891	15,158	18,020	19,657	22,308	20,143	24,730	28,458	25,522	31,573	33,418	37,753	45,089

[a]Not included elsewhere.
. . . denotes that data are not available, unobtainable or not separately available but included in another category.

expected that the traditional non-acceptance of tilapia in some Asian countries, such as India, Pakistan and Bangladesh, will change, and the interest in tilapia culture in those countries will increase. Therefore, the production of farmed tilapia in Asia is very likely to bloom further in the future.

4. The production of tilapia for export is also expected to receive considerable attention. In parallel, a new focus on consumer's preference and product quality will emerge (Dey, 2001). Quality will become a limiting factor for the competition among tilapia producers (see Chapter 9 for further details on tilapia export and import).

5. The availability of huge freshwater resources, environmental conditions suitable for tilapia culture and culture inputs in many countries in the Americas (Brazil and Mexico, for instance) make the future of tilapia culture in those regions very bright. Thus, it is no surprise that tilapia production in those regions is expected to reach 500,000 Mt by 2010 and 1,000,000 Mt by 2020 (Fitzsimmons, 2001a). Brazil and Mexico are very likely to become the major tilapia producers in the Americas. Nile tilapia (mainly sex-reversed, all-males) will dominate the production in the future. More tilapia processing and filleting, in addition to new tilapia products, will also occur globally, especially in the Americas, for domestic markets as well as for international markets.

6. Unlike Asia and the Americas, the future of tilapia culture in Africa is not clear. The shortage of freshwater resources, culture inputs and technical experience are the main handicaps for tilapia culture development in Africa. These factors may limit the future expansion of tilapia culture in the continent. However, tilapia culture may still play a significant role in rural development in Africa if it is integrated with other plant/animal farming systems.

1.8. Constraints

Despite the bright future of tilapia culture in many parts of the world, there are a number of constraints that might restrict the development of this industry:

- The inadequate knowledge of tilapia farmers in many developing countries of biological, technical, environmental and social factors related to tilapia culture. Understanding these factors is critical for improving farmers' skills and increasing farm productivity and sustainability.
- Lack or shortage of funds and technical assistance in many developing countries, especially in Africa and Latin America.
- Poor fingerling quantity and quality, especially in Asia and the Americas, where most farmed tilapia are derived from very small founder stocks.
- Poor extension service programmes, especially in remote rural areas, where tilapia culture is expected to expand.
- The continuous increase in the cost of culture inputs, such as fingerlings, feeds, fertilizer, fuel, labour, etc., which reduces the profitability of tilapia culture enterprises.

2

Basic Biology and Ecology

2.1. Introduction

Apart from the great economic importance of tilapias for aquaculture and fisheries, they play a significant role in tropical aquatic ecosystems. This means that unplanned introductions and/or culture of tilapia are likely to cause severe ecological impacts. Therefore, the study of the basic biology and ecology of tilapia is necessary prior to their introduction into and/or culture in different geographical regions in order to avoid, or at least minimize, these impacts. This chapter discusses, from a broad perspective, the major aspects of tilapia biology and ecology, including taxonomy, external morphology, geographical distribution, introductions and transfers and feeding habits. Detailed information on these aspects is available in *The Biology and Culture of Tilapias* edited by Pullin and Lowe-McConnell (1982), *Cichlid Fishes* by Keenleyside (1991) and the recent book *Tilapias: Biology and Exploitation* edited by Beveridge and McAndrew (2000). More information on the ecological requirements of tilapia is also provided in Chapter 3.

2.2. Taxonomy

The name 'tilapia' was derived from the African Bushman word meaning 'fish' (Trewavas, 1982). Tilapias represent a large number of freshwater fish species within the family Cichlidae. According to Kaufman and Liem (1982), the family Cichlidae is one of the four families (Cichlidae, Embiotocidae, Pomacentridae and Labridae) included in the suborder Labroidei. Despite the fact that over 70 species of tilapia have been described, there is as yet considerable argument over whether these species are truly separate species. The taxonomic classification of tilapia is still confusing and the subject of continuous changes. This is mainly because of the similarity and overlap of their morphological characteristics, and also due to the fact that many species of tilapia freely hybridize in nature.

The genus *Tilapia* was first described by Smith (1840). It was later split, based on breeding behaviour and feeding habits, into two subgenera: *Tilapia* (substrate spawners) (Fig. 2.1) and *Sarotherodon* ('brush-toothed') (mouthbrooders). Mouthbrooders incubate the fertilized eggs and hatched fry in the mouth of the male or female parents or both male and female. Later, the subgenus *Sarotherodon* was raised to a genus and further subdivided into two genera, *Oreochromis* (mountain cichlids) (Figs 2.2 and 2.3) and *Sarotherodon* (Fig. 2.4), based on whether parental females (*Oreochromis*), males (*Sarotherodon*) or both parental sexes (*Sarotherodon*) perform the mouthbrooding behaviour. In the early 1980s, two other alternative classifications of tilapia were proposed by E. Trewavas (cited in Fishelson and Yaron, 1983):

1. The first classification includes five genera: *Tilapia*, *Sarotherodon*, *Oreochromis*, *Tristromella* and *Danakilia*.
2. The second classification includes only one genus, *Tilapia*, with seven subgenera: *Heterotilapia*, *Pelmatilapia*, *Sarotherodon*, *Oreochromis*, *Nyasalapia*, *Alcolapia* and *Neotilapia*.

All these revisions and changes in the taxonomic classification of tilapia did not eliminate or resolve the current confusion. Therefore, many taxonomists

Fig. 2.1. Redbelly tilapia (*Tilapia zillii*), a typical substrate spawner.

Fig. 2.2. Nile tilapia (*Oreochromis niloticus*) is the most widely cultured tilapia species.

Fig. 2.3. Blue tilapia (*Oreochromis aureus*) is another maternal mouthbrooder. It is also widely cultured.

Fig. 2.4. Galilee tilapia (*Sarotherodon galilaeus*) is a biparental mouthbrooder, where both males and females incubate and defend fertilized eggs and hatched fry.

and researchers still prefer to use the old genus *Tilapia* for all tilapia species.

2.3. Body Shape and External Morphology

Tilapias have fairly conventional, laterally compressed, deep body shapes (see Figs 2.1–2.4). The body is covered with relatively large, cycloid scales, which are not easily dislodged (Ross, 2000). The dorsal and anal fins have hard spines and soft rays. The pectoral and pelvic fins are large and more anterior in an advanced configuration. This character provides the fish with great control over swimming and manoeuvring. The fins are also used for locomotion, and this is why cichlid fishes have red muscles designed for relatively low-speed but continuous movements (Ross, 2000). The numbers of scales, vertebrae, gill rakers and fin rays and spines are widely used for species distinction and identification. The fin formulas of different tilapias are presented in Table 2.1. It should be mentioned, however, that the number of fin spines and/or rays of the same species may vary from one strain to another and from one aquatic environment to another (see Table 2.1).

Tilapia bodies are generally characterized by vertical bars, with relatively subdued colours and with little contrast over the body colours. This provides the fish with a modest ability to change their colours, in response to stress, by controlling skin chromatophores. Tilapia have well-developed sense organs, represented by prominent nares and a clearly visible lateral line. The eyes are also relatively large, providing the fish with an excellent visual capability.

2.4. Geographical Distribution

Tilapias are a freshwater group of fish species originating exclusively from Africa (excluding Madagascar) and from Palestine (Jordan Valley and coastal rivers) (Philippart and Ruwet, 1982). They are distributed all over Africa, except the northern Atlas Mountains and south-west Africa (McAndrew, 2000). Outside Africa, they are also widely distributed in South and Central America, southern India, Sri Lanka (Philippart and Ruwet, 1982) and Lake Kinneret, Israel. Tilapias also

Table 2.1. Fin formulas of various tilapia species.

Species	Location	Dorsal fin	Pectoral fin	Anal fin	Reference
Tilapia mariae	Umuoseriche Lake, Nigeria	XV–XVII, 10–14	11–15	9–12	Anene (1999)
Tilapia zillii	Umuoseriche Lake, Nigeria	XIII–XVII, 11–13	12–14	8–10	Anene (1999)
	Lake Edku, Egypt	XIV–XVI, 10–13	12–14	8–10	Abdalla (1995)
	Lake Mariut, Egypt	XIV–XVI, 9–13	12–14	7–9	Akel (1989)
Oreochromis niloticus	Lake Mariut, Egypt	XV–XVIII, 10–13	12–15	8–10	Akel (1989)
	Lake Edku, Egypt	XVI–XVIII, 12–14	12–15	9–10	Bakhoum (2002)
Oreochromis aureus	Lake Mariut, Egypt	XVI–XVII, 10–14	11–14	8–11	Akel (1989)
	Lake Edku, Egypt	XV–XVII, 12–13	12–14	9–10	Bakhoum (2002)
Sarotherodon galilaeus	Lake Mariut, Egypt	XV–XVIII, 11–13	11–14	9–11	Akel (1989)

inhabit a wide range of ecosystems (see Chapter 3 for details). They seem to have evolved as riverine fishes living in marginal waters and flood-plain pools, but they have adapted to lacustrine conditions. This explains why they currently live in various ecological water systems, including slow-moving rivers and their flood-plain pools and swamps, small shallow lakes, large deep lakes, impounded water bodies, isolated crater lakes, soda lakes, thermal springs and brackish-water lakes (Philippart and Ruwet, 1982; Lowe-McConnell, 2000). These fish are also highly adaptable to their environments, as reflected by their tolerance to a wide range of environmental conditions, such as temperature, salinity, dissolved oxygen and ammonia.

Philippart and Ruwet (1982) suggested that the natural distribution of the two tilapia genera, *Tilapia* (substrate-spawners) and *Oreochromis* (mouthbrooders), is a reflection of two types of factors:

1. Historical–geological factors, which have led to geographical isolation and speciation (endemic species in lakes or stretches of rivers).
2. Ecological factors, which represent the requirements of and preferences for various environmental conditions, such as temperature, salinity and water composition, in addition to the behavioural characteristics that reflect feeding and reproduction patterns.

Based on the historical factors, the genus *Tilapia* is widely distributed in west and central Africa, but not in the eastern slope of the eastern Rift Valley and the river basins flowing into the Indian Ocean north of Zambesi River. These fish species are separated by ecological or behavioural barriers rather than by geographical or hydrographic barriers (Philippart and Ruwet, 1982).

On the other hand, all species of the genus *Sarotherodon*, except *S. galilaeus*, are restricted to West Africa. *Sarotherodon galilaeus* has spread eastwards towards the Nile and the first Rift lakes. Meanwhile, the genus *Oreochromis* is widely distributed in the Rift Valley lakes and rivers and the rivers that drain into the Indian Ocean, but it is rare in western Africa. *Oreochromis niloticus* and *Oreochromis aureus* are distributed in the Nilo-Sudanian region. Moreover, *O. niloticus* is spreading eastwards into the Ethiopian Rift Valley and has moved southwards, colonizing all the western Rift lakes (Lake Albert, Lake George, Lake Edward, Lake Kivu and Lake Tanganyika) and Lake Turkana in the eastern Rift Valley. This species is also spreading in central and western Africa, via the Chad and Niger basins.

It appears from the above discussion that *Tilapia* and *Sarotherodon* species are more localized in West Africa, while *Oreochromis* species are more distributed in the central and eastern African regions. However, some species, such as *Tilapia zillii*, *S. galilaeus*, *O. niloticus* and, to a lesser extent, *O. aureus*, have a larger and overlapping distribution. *Tilapia zillii* and *S. galilaeus* are distributed far south to Lake Albert, suggesting that the Chad–Nile connection that enabled them to inhabit the Nile River must have occurred after *O. niloticus* had spread southwards and after the disappearance of the connection between Lake Albert and southerly lakes (McAndrew, 2000).

2.5. Factors Affecting Tilapia Distribution

2.5.1. Habitat diversity

Habitat diversity is one of the major factors behind the wide distribution of tilapia. In Africa, for example, tilapia inhabit a wide range of ecologically and geographically different habitats, such as permanent and temporary rivers, rivers with rapids, large equatorial rivers (Zaïre), tropical and subtropical rivers (Nile, Niger, Senegal, Zambezi, Limpopo), shallow, swampy lakes (Bangweulu, Victoria, Kyoga, Rukwa, Chad), deep lakes (Albert, Kivu, Tanganyika, Malawi), human-made lakes (Nasser, Nobia), alkaline and saline lakes (Magadi, Natron, Manyara, Mweru Wantipa, Chilwa, Chiuta, Turkana, Tana, Qarun), hot springs (Magadi), crater lakes (Chala, Barombi, Mbo, Barombi ba Kotto, Bosumtwi), lakes with low mineral contents (Bangweulu, Nabugabo), acidic lakes (Tumba), open and closed estuaries, lagoons, coastal brackish lakes (Maryut, Edku, Borullos, Manzalla), marine habitats (Gulf of Suez, Alexandria West Harbour, Abu Qir Bay).

2.5.2. Environmental conditions

The above-mentioned habitat diversity also represents a wide range of environmental conditions, including physical parameters (temperature, photoperiod, depth, current velocity, turbidity, etc.), chemical parameters (salinity, pH, dissolved oxygen, mineral and gas contents) and biological factors (competition, food availability, productivity, etc.). It is no surprise, therefore, that tilapia are able to tolerate an extraordinarily varied range of these environmental conditions. The environmental requirements of tilapia are covered in detail in Chapter 3.

2.6. Introductions and Transfers

2.6.1. Introductions in Africa

Several tilapia species have been introduced into various African regions, outside their natural geographical habitats, either within the same basin or from one basin to another. These introductions have been into natural habitats and/or reservoirs. Philippart and Ruwet (1982) summarized the objectives of tilapia introductions as follows:

- Stocking natural lakes which are not inhabited naturally by tilapia (such as *Sarotherodon alcalicus grahami* into Lake Nakuru, *Oreochromis spilurus niger* and *T. zillii* into Lake Naivasha, *Oreochromis macrochir* and *Tilapia rendalli* into Lake Lusiwashi).
- Introductions into natural habitats to fill ecological niches that are not occupied by any of the tilapias present, to increase fisheries yields (such as *T. zillii* and *O. niloticus* into Lake Victoria and Lake Kyoga).
- Introductions into natural water bodies to develop new tilapia-based fisheries (such as *Oreochromis mossambicus* and *O. niloticus* in the reservoir of a southern Tunisian oasis).
- Introductions into natural water bodies to consume plankton production (*O. macrochir* into Lakes Kariba and McIlwaine (Zimbabwe), *O. mossambicus* and *Sarotherodon mortimeri* into the lakes of the Zimbabwe eastern highlands).
- Biological aquatic weed control (such as *T. rendalli* into Sudanese irrigation canals) and the control of mosquitoes.
- Tilapia culture in rice fields, cages and ponds.
- Accidental introductions occurring during the deliberate introduction of other tilapia species (such as *T. rendalli* into Lakes Victoria and Kyoga) or because of the confusion between sympatric species that have begun to differentiate (such as *O. niloticus* and *O. aureus*).

It should be emphasized that, because of the old, unrecorded introductions of tilapia, it is almost impossible in several cases to tell whether the presence of a given species is natural or introduced by humans, accidentally or deliberately. More details on the introduction of tilapia in Africa and their environmental impacts are provided in Chapter 12.

2.6.2. Introductions outside Africa

Tilapias have been introduced into many countries in Asia, Europe, the Americas, Australia and Oceania. The first introduction of tilapia outside Africa was probably the unintentional introduction of *O. mossambicus* into the Indonesian island of Java prior to 1939, where a few individuals were caught in the Serang River (Atz, 1954). Subsequently, the species was further introduced to many other countries in South and South-east Asia. Other economically important tilapia species, such as *O. niloticus*, *O. aureus*, *Oreochromis hornorum*, *S. galilaeus*, *T. zillii* and *T. rendalli*, have also been introduced into various countries (Welcomme, 1988; Tan and Tong, 1989). Various tilapia species currently inhabit natural waters throughout the tropical and subtropical regions. They have also been established in brackish water (China, Philippines), marine waters (China), salt lakes (Salton Sea in California, USA), mangrove swamps (Madagascar, South Pacific and Micronesia) and quarry lakes (Puerto Rico). There are overwintering populations established in power-plant cooling lakes in Texas and *O. niloticus* is established in Pearl Harbour, Hawaii, USA. In Hawaii, they are thought to be escapees from introductions to that state for aquaculture. *Tilapia rendalli* are used for weed control in California (R. Stickney, Texas, 2004, personal communication).

As in the case of Africa, the introduction of tilapia outside Africa was mainly for aquaculture, the stocking of natural habitats, human-made lakes and reservoirs and mining pits and the control of aquatic vegetation (Philippart and Ruwet, 1982). They have also been introduced to stock cooling waters and geothermally heated waters in a number of temperate regions (Kuroda, 1977).

In many cases, the introduction of tilapia, both in and outside Africa, was successful, due to the existence of empty ecological niches, the presence of ecologically complementary species and good acclimatization. The following cases are examples of the successful introduction of tilapia: *T. rendalli* and *O. macrochir* into Lake Lusiwashi (Zambia), *O. niloticus* into Koki lakes (Uganda), *O. mossambicus* throughout South and South-east Asia and several South American countries, *O. niloticus* into many South-east Asian countries (Indonesia, Thailand, Bangladesh, the Philippines), *Tilapia* and *Oreochromis* in the lakes, reservoirs, streams, rivers, mangroves and shallow lagoons on many islands of the South Pacific and Micronesia and *T. zillii* into Madagascar. In most of these cases, tilapia introductions have led to dramatic improvements of inland fisheries and significant development of subsistence and commercial aquaculture, in addition to mosquito control and aquatic weed control.

In many other cases, tilapia introductions have been unsuccessful, or have been successful but have caused tremendous negative ecological impacts, ranging from habitat destruction and hybridization with endemic species to the disappearance of native species. For example, the introduction of *O. spilurus niger* into Lake Bunyoni (Uganda) and *Oreochromis esculentus* into Uganda was a complete failure, while the introduction of *O. niloticus* was successful, but hybridization of *O. niloticus* with the two other introduced tilapia species occurred (Lowe-McConnell, 1958). *Oreochromis niloticus* populations were also subject to retarded growth, infestation by parasites and poor fishery yields (Beadle, 1981). Similarly, the introduction of *T. rendalli* into Madagascar and Mauritius has caused serious disturbance to the indigenous fauna and flora in the lakes, rivers and reservoirs (George, 1976). The environmental impacts of tilapia are further explored in Chapter 12.

2.7. Feeding Habits

The increasing importance of tilapia as an aquaculture candidate makes it necessary to understand their food preferences and feeding regimes in their natural habitats, in order to prepare suitable diets for them and adopt appropriate feeding regimes under culture conditions. Tilapia are generally herbivorous/omnivorous (i.e. they are low on the aquatic food chain) (Table 2.2). This characteristic is extremely important for the economics of tilapia culture. Although they feed at low trophic levels and feed costs are lower than for carnivorous fishes, tilapia are still a source of high-quality protein suitable for human consumption, at a relatively low cost. The feeding habits and dietary preferences of tilapias depend, among other factors, on tilapia species and size, time of day, photoperiod, water depth and geographical location. The feeding habits of tilapia in relation to these factors are briefly discussed in this section, with emphasis on economically important species.

Table 2.2. Feeding habits of some economically important tilapias.

Species	Diet	Remarks	Reference
O. niloticus	Phytoplankton, zooplankton, benthic detritus	Little selectivity of food items. The fish change food and feeding habit with size. Increased preference for debris with increasing size	Moriarty and Moriarty (1973), Saha and Dewan (1979), Harbott (1982), Northcott *et al.* (1991)
O. aureus	Phytoplankton, zooplankton, detritus, vascular plant residues	Food preferences were detected in juveniles and adults	Spataru and Zorn (1978), Jimenez Badillo and Nepita Villanueva (2000)
O. mossambicus	Macrophytes, benthic algae, phytoplankton, zooplankton, detritus, periphyton, fish larvae	Intraspecific variations in diet between environments and over time	Bowen (1979), De Silva *et al.* (1984), de Moor *et al.* (1986)
O. mossambicus × *O. niloticus*	Periphytic detritus, filamentous and colonial algae (periphytic epipelon)	Intraspecific variations in diet between environments	Haroon *et al.* (1998)
Florida red tilapia	Macroalgae, phytoplankton	Variations in diet between environments	Grover *et al.* (1989)
S. galilaeus	Phytoplankton (dinoflagellates)	Fish are highly selective	Corbet *et al.* (1973), Spataru (1976)
Sarotherodon melanotheron	Macrophytes, blue-green and green algae, organic debris, aquatic insects and insect larvae, rotifers, diatoms, benthic invertebrates	Plant materials more preferable than animal sources	Fagade (1971), Oben *et al.* (1998)
T. zillii	Macrophytes, blue-green and green algae, organic debris, zooplankton, aquatic insects and insect larvae and pupae, rotifers, diatoms, benthic invertebrates, arthropods	Plant materials more preferable than animal sources. Variations in diet between seasons	Spataru (1978), Buddington (1979), Oben *et al.* (1998)
T. rendalli	Macrophytes, attached periphyton, zooplankton (cladocerans and copepods), insect larvae, fish eggs, detritus	Feeding behaviour depends on fish size. Small fish are visual feeders, large fish are pump-filter feeders	Caulton (1976), Lazzaro (1991)
T. guineensis	Algae, detritus, sand, invertebrates		Fagade (1971)

During larval stages, tilapia feed initially on zooplankton, especially crustaceans (copepods) (Bowen, 1982; Harbott, 1982). When Florida red tilapia fry were stocked in fertilized seawater pools containing different food resources, copepods were selectively ingested (Grover *et al.*, 1989). Frequent consumption of phytoplankton also takes place during early larval stages of some tilapia species.

The transition period from planktivore stage to a typical, specialized diet(s) is usually short (Bowen, 1976), but in some cases it occurs gradually over an extended period of a year or more (Whitefield and Blaber, 1978). The food of juvenile and adult tilapias consists of a considerable variety of aquatic vegetation, phytoplankton, zooplankton, periphyton and detritus of plant origin, depending on tilapia species, as mentioned earlier. Many research studies have been carried out on the feeding habits of different tilapias in their natural habitats, under a variety of environmental conditions. Generally speaking, tilapia can be classified into one of the following broad categories according to their feeding patterns:

1. Tilapia of the genus *Oreochromis* are primarily microphagous, feeding mainly on phytoplankton, periphyton and detritus. *Oreochromis niloticus*, *O. aureus* and *O. mossambicus* are examples of this category. These species can efficiently ingest the food sources mentioned through 'filter-feeding'. For example, Harbott (1982) reported that all size classes of *O. niloticus* in Lake Turkana are primarily phytophagous, grazing on the dense algal plankton found in sheltered, inshore waters, while little selectivity of food items was recorded. Similarly, Spataru and Zorn (1978) found that *O. aureus* in Lake Kinneret are mainly zooplankton feeders, while vegetable detritus, mixed with plankton and benthos forms, served as additional and alternative foods.

2. Tilapia of the genus *Sarotherodon* are also primarily phytoplankton feeders, but they are more selective. For example, Spataru (1976) found that the dinoflagellate *Peridinium cinctum* was the most abundant food in the stomachs of *S. galilaeus* in Lake Kinneret, sometimes comprising > 95% of the phytoplankton biomass, especially during the blooming season of these dinoflagellates (March–April).

3. Tilapia of the genus *Tilapia* are generally macrophyte feeders (Abdel-Malek, 1972; Caulton, 1976; Lowe-McConnell, 1982). This explains why they are used for biological control of aquatic weeds. However, they cannot avoid ingesting the algae, phytoplankton, zooplankton, bacteria, benthic invertebrates, insect larvae, fish and vertebrate eggs and detritus that are attached to the macrophytes they feed on. These attached materials are therefore an important food component for *Tilapia* species (Bowen, 1982). Fagade (1971) found that the principal components in the stomachs of *Tilapia guineensis* and *Tilapia melanotheron melanotheron* in Lagos Lagoon (Nigeria) were algal filaments, diatoms, sand grains and unidentified organic material, and these components were very similar in both species. Adult *T. rendalli* have also been found to consume a wide variety of food items, including macrophytes, phytoplankton, zooplankton, insect larvae, fish eggs and larvae and detritus (Caulton, 1976; Lazzaro, 1991).

It should be realized that the above-mentioned categories of feeding habits are not rigid, but extensive overlap may also occur in tilapia diets. Intraspecific variations may also occur in these diets between environments over time, as has been reported with *O. mossambicus* in Sri Lankan reservoirs (De Silva, 1985a).

Since light penetration into the water column is affected by water depth and turbidity, natural food productivity in the water is likely to vary between different depths. Accordingly, the feeding patterns of tilapia may vary with varying water masses. Bowen (1982) stated that deposit feeders that feed in the littoral zone frequently ingest algae, detritus and bacteria, which cannot be distinguished from the material attached to macrophytes. On the other hand, deposit feeders that feed in deeper waters ingest a mixture of precipitated phytoplankton and detritus that are indistinguishable from the diets of suspension feeders. By this mechanism, tilapia can switch from one source of food to another with little change in diet composition (Moriarty *et al.*, 1973).

Tilapia feeding patterns and efficiency may also be subject to diurnal and seasonal changes. Extensive studies have considered the diel feeding patterns of tilapia, both in their natural habitats and under aquaculture conditions. Harbott (1982) found that all size classes of *O. niloticus* in Lake Turkana followed a daytime feeding cycle, but fish fry infrequently consumed invertebrates during the hours of darkness. The feeding activity of *O. niloticus* adults reared under conditions of self-feeding (using demand feeders) was also observed exclusively during the light phases (Toguyeni *et al.*, 1997). Moreover, Nile tilapia in Lake Rudolf exhibited a regular diurnal feeding rhythm, commencing between 05.00 h and 08.00 h and ceasing between 14.00 h and 18.00 h, and the largest fish individuals appeared to feed longer (Harbott, 1975). On the other hand, Nile tilapia in Bangladeshi ponds are reportedly continuous feeders, but they increase feeding activity from noon to midnight, with a peak immediately after dusk and reduced feeding activity after midnight (Dewan and Saha, 1979).

Fish size may also affect feeding rhythms in tilapia. For example, Richter *et al.* (1999) analysed the diel feeding patterns and daily ration of *O. niloticus* in cages in Laguna de Bay, Philippines. They found that, in May, larger fish (31.5 g) feeding on natural food alone fed continuously from dawn to dusk, whereas smaller fish (9.8 g) exhibited two feeding periods per day, from sunrise to mid-morning and again from mid-afternoon until after sunset. In another study, Haroon *et al.* (1998) investigated the diel feeding patterns and ration of tilapia hybrids (*O. mossambicus* × *O. niloticus*) at two sizes (4.3–9.3 and 9.5–13.8 cm total length) in pond and paddy field in Bangladesh. Both small and large tilapia exhibited a single feeding peak around afternoon–dusk in the pond, but irregular feeding peaks in the rice field. However, small fish were more active feeders than large fish, and both feeding intensity and food consumption decreased with increasing fish size. Periphytic detritus was the principal food in the paddy field, while filamentous and colonial algae attached to sediments and mud (periphytic epipelon) were the main food in the pond. On the other hand, zooplankton was an insignificant dietary component in both habitats.

The feeding habits of *T. rendalli* in southern Brazilian lakes and reservoirs in relation to fish size has also been studied (Lazzaro, 1991). Fish smaller than 30 mm in length were visual feeders, while fish with a size ranging from 30 to 50 mm were either visual feeders or pump–filter feeders, depending on zooplankton size. Fish larger than 70 mm were pump–filter feeders. Electivity and feeding rate increased with prey length. A difference in food preferences between juvenile and adult *O. aureus* in Mexico was also detected, in addition to a variation in the consumption of some food items during the rainy and dry seasons (Jimenez Badillo and Nepita Villanueva, 2000).

The time of year may also have an influence on tilapia feeding habits. Spataru (1978) found that *T. zillii* in Lake Kinneret have great variations in their food in relation to season of the year. In winter and spring, *Chironomida* pupae (Diptera) were prevalent in fish stomachs, while zooplankton forms were dominant in summer and autumn. Spataru and Zorn (1978) also found that *O. aureus* in the same lake feed more intensely on zooplankton in spring, when zooplankton forms are more abundant. Nile tilapia also showed seasonal variation in feeding activity in Bangladeshi fish ponds (Dewan and Saha, 1979). The feeding activity was greater in summer than in winter. Greater amounts of phytoplankton were recorded in fish stomachs during winter, while debris was consumed more during summer.

From the above discussion, it appears that tilapia have irregular feeding patterns, depending, as mentioned earlier, on fish species and size, season of the year, time of the day, photoperiod, water depth, geographical location and type of habitat. It is crucially important that tilapia farmers, researchers and farm managers consider these factors in order to adopt the most suitable fertilization regimes, prepare the best pelleted diets and select the most convenient feeding regimes in terms of feeding frequency, timing and quantity.

2.8. Gut Morphology

The digestive system of tilapias is characterized by certain modifications reflecting the types of food they ingest. It is relatively simple and unspecialized, consisting of a very short oesophagus connected to a small, sac-like stomach, and a very long, coiled intestine, which can reach up to seven to 13 times the total fish length (Caulton, 1976; Balarin and Hatton, 1979). Such long intestines reflect the herbivorous feeding habits of these fish, perhaps because vegetable foods are digested less readily than animal sources. Therefore, herbivorous fish have much higher amylase activities in the gut than carnivorous species.

Tilapia have two types of teeth: jaw teeth and pharyngeal teeth. Both types of teeth are varied in configuration from one tilapia species to another, to suit the different diet preferences. The jaw teeth are small, unicuspid, bicuspid or tricuspid structures, arranged in one to five rows and flattened distally to form blades that can be used as scrapers

(Fryer and Iles, 1972a; Lanzing and Higginbotham, 1976). The pharyngeal teeth of the phytoplanktivorous tilapia, such as *Sarotherodon esculantus*, are fine, thin and hooked on the pharyngeal bones, whereas those of macrophyte feeders, such as *T. rendalli*, are coarse and robust (Caulton, 1976). Microphagous tilapia are also characterized by numerous, long, thin, closely spaced gill rakers, while fish species that feed on large particles have fewer, large gill rakers. While gill rakers are specialized for feeding in many tilapia species, their role in feeding mechanism in some other tilapia species may not be significant. Drenner *et al.* (1987) found that the surgical removal of gill rakers did not affect phytoplankton ingestion rates of *S. galilaeus*. However, it may be unwise to adopt a firm conclusion based on only one study. Clearly, more work should be conducted on different tilapia species that employ different feeding strategies before a concrete conclusion can be drawn on the role of gill rakers in feeding mechanisms of tilapia.

2.9. Closing Remarks

1. The taxonomic classification of tilapia is still confusing and subject of continuous modification. Currently, tilapia are divided into three genera, *Tilapia* (substrate spawners), *Sarotherodon* (maternal/paternal mouthbrooders) and *Oreochromis* (maternal mouthbrooders).

2. Tilapias are freshwater groups of fish species originated exclusively from Africa (excluding Madagascar) and from the Middle East (Jordan Valley and coastal rivers). They are distributed all over Africa, except the northern Atlas Mountains and south-west Africa.

3. The genus *Tilapia* is widely distributed in West and Central Africa, but not in the eastern slope of the eastern Rift Valley and the river basins flowing into the Indian Ocean north of the Zambezi River.

4. The genus *Sarotherodon*, except *S. galilaeus*, is restricted to West Africa, while *S. galilaeus* spread eastwards towards the Nile and the first Rift lakes.

5. Fishes of the genus *Oreochromis* are distributed more in the Central and Eastern African regions, but they are rare in western Africa.

6. Outside Africa, tilapia are widely distributed in South and Central America, southern India, Sri Lanka and Lake Kinneret (Israel).

7. Several tilapia species have been introduced into many countries, both in Africa and outside Africa, for various objectives. Many introductions were successful, but many others were unsuccessful or successful but had negative side effects.

8. Feeding and dietary preferences of tilapias depend on tilapia species and size, time of day, photoperiod, water depth and location, with little selectivity.

9. Tilapia of the genus *Oreochromis* are primarily microphagous, feeding mainly on phytoplankton, periphyton and detritus.

10. Tilapia of the genus *Sarotherodon* are also primarily phytoplanktivorous, but they are more selective.

11. Tilapia of the genus *Tilapia* are generally macrophyte feeders. They also ingest algae, phytoplankton, zooplankton, invertebrate larvae, fish eggs, insects, bacteria and detritus that are attached to the macrophytes.

12. Feeding patterns and efficiency of tilapia may be subject to diurnal and seasonal changes. Generally, tilapia feed intensively during early morning and late afternoon and reduce feeding at midday and at night.

13. The digestive system of tilapias is relatively simple and unspecialized, consisting of a very short oesophagus, a small, sac-like stomach and a very long intestine. Such a digestive tract reflects the herbivorous feeding habits of tilapia.

14. Tilapia have two types of teeth, jaw teeth and pharyngeal teeth. Both types of teeth vary from one tilapia species to another to suit the different diet preferences.

15. Microphagous tilapia have numerous, long, thin, closely spaced gill rakers, while tilapia that feed on large particles have fewer, large gill rakers.

3
Environmental Requirements

3.1. Introduction

Tilapia are generally cultured semi-intensively or intensively, under different culture and environmental conditions, stocking densities and management strategies. Water quality is very likely to be subjected to continuous changes, especially in intensive or super-intensive culture systems. Thus, water quality management becomes a key factor for successful aquaculture practices. It is necessary therefore to understand the major water quality parameters and their interrelationships, which affect fish growth and health and determine the failure or success of overall culture practices. Such understanding will enable fish farmers and farm managers to determine the chemical, biological and physical processes that take place in tilapia farms, and, in turn, to take the proper and correct supporting managerial decisions. In the meantime, tilapia have been introduced to many countries across the globe, mainly for aquaculture. The introduction of tilapia into many countries where the environmental conditions are outside their tolerance limits has made the traditional land-based systems unsuitable for their culture. Tilapia culture will certainly be affected by the varying environmental conditions in these geographically different regions.

This chapter discusses and analyses the major environmental factors affecting tilapia in the wild or under aquaculture conditions. Special emphasis will be given to water temperature, salinity, dissolved oxygen, ammonia and nitrites, pH, photoperiod and water turbidity.

3.2. Temperature

Temperature is one of the most important factors affecting the physiology, growth, reproduction and metabolism of tilapia. Temperature is of prime importance in temperate and subtropical regions, which are characterized by seasonal fluctuations in water temperature. Tilapia are thermophilic fish and known to tolerate a wide range of water temperatures. Extensive research has been conducted on the effects of water temperature on tilapia performance. The temperature range for the normal development, reproduction and growth of tilapia is about 20 to 35°C, depending on fish species, with an optimum range of about 25–30°C (Balarin and Haller, 1982; Chervinski, 1982; Philippart and Ruwet, 1982). However, a tremendous difference in the growth and feed efficiency of tilapia may occur even within this narrow range of water temperatures. In a recent study, A.-F.M. El-Sayed and M. Kawanna (unpublished results) evaluated the effects of three water temperatures (24, 28 and 32°C) (lying within the optimum range of tilapia tolerance) on the growth and feed conversion of Nile tilapia fry, reared in an indoor, recirculating system. The growth of the fish at 28°C was almost double the growth at 24 and 32°C (Fig. 3.1). Fish performance at 24 and 32°C was not significantly different. It is clear that optimum water temperature (not only optimum temperature range) is essential for maximum fish growth.

Tilapia can also tolerate temperature as low as 7–10°C, but only for brief periods (Balarin and Haller, 1982; Chervinski, 1982; Jennings, 1991; Sifa *et al.*, 2002). Longer exposure of tilapia to

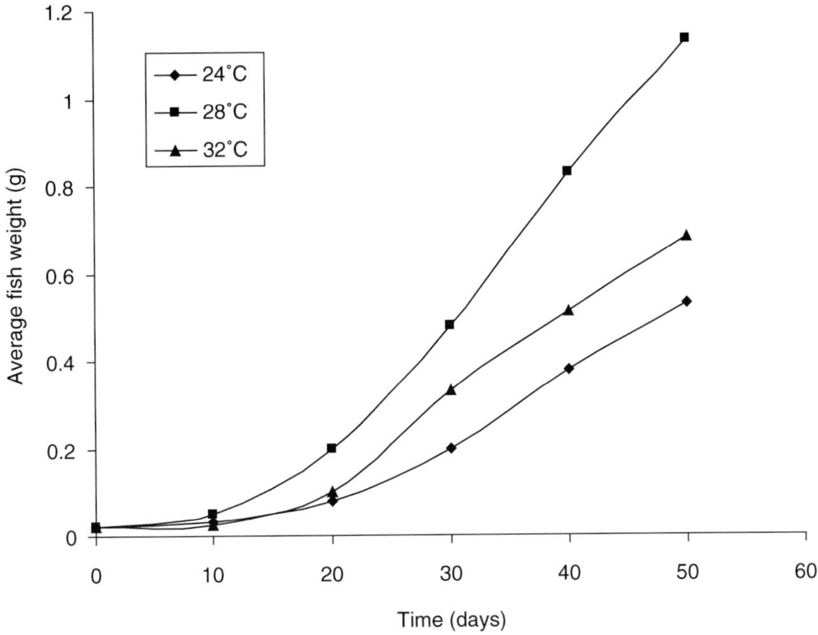

Fig. 3.1. Effect of water temperature on weight gain of Nile tilapia fry reared in an indoor, closed system (A.-F.M. El-Sayed and M. Kawanna, unpublished results).

these low temperatures will certainly lead to mass mortality. Tilapia feeding is sharply reduced below 20°C, and they stop feeding at about 16°C, while severe mortality occurs at 12°C (Balarin and Haller, 1982; Chervinski, 1982). Of course, there are some variations among tilapia species in response to these temperature limits, as shown in Table 3.1.

Contrary to the limited lower temperature tolerance of tilapia, they can tolerate relatively high water temperatures. The upper lethal temperature limits for tilapia vary from one species to another, but it has been reported that most tilapias cannot tolerate water temperatures above 40–42°C for a long time (Morgan, 1972; Hauser, 1977; Balarin and Haller, 1982; Chervinski, 1982). The response of tilapia to varying water temperatures has been reviewed by Balarin and Hatton (1979) and Chervinski (1982). The influence of temperature on tilapia depends on species, strain and size, duration of exposure, other environmental factors, culture systems and geographical location. These factors will be discussed in this section.

Since different tilapia species and strains have been introduced to many different geographical regions of the world, their response to water temperature in their new habitats requires prime attention. It has been realized that the further the geographical location from the equator, the more tolerant are Nile tilapia strains to cold (Khater and Smitherman, 1988; Sifa *et al.*, 2002). Sifa *et al.* (2002) related this phenomenon to 'natural selection' processes. These authors also found that Sudanese and Egyptian strains of Nile tilapia had better cold tolerance in China than genetically improved farmed tilapia (GIFT). They attributed this behaviour to the following:

1. GIFT were introduced into China recently and had not been exposed to selective pressures for low water temperature for enough time.
2. The Sudanese strain, introduced into China several years ago, had been subjected to cold tolerance and in turn had gone through some natural selection response to low water temperature.
3. GIFT might have been contaminated with *Oreochromis mossambicus*, which have poor cold tolerance.

It is evident therefore that newly introduced tilapia have lower cold tolerance than those which have been introduced and adapted to local cold temperature for a longer time.

Table 3.1. Temperature tolerance ranges of tilapia.

Species	Tolerance (°C) Lower lethal	Upper lethal	Optimum	Reference
Oreochromis niloticus	10.5	42	27–30	Denzer (1968), Fukusho (1968), Beamish (1970)
O. niloticus			28	A.-F.M. El-Sayed and M. Kawanna (unpublished data)
O. niloticus	8.28		28–32	El Gamal (1988)
O. niloticus:				Sifa *et al.* (2002)
GIFT	8.4–11			
Sudan strain	7.4–9.8			
Egypt strain	7.4–11			
O. niloticus:				Khater and Smitherman (1988)
Egypt strain	10			
Ivory coast strain	12.2			
Ghana strain	14.1			
Oreochromis aureus	6–8	41–42		Yashouv (1960), Philippart and Ruwet (1982), El Gamal (1988)
Oreochromis mossambicus	8–9.5	42		Kirk (1972), Chervinski (1982), Shafland and Pestrak (1982)
			28–30	Job (1969a)
Sarotherodon melanotheron	18	33		Pauly (1976)
	6.9–10.3			Shafland and Pestrak (1982), Jennings (1991)
			32	Philippart and Ruwet (1982)
Tilapia zillii	6.5	42.5	28.8–31.4	Kirk (1972), Platt and Hauser (1978), Philippart and Ruwet (1982)
Tilapia rendalli	11	41		Philippart and Ruwet (1982)
O. aureus × *O. mossambicus*	10			Behrends and Smitherman (1984)
Florida red tilapia			27 (at 0‰) 32 (at 18 or 36‰)	Watanabe *et al.* (1993a)

GIFT, genetically improved farmed tilapia.

The conditioning of tilapia at an intermediate temperature before their exposure to lower temperatures may improve fish survival and tolerance to cold stress. Hofer and Watts (2002) reared genetically male tilapia (GMT) at 28°C and exposed them to reduced temperatures of 20, 18, 16, 14, 13 or 12°C for 24 h and then returned them to 28°C. They found that 43% of the fish died at 14°C and all of them died at 12°C. When the fish were preconditioned at 20°C, their survival was 72 and 23% when they were exposed to 14 and 12°C, respectively. In a similar study, Chervinski and Lahav (1976) found that *Oreochromis aureus* acclimatized to 28°C for 2 weeks started to die at 11°C, while fish acclimatized to 18°C for the same period began to die at 9°C.

Fish species and size play a significant role in tilapia response to water temperature. Small fingerling tilapia were more susceptible to low temperature than larger fish (Hofer and Watts, 2002). The survival of large monosex Nile tilapia fry during overwintering has also been reported to be higher than that of smaller fry (Dan and Little, 2000a). This means that large fingerlings are preferred for transportation and culture if water temperature is expected to drop, because of their higher cold tolerance. However, Cnaani *et al.* (2000) found that cold tolerance and fish size (2.3–10.5 cm) of *O. mossambicus* and *O. aureus* and their hybrids were not correlated, but *O. mossambicus* was the most cold-sensitive group, and genetic variations in cold tolerance had a large dominance component. Behrends *et al.* (1996) also found that *O. aureus* and their hybrids were more cold-tolerant and had better growth than Nile tilapia when raised at suboptimal temperatures (17.2–21.2°C). In addition, Chen *et al.* (2002) found that cold stress in *O. aureus* led to an increase in the secretion of plasma noradrenaline, adrenaline and cortisol and a decrease in both phagocytic activity of leucocytes and plasma immunoglobulin. These fish can tolerate temperature as low as 6–7°C, but for only a few hours. Long exposure to such low temperature renders the fish unable to maintain body position (Yashouv, 1960).

The effects of culture systems and environmental conditions on tilapia response to water temperature are also evident. El-Sayed *et al.* (1996) found that the growth and survival of Nile tilapia reared in earthen ponds were significantly affected by pond depth and water temperature. Fish growth was poorest and mortality highest at a pond depth of 50 cm, whereas the best performance and survival were achieved at 100–200 cm depth. At 50 cm pond depth, 85% of the fish died, while increasing the depth to 200–300 cm provided the fish with a refuge for escaping extreme temperatures during summer (escaping towards the bottom) and winter (escaping towards the surface) seasons, leading to a significant improvement in fish growth and survival. Similarly, Dan and Little (2000b) overwintered Thai, GIFT, Egyptian and Vietnamese strains of Nile tilapia broodstock in deep and shallow ponds and in shallow and deep hapas suspended in deep ponds. They found that the survival was 99.6 and 100% in deep and shallow hapas in ponds, which was significantly higher than that of fish stocked in deep and shallow ponds (74.4–90%).

The effects of water salinity on the response of tilapia to water temperature have also been investigated by a number of authors. Allanson *et al.* (1971) found that *O. mossambicus* tolerated 11°C at 5‰, while fish reared in fresh water did not survive at that temperature. These authors suggested that the ability of *O. mossambicus* to tolerate low water temperature is associated with the maintenance of high plasma sodium and chloride concentration. The effect of salinity on the growth and survival of tilapia at different temperatures is also evident. Watanabe *et al.* (1993a) found that sex-reversed Florida red tilapia attained their best performance at 27°C when water salinity was 0‰. At 18 and 36‰, the best growth and feed efficiency were reached at 32°C. These results suggested that increasing water temperature above 27°C is only justifiable for tilapia reared in brackish water, not in fresh water. Similar trends have been reported in Nile tilapia (Likongwe *et al.*, 1996), where the growth of fish reared at 8, 12 and 16‰ increased with increasing water temperature from 24 to 32°C at all salinities. On the other hand, Jennings (1991) found that the cold tolerance of *Sarotherodon melanotheron* did not differ with water salinity ranging from 5 to 35‰.

Tilapia are often subjected to abrupt or gradual transfer from fresh water to brackish water and vice versa, during transportation and culture practice, presumably because of their eurythermal and euryhaline nature. Understanding the response of these fish to thermal and thermohaline shocks would be necessary for tilapia culture and farm managers in order to minimize thermal and thermohaline stresses, particularly during transportation. Al-Amoudi *et al.* (1996) studied the effects of thermal and thermohaline shocks on the tilapia *O. mossambicus* and *O. aureus* × *O. niloticus* hybrids transferred abruptly from 25°C fresh water to fresh water or salt water (26‰) maintaining temperatures of 15, 20, 30 and 35°C. In fresh water, fish survival was not affected by thermal shock, but they suffered from coma and fungal infection at 15°C. In salt water, the survival of *O. mossambicus* was not affected by thermohaline shocks, while tilapia hybrids maintained 7.6, 100, 100, 70 and 59% survival at 15, 20, 25, 30 and 35°C, respectively. Plasma osmotic concentrations of *O. mossambicus* and tilapia hybrids transferred to 15 and 35°C salt water were sharply increased, but gradually decreased to normal levels in *O. mossambicus*, while never returning to the initial low levels in tilapia hybrids. This study demonstrated that these fish were more tolerant to high rather than low temperature shocks, and *O. mossambicus* are more resistant to thermohaline shocks than tilapia hybrids.

One of the factors that tilapia farmers, farm managers and researchers may not be aware of is the significant effects of water temperature on sex differentiation and morphological development during early larval stages of tilapia. Increased water temperature may limit larval growth and induce body deformation. Wang and Tsai (2000) studied the effects of water temperature on deformity and sex differentiation of *O. mossambicus* larvae. Zero (day of hatch), 5- and 10-day old larvae were reared at 20, 24, 28 and 32°C. The authors found that the percentage of deformity increased at higher water temperatures (28 and 32°C) before 5 days old, while low temperature (20°C) had no effects. The exposure of the larvae to low temperature before 10 days old induced a high proportion of females, while elevated temperature induced a high percentage of males after 10 days old. The sex ratio of *O. niloticus* has also been found to be affected by water temperature. Baroiller *et al.* (1995) found that increasing water temperature to 34–36°C significantly increased the proportion of Nile tilapia males (69–91%), while low temperature (19–23°C) had no effects on sex ratio. The same authors reported similar results with Florida red tilapia. These results suggest that a temperature–sex determination pattern could exist in tilapia with strong genotype–temperature interaction. It is also essential that water temperature during early larval stages be carefully monitored and optimized.

The effects of temperature on survival, growth and phenotypic sex of mixed (XX–XY) progenies of Nile tilapia have been further investigated (Baras *et al.*, 2001). Fish larvae were exposed to different temperatures (20.4–39°C) during the first 28 days of their life. About 90% masculinizing efficiency was attained at 37.8–39.2°C, which was very close to the upper lethal temperature and in turn resulted in high mortality (> 70%). This result suggested that producing faster-growing Nile tilapia males at high temperature may not compensate the high loss of production incurred during masculinizing processes. Further work is needed to select the most thermosensitive breeders or strains.

3.3. Salinity

The competition for fresh water with agriculture and urban activities has increased the pressure to develop aquaculture in brackish water and seawater. The first candidate to think of for aquaculture in brackish water is tilapia. These fish, despite being freshwater fish, are believed to have evolved from marine ancestors (Kirk, 1972). This may explain the ability of most tilapia species and hybrids to tolerate a wide range of water salinity. They can normally grow and reproduce in brackish water. Some species can even grow and reproduce at very high water salinity. But salt tolerance does not necessarily indicate suitable conditions for maximum production. Extensive research has been conducted on the tolerance and adaptability of tilapia to water salinity. Available information indicates that salt tolerance depends on tilapia species, strains and size, adaptation time, environmental factors and geographical location (Balarin and Haller, 1982; Chervinski, 1982; Philippart and Ruwet, 1982; Suresh and Lin, 1992a). The results of these studies will just be outlined in this section (Table 3.2), while more detailed discussion will be devoted to the recent studies on this subject.

SPECIES, SEX AND SIZE. As indicated in Table 3.2, *O. mossambicus*, *O. aureus* and *Tilapia zillii* are the most salinity-tolerant tilapia species. *Oreochromis mossambicus* have been reported to tolerate up to 120‰ water salinity (Whitefield and Blaber, 1979). Moreover, they can grow normally and reproduce at a water salinity of 49‰, and their fry live and grow reasonably well at 69‰ (Whitefield and Blaber, 1979). *Oreochromis aureus* are less salinity-tolerant, but can grow well at a salinity of up to 36 to 44‰, while reproduction occurs at 19‰. With gradual acclimatization, they can tolerate a salinity of up to 54‰ (Balarin and Haller, 1982). However, McMahon and Baca (1999) found that *O. aureus* reared at a salinity of 30‰ developed a toxic algae bloom and suffered from epithelial erosion, caused by the dinoflagellate *Pfiesteria piscicida*, and brighter flashing on the dorsal and pectoral fins and tail. The authors also found that higher salinity severely restricted fish reproduction

Table 3.2. Salinity tolerance of tilapia (‰).

Species	Upper limit		Optimum limit	Remarks	Reference
	Direct transfer	Gradual transfer			
O. niloticus	18[1]	36[1]	5–10[2], 15[3]	Reproduce at 13.5–29‰[4]	[1]Al-Amoudi (1987a), [2]Payne and Collinson (1983), [3]Alfredo and Hector (2002), [4]Balarin and Haller (1982)
O. mossambicus	27[5]	120[6]	17.5[7]	Spawn at up to 49‰[8]	[5]Al-Amoudi (1987b), [6]Whitefield and Blaber (1979), [7]Canagaratnam (1966), [8]Popper and Lichatowich (1975)
O. aureus	27[5]	54[4]	10–15[9]	Reproduce at 5–20‰, low growth and high mortality at 36‰[10]	[9]Perry and Avault (1972), [10]McGeachin *et al.* (1987)
O. spilurus	33[4]	40[11]	3–8[11]	Good growth and survival in seawater, but low fecundity[11]	[11]Al-Ahmed (2001)
Sarotherodon galilaeus		29[4]	19[4]	Reproduce in the wild at 29‰[4]	
Tilapia rendalli		13–19[4]	0[12]		[12]Likongwe (2002)
T. zillii		45[13]		Grow and reproduce naturally at 10–> 30‰[14]	[13]Chervinski (1982), [14]A.-F.M. El-Sayed (personal observation)
O. niloticus × *O. mossambicus*		35[15]	15[15]	At 35‰, the fish failed to adapt	[15]Alfredo and Hector (2002)
Florida red tilapia			17.8[16]	Grow normally at up to 36.2‰	[16]El-Ebiary *et al.* (1997)

The superscript numbers in columns 2, 3, 4 and 5 correspond to the superscript numbers in column 6 to indicate from which reference the data are derived.

despite increasing somatic growth. However, the authors did not mention whether the algae bloom occurred as a coincidence of suitable conditions, with or without the presence of the fish, and whether it caused the problems that were observed. Similarly, McGeachin *et al.* (1987) found that *O. aureus* reared in seawater cages (36‰) showed a sharp reduction in growth rates and were infected with *Bacillus* sp., which led to severe mortality.

Tilapia zillii are among the most salinity-tolerant tilapia species. They are found in Bardaweel Lagoon (North Sinai, Egypt), Gulf of Suez, Lake Qarun (Fayyum Governorate, Egypt) and Abu Qir Bay (Alexandria, Egypt), which are highly saline water bodies (36–45‰) (Balarin and Hatton, 1979; A.-F.M. El-Sayed, Alexandria, Egypt, 1996, personal observation). *Tilapia zillii* can also reproduce at 29‰, and at even higher salinities in Lake Qarun (> 30‰). *Oreochromis spilurus* has also been successfully cultured in full seawater in Kuwait, and trials are going on in Israel for their commercial culture in seawater cages and tanks, with promising results (Carmelo, 2002).

Other tilapias are generally less euryhaline and can tolerate water salinities ranging from about 20 to 35‰ (see details in Table 3.2). Most of these tilapia grow, survive and reproduce at 0–29‰, depending on the species and acclimatization period. On the other hand, some other tilapia species are considered 'stenohaline', since they tolerate only a narrow range of water salinity. *Tilapia rendalli*, *Tilapia sparrmanii* and *Oreochromis macrochir*, for example, have a maximum salinity tolerance of 19, 18 and 20‰, respectively (Whitefield and Blaber, 1976; Balarin and Haller, 1982; Philippart and Ruwet, 1982).

The salinity tolerance of tilapia is also affected by fish sex and size. Perschbacher and McGeachin (1988) evaluated the salinity tolerance of red tilapia (*O. mossambicus* × *Oreochromis urolepis hornorum*) fry, juveniles and adults. Adult fish were more salt-tolerant than fry and juveniles. Fry and juveniles tolerated direct transfer to 19‰, without apparent stress and mortality, but 100% mortality occurred at 27‰. On the other hand, adult fish tolerated a direct transfer to 27‰, with 100% mortality at 37‰. Similarly, Watanabe *et al.* (1985) studied the ontogeny of salinity tolerance in Nile tilapia, blue tilapia and hybrid tilapia, *O. mossambicus* female × *O. niloticus* male. The median lethal salinity at 96 h (MLS-96) for Nile tilapia and blue tilapia over an age of 7–120 days post-hatching (dph) was 18.9 and 19.2‰. In contrast, the MLS-96 of tilapia hybrids changed with age and increased from 17.2‰ at 30 dph to 26.7‰ at 60 dph. The authors attributed these ontogenetic changes in salinity tolerance to body size rather than to chronological age. Watanabe *et al.* (1985) also reported that male tilapia tend to be more salt-tolerant than females.

It is also believed that tilapia hybrids descended from salt-tolerant parents (such as *O. mossambicus* and *O. aureus*) are highly salt-tolerant (Suresh and Lin, 1992a; Romana-Eguia and Eguia, 1999). This may explain why Taiwanese red tilapia (Liao and Chang, 1983) and Florida red tilapia (Watanabe *et al.*, 1988a, b) grow faster in seawater (SW) and brackish water (BW) than in fresh water (FW). Moreover, the tolerance of tilapia hybrids to salinity may be influenced by fish strain, as has been reported by Romana-Eguia and Eguia (1999). The authors studied the growth of five Asian red tilapia strains (BFS, NIFI, FAC, PF and HL) at 0, 17 and 34‰, for 10 weeks. The Philippine strain (PF) grew best in SW, while the Thai strain (NIFI) grew well in BW. The authors concluded that some Asian tilapia strains can be developed for use in sustainable brackish- and seawater culture systems. Tayamen *et al.* (2002) also evaluated progenies from 27 cross combinations of four *Oreochromis* species (five pure-bred and 22 cross-bred) for saline waters in the Philippines, in ten environments with different water salinities. They found that *O. aureus* × *O. spilurus* gave the highest body growth, while *O. mossambicus* × *O. spilurus* produced the best survival.

Water salinity has also been reported to affect the reproduction of tilapia. Gonadal development and spawning of Nile tilapia occurred at salinities of 17–29‰, while the onset of reproduction was delayed with increasing water salinity from 25 to 50‰, and reproduction stopped completely at salinity above 30‰ (Fineman-Kalio, 1988). More recently, El-Sayed *et al.* (2003) found that spawning performance and protein requirements of Nile tilapia broodstock were significantly affected by water salinity. The fish were fed test diets containing 25, 30, 35 and 40% crude protein at three water salinities (0, 7 and 14‰). The size at first maturation increased with increasing dietary protein at all salinities. At 25 and 30% protein levels, broodstock reared at 0‰ reached their sexual maturity at bigger sizes than those reared at 7 and 14‰. At 0‰, spawning intervals were not significantly affected by dietary protein levels, while, at 7 and 14‰, spawning intervals significantly decreased with increasing dietary protein levels. Spawning frequency and number of eggs per spawn were increased with increasing dietary protein levels. The total number of spawnings per female and absolute fecundity were better in fish fed 40% protein in fresh water than at 7 and 14‰ salinity. On the other hand, Watanabe and Kuo (1985) found that the total number of spawnings of Nile tilapia females was greater in brackish water (5–15‰) than in either full strength seawater (32‰) or fresh water.

ACCLIMATIZATION. Pre-acclimatization to salt water and gradual transfer to high salinity have a significant effect on tilapia growth and survival, as has been reported by Al-Amoudi (1987b). The author found that *O. aureus*, *O. mossambicus* and *O. spilurus* required shorter acclimatization time (4 days) for a transfer to full-strength seawater than *O. niloticus* and *O. aureus* × *O. niloticus* hybrids (8 days). These results indicated that the former tilapia group is more euryhaline than the latter group. Al-Amoudi *et al.* (1996) also found that *O. mossambicus* are more resistant to thermohaline shocks than *O. aureus* × *O. niloticus* hybrids. Similarly, the physiological and respiratory responses of *O. mossambicus* to salinity acclimatization have been evaluated by Morgan *et al.* (1997). Fish reared in FW were transferred to FW, isotonic salinity (ISO, 12‰) and 75% seawater (SW, 25‰), and a number of physiological parameters were measured. The authors found that plasma Na^+ and Cl^- were elevated 1 day after transfer to SW, but returned to FW levels on day 4. Plasma cortisol and glucose levels were higher; while growth hormone, Na^+, K^+ adenosine triphosphatase (ATPase) activities and prolactins were lower in FW and ISO than in SW. These results suggested that the physiological changes associated with SW acclimatization in tilapia are short-term, are energy-demanding and may account for as much as 20% of total body metabolism after 4 days in SW. The increase in the metabolic energy diverted into osmoregulation with increasing water salinity has also been reported in *O. mossambicus* and *O. spilurus* (Payne *et al.*, 1988), *O. niloticus* × *O. aureus* and common carp (Payne, 1983).

Feeding tilapia broodstock with diets containing higher salt levels may produce seed with better adaptability to water salinity. Turingan and Kubaryk (1992) studied this assumption by feeding Taiwanese red tilapia (*O. mossambicus* × *O. niloticus*) broodstock diets containing 0.8, 3, 6, 9 or 12% salt for 2 months prior to spawning. They found that egg hatchability was higher in SW than in FW. The hatchability and larval growth were highest in fish fed 12% salt in SW and lowest in FW. In another study, Watanabe *et al.* (1985) found that the survival of fry produced from fertilized eggs of Nile tilapia spawned in FW and incubated at elevated salinities of 0, 5, 10, 15, 20, 25 and 32‰ was 85.5, 84.4, 82.5, 56.3, 37.9, 20.0 and 0%, respectively. Fry salinity tolerance also increased with increasing the salinity of spawning, hatching or acclimatization. In addition, at equivalent salinity, early exposure of tilapia broodstock to high salinity produced progeny with higher salinity tolerance than those spawned in FW and hatched at high salinity.

Steroid hormones may reduce the routine metabolism of euryhaline tilapia reared at high salinity and, in turn, improve fish growth. The growth of *O. mossambicus* continuously treated with 17α-methyltestosterone (MT) was faster than early or delayed MT-treated fish in FW and SW. The growth of continuously treated fish was five to seven times higher in SW than in FW (Kuwaye *et al.*, 1993). Similar results were reported by Ron *et al.* (1995), who studied the effects of MT treatment and feeding rate on the growth of *O. mossambicus* reared in FW and SW. The best growth was observed in MT-treated fish fed double ration in SW. Oxygen consumption (routine metabolism) was also much lower in SW than in FW. The authors suggested that the reduction in routine metabolism in fish reared in SW may have accounted for the increase in growth rates compared to fish reared in FW.

3.4. Dissolved Oxygen

Dissolved oxygen (DO) is one of the limiting environmental factors affecting fish feeding, growth and metabolism. DO fluctuation is affected by photosynthesis, respiration and diel fluctuation. These factors must be fully considered where DO is concerned. Ambient DO range produces the best fish performance, while low DO levels limit respiration, growth and other metabolic activities of fish (Tsadik and Kutty, 1987).

Tilapia are known to withstand very low levels of DO. Most tilapias can tolerate DO levels as low as 0.1–0.5 mg/l for varying periods of time

(Magid and Babiker, 1975; Tsadik and Kutty, 1987). They can even survive at zero DO concentration, if they are allowed access to surface air. But tilapia usually suffer from high mortality if they fail to reach surface air. On the other hand, tilapia can tolerate conditions of high oxygen supersaturation (up to 400%), which usually occurs because of high photosynthesis resulting from phytoplankton and macrophytes blooming (Morgan, 1972).

It is well known that increasing water temperature reduces the rate of DO in the water. This leads to increased respiration rate and oxygen consumption in tilapia, because under increased water temperature the rate of metabolism and, in turn, the tissue demand for oxygen increase. Franklin *et al.* (1995) found that the rate of oxygen consumption in tilapia increased from 0.74 to 0.97 mg/l/h with increasing water temperature from 37 to 42°C. Similar results were reported with a number of tilapia species and their hybrids (Becker and Fishelson, 1986). Moreover, early studies of Job (1969a, b) indicated that the respiration of tilapia was independent of DO at oxygen saturation of 25–32% (at temperature ranging from 15 to 30°C), while below these saturation levels the metabolic rate became dependent on oxygen availability, and mortality occurred when DO remained below 20% saturation for more than 2–3 days. On the contrary, Teichert-Coddington and Green (1993) found that reducing the aeration in tilapia ponds in Honduras from 30% to 10% of oxygen saturation did not affect fish growth and water quality, and both aeration levels produced better growth than unaerated ponds. Nevertheless, the authors reported that aeration leads to the suspension of settled particulate matter and causes water turbidity and suggested that more research is needed to improve aeration techniques in order to reduce such turbidity.

Handling stress has a significant effect on oxygen consumption in tilapia. Ross and Ross (1983) found that handling stress in Nile tilapia increased oxygen consumption rate from +150% to +300% of the resting value, and did not fully return, in many cases, to the resting value after 3 h. The authors suggested that, following handling stress, the fish should be returned to water containing high levels of DO and should not be fed for at least 1 h. The message from this result is clear: 'handling stress should be minimized'. The respiration rate of the fish also increased curvilinearly with increasing body weight.

Increases in carbon dioxide tension in the water lead to decreased oxygen consumption (but increased ventilation rate). However, in addition to their tolerance of low levels of DO, tilapia can also withstand very high levels of CO_2, ranging from 50 to 72.6 ppm (Fish, 1956). It has been suggested that high CO_2 water supplies caused nephrocalcinosis in tilapia (R.J. Roberts, cited in Chervinski, 1982). This means that CO_2 concentration should also be kept minimal in tilapia culture water.

3.5. Ammonia and Nitrite

3.5.1. Ammonia

Most of the nitrogenous wastes of fish are excreted via gills in the form of ammonia. Excreted ammonia exists in un-ionized NH_3 form (UIA-N), which is toxic to fish, and ionized NH_4^+, which is non-toxic (Chervinski, 1982). The toxicity of ammonia depends on DO, CO_2 and pH. The toxicity increases with decreasing DO and decreases with increasing CO_2 (Chervinski, 1982). Fish species and size, acclimatization time and culture systems also affect the toxicity of ammonia to fish.

The toxicity of ammonia to tilapia has been related to different physiological changes. Ahmed *et al.* (1992) found that Nile tilapia exposed to ammonia had a lower number of red blood cells and haemolytic anaemia, leading to a significant reduction in blood oxygen content, which enhances ammonia toxicity. Oxygen saturation of the arterial blood of *T. zillii* was also decreased after exposure to a sublethal ammonia concentration (1.1–3.3 mg/l) for 2 weeks (El-Shafey, 1998). El-Shafai *et al.* (2004) also evaluated the effect of prolonged exposure to sublethal UIA-N on the growth performance of Nile tilapia fed on fresh duckweed. They found that the toxic level of NH_3-N and its negative effect on the growth performance ranges from 0.07 to 0.14 mg UIA-N/l. They further recommended that the NH_3-N concentration should be maintained below 0.1 UAI-N mg/l.

The median lethal concentration (LC_{50}) of NH_3 has also been studied in a number of tilapia species and hybrids. Redner and Stickney (1979) found that the 48 h LC_{50} for *O. aureus* was 2.46 ppm. When the fish were acclimatized to the sublethal concentration of 0.43–0.53 ppm for 35 days, they were able to withstand a concentration of 3.4 ppm NH_3 without any mortality for 48 h. Daud *et al.* (1988) also reported that the LC_{50} of un-ionized ammonia at 48 h, 72 h and 96 h in red tilapia (*O. mossambicus* × *O. niloticus*) fry was 6.6, 4.07 and 2.88 ppm, respectively. The authors related the relatively high resistance of fish fry to un-ionized ammonia to the high levels of DO (7–10.1 mg/l). The threshold lethal concentration was 0.24 ppm. Prior to death, fry swam erratically and exhibited haemorrhage of gill filaments.

The effect of ammonia on tilapia performance is also related to water pH and exposure period, in addition to ammonia concentration. Hargreaves and Kucuk (2001) found that brief daily exposure of *O. aureus* to up to 0.91 mg/l NH_3-N at pH 9 did not affect fish growth and feed utilization, while brief daily exposure to 1.81 mg/l NH_3-N reduced specific growth rate (SGR). At a concentration of 3.23 mg/l NH_3-N at pH 9.5, SGR was only 35% that of the control fish. These results suggested that brief sublethal ammonia exposure in aquaculture has little effect on tilapia performance.

3.5.2. Nitrite

Ammonia is oxidized into nitrite (NO_2) and then into nitrate (NO_3) through nitrifying bacteria grown on suspended organic matter. The bacteria remove the organic matter from the culture system by using it as food, while the bacteria themselves can be used as natural food for filter-feeding fish such as tilapia and carp. Nitrate is relatively non-toxic to tilapia; however, prolonged exposure to elevated levels of nitrate may decrease immune response and induce mortality (Plumb, 1997). Nitrite is highly toxic to fish, including tilapia, because it disturbs the physiological functions of the fish and leads to growth retardation. Sudharsan *et al.* (2000) studied the effect of nitrite toxicity on the dehydrogenases in Mozambique tilapia (*O. mossambicus*). The activity of dehydrogenases, lactic dehydrogenase (LDH), succinate dehydrogenase (SDH) and glutamate dehydrogenase (GDH), were estimated in various tissues of the fish exposed to sublethal concentrations of nitrite for 24, 48, 72 and 96 hours. The activities of these dehydrogenases were tissue-specific and time-dependent. A significant decrease in the activity levels of LDH occurred, while SDH and GDH showed an increase in their activity when compared with the control.

The tolerance of tilapia to nitrite is also influenced by fish size. Atwood *et al.* (2001) found that small-sized Nile tilapia (4.4 g) were more tolerant to nitrite than large fish (90.7 g). The 96 h LC_{50} of NO_2-N were 81 and 8 mg/l in small and large fish, respectively. The addition of a chloride source (500 mg/l $CaCl_2$ or NaCl) to culture water protected both small and large fish from nitrite toxicity. Such protection was achieved at a chloride : nitrite ratio of 1.5 : 1 (by weight).

3.6. pH

Some tilapias are known to tolerate a very wide range of water pH. For example, the tilapia *Oreochromis alcalicus grahami* live successfully in Lake Magadi (Kenya), which has a pH of 10.5 (Randall *et al.*, 1989). These fish were reported to tolerate a pH range of 5 to 11 for at least 24 h, but they die at pH < 3.5 and > 12 (Reite *et al.*, 1974). They overcome the problem of ammonia excretion by excreting about 90% of their nitrogenous wastes as urea. This process is facilitated by seawater-type gill chloride cells (Wilkie and Wood, 1996). Nile tilapia can also survive at a pH range of 4–11 (Balarin and Hatton, 1979; Wangead *et al.*, 1988). In the meantime, *Tilapia congica* live in the very acidic water (4.5–5.0) of Lake Tumba (Zaire).

Wangead *et al.* (1988) studied the effects of acid water on the survival, behaviour and growth of Nile tilapia fingerlings (0.4–1.0 g) and adults (45.4–46.3 g). They found that both fingerlings and adults died at pH 2–3 within 1–3 days. Both size groups tolerated pH 4–5 very successfully, and attained survival and growth rates similar to the control group (pH 7) after 60–70 days. However, adult fish were more resistant to low pH, with a survival rate of 86.6, 100 and 100% at pH 4, 5 and 7, respectively, whereas the survival of fingerlings was 57.8, 82.2 and 84.5%, respectively, at the same pH values.

Other tilapias are less tolerant to water pH and may develop physiological changes following transfer from neutral water to acidic water. Yada and Ito (1997) studied the effects of transferring *O. niloticus* and *O. mossambicus* from neutral water to acidic water with a pH of 4.5, 4.0 and 3.5. After 3 days, *O. niloticus* showed the lowest level of plasma Na^+ at a pH below 4.5, while the lowest level of plasma Na^+ in *O. mossambicus* occurred only at pH 3.5. Plasma Na^+ of *O. mossambicus* tended to get back to normal, while that of *O. niloticus* continued to decrease after the exposure to pH 3.5. In both species, gill Na^+, K^+ ATPase activity increased after the exposure to acidic water. These results indicated that *O. mossambicus* have a greater ability to maintain plasma Na^+ in acidic water than *O. niloticus*. On the other hand, low or high water pH may lead to behavioural changes, damage of gill epithelial cells, reduction in the efficiency of nitrogenous excretion and increased mortality. In support, Wangead *et al.* (1988) reported that fingerling and adult Nile tilapia exposed to pH 2–3 showed rapid swimming and opercular movements, surfacing and gulping of air, lack of body position and mass mortality within 1–3 days. Similarly, Chen *et al.* (2001) found that *O. mossambicus* exposed to high pH for 7 days decreased ammonia nitrogen excretion, but increased urea nitrogen excretion. This situation was reversed at low pH (i.e. increased ammonia excretion and decreased urea excretion). The authors also reported that the median lethal dissolved oxygen (LDO_{50}) increased significantly at low pH (7.14, 4.02, 3.36, 0.84 and 0.32 mg/l at pH 4, 5, 6, 8.3 and 9.6, respectively).

Bonga *et al.* (1987) found that the short-term effects of acid exposure on prolactin secretion in *O. mossambicus* depend on pH and rate of acidification. Rapid acidification to pH 4 resulted in necrosis of the integumental epithelium and caused damage of the skin and high plasma electrolyte loss and impaired the response of prolactin cells. At a lower acidification rate (0.5 pH units/min), these symptoms hardly occurred and branchial permeability to water and ions increased and prolactin secretion increased threefold. Slow acclimatization of tilapia to low pH may enable the fish to withstand long-term exposure to low environmental pH and prevent epithelial cell damage. Van Ginneken *et al.* (1997) exposed *O. mossambicus* to water gradually acidified to pH 4 for 37 days. The fish successfully withstood this low pH without any mortality or significant changes in energy-rich compounds and plasma Na^+, Cl^-, cortisol and glucose between the control and acidified groups. This result indicated that the fish were able to maintain ionic balance and there was no activation in the pituitary–interrenal axis. The effects of gradual acidification on metabolic rates of *O. mossambicus* and common carp (*Cyprinus carpio*) have also been investigated by Van Dijk *et al.* (1993). They found that standard metabolic rate, maximum metabolic rate and oxygen consumption in tilapia decreased in acidic water (pH 4), whereas these metabolic parameters did not decrease in carp. Wokoma and Marioghae (1996) also studied the median lethal time (LT_{50}) and physiological changes in *Tilapia guineensis* in relation to varying water pH. They found that LT_{50} increased from 1.2 h to 62 h with the pH increasing from 2 to 3, while the critical lethal pH was 3.3. They also found that the routine metabolic rate of small fish (1.4 g) was higher than that of larger fish (10.6 g) at high ambient oxygen concentration (6.3 mg/l) and the metabolic rate decreased at low ambient oxygen concentration.

3.7. Photoperiod

Photoperiod is a key artificial factor in regulating the daily activity rhythms of fish (Biswas *et al.*, 2002), promoting fish growth, metabolic rates, body pigmentation, sexual maturation and reproduction (Biswas and Takeuchi, 2002; Biswas *et al.*, 2002; Trippel and Neil, 2002; El-Sayed and Kawanna, 2004). The effects of photoperiod on these physiological functions depend on fish species and size.

Despite the significant effects of photoperiod on fish performance, its influence on tilapia growth, feed efficiency and other physiological functions under culture conditions has not been well studied. Only a few, recent studies have considered the role of photoperiod in the growth and reproduction of Nile tilapia (Ridha and Cruz, 2000; Biswas and Takeuchi, 2002; Biswas *et al.*, 2002; Campos-Mendoza *et al.*, 2004; El-Sayed and Kawanna, 2004).

It has been reported that the response of Nile tilapia to photoperiod cycles depends on fish developmental stage and sex (El-Sayed and Kawanna, 2004). Larval stage was more sensitive to photoperiod than fingerling and juvenile stages. Fish fry subjected to long photoperiods (24 and 18 h) had significantly better performance than those exposed to intermediate or short photoperiods (12 or 6 h). One possible explanation of the decreased larval growth during shorter photoperiods is that during short light phases insufficient time will be available for the establishment of a robust rhythmicity, as has been reported by

Biswas and Takeuchi (2002) and Biswas *et al.* (2002). The effect of photoperiod on synchronizing an endogenous rhythm to the external environment may also require more energy in the shorter photoperiods, leading to a reduction of somatic fish growth (Biswas and Takeuchi, 2002; Biswas *et al.*, 2002).

On the other hand, it has been suggested that the improvement in the growth of fingerling Nile tilapia with increasing photoperiod is related to the reduction of standard metabolic rate (Biswas and Takeuchi, 2002; Biswas *et al.*, 2002). Therefore, the adoption of the optimum photoperiod in tilapia culture systems, especially in closed systems, is essential since it will reduce the amount of energy used for standard metabolism, and in turn increase the energy spared for somatic growth, leading to increasing fish production and profitability.

Very few studies have considered the effects of light intensity and photoperiod on reproductive performance of tilapia. The information reported, despite being limited, indicated that low light intensity leads to low spawning activity (Duarte, 1989; Ridha and Cruz, 2000). Ridha and Cruz (2000) recommended a light intensity of 2500 lux and a photoperiod of 18 h/day for optimum seed production of Nile tilapia. Gonad development, fecundity and spawning frequencies also tend to improve with increasing photoperiod (Behrends and Smitherman, 1983). It appears that the best reproductive performance is achieved under normal day length (12 h light : 12 h dark cycles) (Campos-Mendoza *et al.*, 2004; A.-F.M. El-Sayed and M. Kawanna, unpublished results).

3.8. Water Turbidity

Water turbidity can be a major problem in semi-intensive, fertilized freshwater fish ponds. According to Yi *et al.* (2003a), this turbidity originates from a number of sources, namely: (i) turbid source water; (ii) rainwater runoff from pond dykes containing clay materials; and (iii) resuspension of bottom mud by water and fish movements. The high concentrations of suspended colloidal particles reduce fertilizer effect, cause water acidity and inhibit light penetration, and in turn adversely affect primary productivity in fish ponds (Boyd, 1990). The influence of turbidity on growth, feed conversion and survival of all-male, sex-reversed Jamaica red tilapia fry has been studied by Ardjosoediro and Ramnarine (2002). The fish were reared in water at clay turbidity levels of 0, 50, 100, 150 and 200 mg/l for 7 weeks. Fish growth, feed efficiency and survival were all significantly reduced with increasing water turbidity, but the differences at the higher levels of 100–200 mg/l were not significant. The authors suggested that clay turbidity levels in earthen ponds should be kept below 100 mg/l.

It is essential that proper mitigation techniques be applied in order to minimize the adverse effects of turbidity on pond productivity and fish growth. In a recent study, Yi *et al.* (2003a) described the source of clay turbidity and evaluated different clay turbidity mitigation techniques and their effects on the growth of sex-reversed all-male Nile tilapia and on water quality, in order to find a suitable mitigation method for controlling turbidity during the rainy season, for 149 days. Five mitigation treatments were adopted: (i) control; (ii) covering upper 50 cm of pond dykes with black plastic material to prevent turbidity from runoff (edge-covered); (iii) covering pond bottoms with green manure (terrestrial weeds) to alter soil texture (weed-covered); (iv) covering pond bottoms with small mesh (1 cm) net to prevent turbidity from fish disturbance (bottom-covered); and (v) covering pond dykes with rice straw (straw-covered). The ponds were fertilized with chicken manure at a rate of 500 kg/ha/week, in addition to supplemental urea and triple superphosphate (TSP) to provide 28 kg N/ha/week and 7 kg P/ha/week. The authors found that the runoff from pond dykes represented the main clay turbidity and covering pond dykes was effective in mitigating this turbidity. The straw- and weed-covered treatments produced the best growth rates and yield. The authors attributed the better performance and effective control of turbidity in straw- and weed-covered treatments to the following: (i) the increased natural food availability; (ii) metabolic and decomposition products from plant materials cleared pond water and reduced the shading effect of turbid water; (iii) the aggregation of clay particles and algal cells may have led to the sedimentation of larger particles; and (iv) the adsorption of clay on weed and straw. However, covering pond dykes with rice straw significantly reduced clay turbidity caused by runoff and enhanced fish growth through the microbial biofilm developed on the rice straw. Therefore, the authors suggested that covering pond dykes with rice straw can be a cost-effective tool for clay turbidity mitigation in tilapia ponds during rainy seasons.

3.9. Closing Remarks

1. Water quality management has become a key factor for successful aquaculture practices. It is therefore necessary to understand the major water quality parameters and their interrelationships, which affect fish growth and health and determine the failure or success of overall culture practices.

2. A temperature of about 28–30°C appears optimal for growth performance of Nile tilapia under recirculating culture systems.

3. The further the geographical location from the equator, the more tolerant are Nile tilapia strains to cold.

4. The conditioning of tilapia at an intermediate temperature before exposing them to lower temperatures may improve fish survival and tolerance to cold stress.

5. Small fingerling tilapia appear more susceptible to low temperature than larger fish. Large fingerlings must be used if water temperature is expected to drop.

6. The resistance of tilapia to cold stress may be better in brackish water than in fresh water.

7. Increased water temperature may limit larval growth and induce body deformation. The exposure of the larvae to low temperature before 10 days old may induce a high proportion of females, while elevated temperature may induce a high percentage of males.

8. Most tilapia species and hybrids tolerate a wide range of water salinity, and can normally grow and reproduce in brackish-water environments.

9. *Oreochromis mossambicus*, *O. aureus* and *T. zillii* are the most salinity-tolerant tilapia species, while other tilapias are generally less euryhaline and can grow, survive and reproduce at 0–29‰, depending on the species, size and sex and acclimatization period.

10. Pre-acclimatization to salt water and gradual transfer to high salinity significantly improve tilapia growth and survival. Feeding tilapia broodstock with diets containing higher salt levels may also produce seeds with better adaptability to water salinity.

11. Handling stress increases oxygen consumption in tilapia and therefore handling should be minimized.

12. Ammonia is very toxic to tilapia and therefore the UIA-N concentration should be maintained below 0.1 mg/l.

13. Nitrite is also toxic to tilapia. The addition of a chloride source (500 mg/l $CaCl_2$ or NaCl) (at a chloride : nitrite ratio of 1.5 : 1) to culture water may protect tilapia from nitrite toxicity.

14. Larval stages of tilapia (Nile tilapia) may be more sensitive to photoperiod than fingerling and juvenile stages. Long photoperiods (18 h light : 6 h dark (18L : 6D) and 24L : 0D cycle) produce better performance than short photoperiods (6L : 18D cycle). A 12L : 12D cycle has been recommended for optimum reproduction performance of Nile tilapia.

15. Water turbidity can be a major problem in semi-intensive, fertilized freshwater fish ponds. Covering pond dykes with rice straw (or similar plant materials, such as wheat straw) or covering pond bottoms with terrestrial weeds can be cost-effective tools for clay turbidity mitigation in tilapia ponds, especially during rainy seasons.

4
Semi-intensive Culture

4.1. Introduction

Semi-intensive culture (SIC) is simply the production of fish, and other aquatic animals, using natural food, through pond fertilization, or fertilization and supplemental feeding. In other words, SIC is a means of producing low-cost fish, through low production inputs, which contributes to hunger alleviation and food security, especially in rural areas in developing countries. It is no surprise, therefore, that SIC systems are adopted mainly for herbivorous and omnivorous fish that feed low down in the food chain, such as tilapia and carp. The production of SIC systems is usually lower than that of intensive systems due to the lower stocking density and production inputs. Taking into account the input/output (cost/benefit) ratio, this system could be more convenient and cost-effective for small-scale farmers than intensive farming systems.

Semi-intensive farming of tilapia has been practised for many years in different parts of the world, particularly in South-east Asia, in either monoculture or polyculture systems. Over the past two decades, SIC of tilapia with other herbivorous fish, such as carp, has experienced increased popularity in several countries in the world, particularly among small-scale farmers in Asia. This chapter discusses the different aspects of SIC of tilapia, including pond fertilization, monoculture, polyculture and integrated culture, under different culture and environmental conditions.

4.2. An Overview of Pond Fertilization

The main objective of pond fertilization is to stimulate the primary productivity in fish ponds and enhance autotrophic and heterotrophic food production. Practically, nitrogen (N), phosphorus (P) and carbon (C) are the major inputs in pond fertilization, and the ratio between them is also of prime importance. Potassium (K) is generally not considered a fertilizer element, since earlier studies revealed that K fertilization did not affect pond production (Hickling, 1962). However, K may be required if the K content and alkalinity of culture water are low. The average nutrient composition of phytoplankton in fish ponds is 45–50% C, 8–10% N and 1% P, giving a ratio of about 50 : 10 : 1 (Edwards *et al.*, 2000). Efficient fertilization requires a number of measures that should be taken into account. These are:

- Availability of considerable amounts of carbon.
- Alkalinity should be > 20 mg/l (Boyd, 1974). If the water is relatively soft and acidic, with low alkalinity (< 20 mg/l), liming may be necessary to raise alkalinity before the application of inorganic fertilizer.
- Fertilization history of the pond. Understanding the fertilization history, water quality, bottom composition and overall pond dynamics is extremely important for efficient and sustainable fertilization. For example, available P in fish ponds is rapidly removed

by reaction with divalent cations, such as Fe^{2+}, Ca^{2+} and Mg^{2+}, forming inactive precipitation. Pond bottom mud also adsorbs P and leads to its removal from culture water. The more mud the bottom contains, the more P it adsorbs (Shrestha and Lin, 1996a, b). Therefore, P requirement for pond fertilization depends on the type of bottom soils and their P saturation (Knud-Hansen, 1992). It has been reported that maximum N and P fertilization rates of ponds in acid sulphate soil with high P binding capacity were 4 and 2 kg/ha/day, respectively (Asian Institute of Technology (AIT), unpublished data, cited in Yakupitiyage, 1995). Further increase in N loading may lead to ammonia toxicity. There was a negative relationship between the ability of pond sediments to remove soluble P from water and organic matter and P contents of pond bottoms (Knud-Hansen, 1992).

- Season of the year also has a significant effect on fertilization effectiveness of fish ponds. Green *et al.* (1990) found that the production of Nile tilapia in ponds fertilized with chicken manure in Panama was better in dry seasons than in rainy seasons. Similarly, the production of blue tilapia (*Oreochromis aureus*) in ponds fertilized with chicken manure in addition to 0–20–0 inorganic fertilizer was 48% less in the wet season than in the dry season (Ledgerwood *et al.*, 1978). Low production during the rainy season was attributed to the increased turbidity resulting from bank erosion following heavy rains, which reduces light penetration and decreases the photosynthesis process.
- Cost-efficiency analysis. Fertilization strategies are often evaluated based on their ability to enhance fish growth and pond yield. Nevertheless, a fertilization regime that produces the best yield may not necessarily be the most cost-effective. It is therefore essential that the farmers understand the dynamic processes in the ponds and how fertilization efficiency is affected by light penetration, pond depth, bottom soil composition and nutrient contents of the water. Economic and ecological analyses of fertilization application are also appreciated, so that the farmer can choose the best fertilization strategy (more discussion on the economic consideration is provided in Section 4.7).

4.3. Fertilization of Tilapia Ponds

4.3.1. Organic fertilizers

Organic fertilizers have been widely used in tilapia ponds, especially in Asia (Edwards *et al.*, 1994a, b; Shevgoor *et al.*, 1994), Central America (Green *et al.*, 1990; Watanabe *et al.*, 2002) and Africa (Njoku and Ejiogu, 1999; Abdelghany and Ahmad, 2002). A wide variety of organic fertilizers, including poultry manure, cattle manure, domestic sewage (sludge), green manure and composted agricultural wastes, are currently in use in tilapia pond fertilization. The effectiveness of fertilization depends on type and composition of fertilizer (Table 4.1), culture systems (monoculture vs polyculture), fish species and species combinations, stocking size and density and environmental conditions.

Water quality, particularly alkalinity and ammonia concentration, should be considered when applying organic fertilization in tilapia ponds. McGeachin and Stickney (1982) found that 70–140 kg/ha/day dry, laying hen manure produced the best growth and survival of *O. aureus* reared in simulated sewage lagoons, depending on water alkalinity. About 70 kg/ha/day was suggested at high alkalinity, while 140 kg/ha/day produced the best yield at low alkalinity. The levels of un-ionized ammonia also increased with increasing manuring rate, due to the increase in water alkalinity. A significant amount of this ammonia is also lost from pond water through volatilization to the atmosphere at high alkalinity (> 8.5).

Organic fertilization should also be applied at proper rates, because insufficient fertilizer will result in low yield, while excessive fertilization may lead to significant deterioration in water quality and to dissolved oxygen (DO) deficiency, and in turn will result in a reduction in fish yield. Massive mortality of Nile tilapia has been reported in a number of farms in the Kafr El-Shaikh Governorate in Egypt, due to excessive application of organic fertilizer (A.-F.M. El-Sayed, Kafr El-Shaikh, 2004, personal observation). Anuta (1995) found that the best yield of *Tilapia guineensis* reared in tanks fertilized with 0, 100 and 150 kg chicken manure/ha/day was achieved at 100 kg/ha/day. The NH_4-N, NO_3 and chemical oxygen demand (COD) were all increased with increasing fertilization rates. Organic manure stimulates food production by adding organic detritus, which stimulates

Table 4.1. Feed and fertilization strategies suggested for optimum yield of tilapia in semi-intensive culture systems.

Cultured fish								
Species (country)	Size (g)	Density/ha	Fertilization regime	Feeding	Culture Period (days)	Yield (Mt/ha)	Remarks	Reference
O. n (Honduras)		10,000	Chicken manure, 1000 kg/ha/week		152	1.76	During rainy season	Green *et al.* (1990)
O. n (Honduras)		10,000	Chicken manure, 1000 kg/ha/week		150	1.71	During dry season	Green *et al.* (1990)
O. n (Panama)		10,000	Chicken manure, 1000 kg/ha/week		149	2.07	During dry season	Green *et al.* (1990)
O. n (Panama)		10,000	Chicken manure, 1000 kg/ha/week		141	1.68	During rainy season	Green *et al.* (1990)
O. n (Philippines)		20,000	Chicken manure, 500 kg/ha/week	23% cp diet, 1.5% bw/day	151	4.35	Based on cost/benefit analysis	Green (1992)
O. n (Cameroon)	202–222	7600	Cattle manure, daily, 226 kg dry wt/ha/day	• CSC : 3% bw/day • 6% bw/day	100	4.80/year 6.50/year	Polyculture with *Clarias gariepinus* (1100 fish/ha) as police fish	Middendorp (1995)
♂ *O. n* (Thailand)	15	20,000	280 kg chicken manure + 56.3 kg urea + 17.5 TSP/ha/week		234	4.0–4.2	2.5 m deep ponds, no water addition. Weekly fertilization produced best growth	Diana and Lin (1998)
Red tilapia (Thailand)	33	62.5/m^3	Urea + TSP; 4 kg N + 2.18 kg P/ha/day	Commercial diet, 0–100% satiation	90	13.5 kg/m^3	Cages in fertilized ponds, 50% satiation was economically best	Yi *et al.* (2004a)

Continued

Table 4.1. *Continued.*

Cultured fish								
Species (country)	Size (g)	Density/ha	Fertilization regime	Feeding	Culture Period (days)	Yield (Mt/ha)	Remarks	Reference
♂ *O. n* (Thailand)	23–24	30,000	Urea + TSP; 28 kg N + 7 kg P/ha/week, N : P = 4 : 1	30% cp diet:	160		Fertilization plus feeding was better than fertilization followed by feeding	Thankur *et al.* (2004)
				• Fertilization + feeding at day 80		16.7/year		
				• Fertilization until day 80, followed by feeding		13.0/year		
♂ *O. n* (Philippines)	0.05	40,000	Ammonium phosphate (28 kg N + 5.6 kg P)/ ha/week	28.6% cp diet:	120			Brown *et al.* (2002a)
				• 100% satiation		3.14	Survival 57%	
				• 67% satiation		3.58	Survival 65%	
♂ *O. n* (Kenya)	90	1000	Diammonium phosphate + urea, 20 kg N/ha/week		133	1.72	During warm season (23.5–28.2°C)	Veverica *et al.* (2001a)
♂ *O. n* (Kenya)	16.9	1000	Diammonium phosphate + urea, 20 kg N/ha/week		147	2.95	During cool season (22.5–26.4°C)	Veverica *et al.* (2001a)
♂ *O. n* (Egypt)	1–3	20,000	Chicken manure, 1000 kg/ha/week, for 60 days. 54.4 kg urea + 92.4 kg superphosphate/ ha/week	• 30% cp diet, 3% bw/day, starts day 60	145	7.40/year	African catfish (59 g) were used for seed control	Green *et al.* (2002)
				• No feeding	145	3.20/year	Same regime	
Nile tilapia Silver carp Common carp (Egypt)	13.8 1.9 10.7	30,000	750 kg chicken litter/ha, biweekly application, 100 TSP and 20 kg urea/ha	25% cp diet, 3% bw/day, 6 wks after stocking	133	4.75	Feeding at 6-week delay was comparable to no delay, and better than at 13-week delay	Abdelghany *et al.* (2002)

O. n, *Oreochromis niloticus*; bw, body weight; CSC, cottonseed cake; TSP, triple superphosphate.

heterotrophic food chains and increases primary productivity. Produced phytoplankton provides oxygen for the respiration of aerobic organisms present in the pond.

Fresh manure may be preferred to dry manure for pond fertilization. Hepher and Pruginin (1982) suggested that fresh manure be used for tilapia pond fertilization, and stated that fresh manure disintegrates in water into colloidal particles, which are easily attacked by bacteria and readily incorporated into the food web. Some semi-intensive tilapia farmers in Egypt pile the dry animal manure on farm dykes and spray it with water for a few days before washing it into the ponds (A.-F.M. El-Sayed, Alexandria, 1998, personal observation). This process increases fermentation rate and reduces the time needed for achieving maximal primary production in the ponds, compared with dry manure. However, fermentation may reduce the nitrogen content of the manure and in turn would reduce its potential effectiveness.

Animal manures are also better applied in tilapia ponds in a fine particulate or colloidal state. Small-sized particles stimulate the growth of bacteria by providing enough surface areas for attachment and mineralization, the minerals being used for photosynthesis and phytoplankton production. In this regard, buffalo manure has not been recommended as a major organic fertilizer in tilapia ponds, as it causes a reduction in dissolved oxygen, due to the respiratory demand of the bacteria (Edwards *et al.*, 1994a). The presence of tannin in buffalo manure stains pond water brown, which reduces light penetration and in turn inhibits phytoplankton productivity. Buffalo manure is also characterized by large-size particles, a high C : N ratio (about 26 : 1) and poorly soluble nutrients. Shevgoor *et al.* (1994) reported that only 6% of buffalo manure nitrogen was released as dissolved nitrogen and 35% of total P was released as soluble, reactive P. A fertilization rate higher than 100 kg dry matter (DM)/ha/day caused water quality deterioration, high suspended solids and reduced light penetration and phytoplankton production. Under such conditions, DO declined to near 0 at dawn. Therefore, buffalo manure rate should not exceed 50 kg/ha/day (Edwards *et al.*, 1994b). It is also recommended that buffalo manure be supplemented with urea (Edwards *et al.*, 1996).

On the other hand, poultry manure is a better source of manure in tilapia ponds than that of large ruminants as it has a higher N (and P) content. Indeed, poultry manure is widely used for tilapia pond fertilization, the main reasons being that it is readily available in a convenient dry form for transportation and application. It is also characterized by small-size particles and a better C : N ratio (about 10 : 1). Generally speaking, a fertilizing rate of < 100 kg/ha/day of poultry manure increases phytoplankton productivity and improves fish production. However, contradictory results have been reported on the use of poultry manure for tilapia pond fertilization.

Hopkins and Cruz (1982) reported that the yield of Nile tilapia reared in polyculture with common carp and snakehead murrel, *Channa striata* or *Clarias batrachus*, in the Philippines was better with pig manure or chicken manure than with duck manure. Green *et al.* (1989) compared the effects of chicken manure, dairy cow manure and inorganic fertilizer on the production of male Nile tilapia. They found that chicken litter produced the best net fish production. Similar results have also been reported by Green *et al.* (1990, 2002) and Green (1992). These authors attributed the better tilapia growth and yields in ponds fertilized with organic fertilizer (poultry manure), compared to inorganically fertilized ponds, to increased heterotrophic production and/or direct consumption of organic manure by the fish. Moreover, it has been suggested that diel oxygen stratification is more pronounced in inorganically fertilized ponds, presumably due to the greater water clarity and the more even distribution of primary production in organically fertilized ponds (Diana *et al.*, 1991).

In contrast, Ledgerwood *et al.* (1978) found that chicken manure was not a very effective fertilizer in blue tilapia ponds in El Salvador, and the addition of inorganic P (but not N) increased fish yields. Knud-Hansen *et al.* (1993) found that the addition of urea and triple superphosphate (TSP) into Nile tilapia ponds increased N and P inputs in the ponds and improved primary productivity, while the addition of chicken manure to these inorganic fertilizers did not enhance fish yield. The effectiveness of chicken manure-P was only 10% of TSP-P. The economic analyses in this study discouraged the use of chicken manure as a source of N and P in tilapia ponds in Thailand.

It appears from these discrepancies concerning pond fertilization that this subject is more complex, confusing and wider than we may have imagined. Indeed, the types and quantities of nutrients required and the relative amounts of fertilizers needed to provide such nutrients vary depending on the type of fertilizer used, nutrient contents of the water, types of bottom soil, water quality, environmental climatic conditions and geographical location. This means that the effectiveness of certain fertilizers is probably site-specific, and fertilizer requirements should be evaluated on a farm-by-farm basis, and even on a pond-by-pond basis.

4.3.2. Inorganic fertilizers

As stated earlier, N, P and C are the most important nutrients required for optimum pond management and fish production in semi-intensive systems. Not only is this the case, but the ratios between them are also important. Boyd (1976) studied the effects of N and N : P : K ratios in blue tilapia ponds fertilized with 22.5 kg/ha of 0–20–5, 5–20–5 and 20–20–5 (N–P–K). He found that the best fish yield was obtained at 5–20–5 kg/ha. Excessive nitrogen fertilization is not recommended, as it may lead to waste of resources without improving fish yields. Veverica *et al.* (2001a) fertilized sex-reversed Nile tilapia ponds with diammonium phosphate during cool and warm seasons in Kenya for 147 and 133 days. Triple phosphate and sodium carbonate were also applied to the ponds to maintain sufficient P and C levels. The authors found that increasing N levels beyond 20 kg/ha/week did not improve tilapia yield. Gross profit was also higher at 20 kg/ha/week than at 30 kg/ha/week during both cold and warm seasons. Meanwhile, optimal fertilization rate in tilapia ponds in Thailand was reported to be about 28 kg N/ha/week at an N : P ratio of 4 : 1 by weight (Knud-Hansen *et al.*, 1993).

The frequency of fertilization may or may not affect the production of tilapia in fertilized ponds. Knud-Hansen and Batterson (1994) investigated the most efficient frequency of urea and TSP application on phytoplankton productivity in Nile tilapia ponds. The ponds were fertilized daily, twice a week, weekly, twice every 3 weeks or once every 2 weeks, for 5 months. No relationship was found between the frequency of fertilization and either phytoplankton production or fish yields. However, primary productivity and fish yield were linearly correlated. Light availability, nutrients and temperature were the main factors affecting primary productivity. The highest productivity was achieved in water with low turbidity and relatively high alkalinity. This means that nutrient uptake by phytoplankton decreases as inorganic turbidity increases, due to low light penetration.

4.3.3. Periphyton-based pond culture

Periphyton-based pond culture (also known in West Africa as 'acadja') is a simple and cheap way of producing natural food for herbivorous and omnivorous fishes, such as tilapia and carp. In this culture system, woody branches, bamboo poles or any other hard substrates are planted or fixed in shallow waters, such as ponds, lagoons, reservoirs, etc., to allow the growth of sessile autotrophic and heterotrophic aquatic biota, known as periphyton (Fig. 4.1). The periphyton community comprises bacteria, fungi, protozoa, phytoplankton, zooplankton, benthic organisms and a wide range of invertebrates (Milstein, 1997; Azim *et al.*, 2001). This system enhances natural productivity, which provides natural food for herbivorous and omnivorous fish.

The production of fish in periphyton-based systems has been tested in Africa and Asia with considerable success (Azim *et al.*, 2001, 2003, 2004a, b; van Dam *et al.*, 2002). Recent studies (Azim *et al.*, 2001; Keshavanath *et al.*, 2004; Milstein and Lev, 2004) indicated that this system is very suitable for tilapia culture. It has been reported that 10 bamboo poles/m^2 increased fish yield 20% to 100%. The quantity and quality of periphyton produced varied with substrate type, grazing pressure and fertilization level. Azim *et al.* (2001) found that bamboo poles produced more and better periphyton than jute stick and branches of hizol tree.

Periphyton may be a partial or total substitute for supplemental feed in tilapia production. This assumption has recently been tested by Keshavanath *et al.* (2004), who evaluated the growth and production of hybrid red tilapia (*Oreochromis mossambicus* × *Oreochromis niloticus*) fry in mud-bottomed tanks with bamboo poles. Substantial improvement in growth rates was achieved in the substrate-based system compared to substrate-free tanks. The results indicated that periphyton can replace or complement supplemental feeding in tilapia fingerling culture. Similarly, Milstein and Lev (2004) found that blue tilapia reared on natural

Fig. 4.1. Periphyton-based tilapia culture.

periphyton showed similar growth, survival and yield to fish fed on pelleted diets. Moreover, the periphyton-based system resulted in a considerable reduction in production costs.

It is clear that periphyton-based aquaculture can be an appropriate technology to reduce production costs and allow economically viable tilapia production. This approach can be an ideal alternative in resource-limited regions in Asia, Africa and Latin America, where small-scale, rural tilapia culture is commonly practised.

4.4. Supplemental Feeding

4.4.1. Rationale

Pond fertilization in semi-intensive tilapia culture is often associated with supplemental feeding, especially during the later grow-out stages. The adoption and effectiveness of supplemental feeding in semi-intensive culture depend on economic, social, biological and technical factors. It has been reported that tilapia can rely totally on natural food, produced through fertilization, up to a certain size (critical standing crop, CSC). Beyond this CSC, fish growth will slow down because natural food would be insufficient to meet their requirement. At this size, supplemental feed becomes necessary to sustain fish growth (Fig. 4.2). Thus, supplemental feeding at a smaller fish size will lead to a waste of resources and an unnecessary increase in

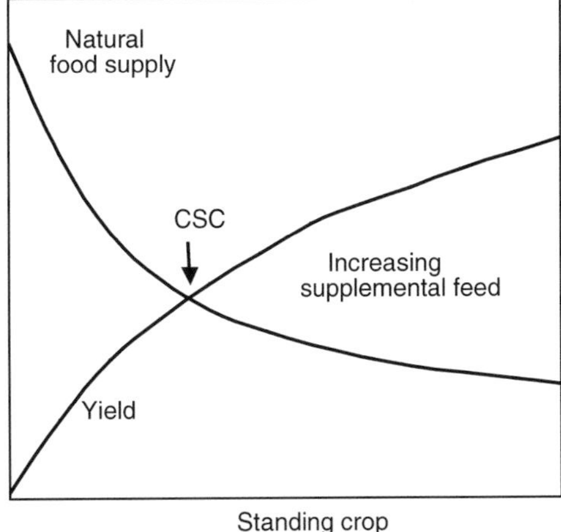

Fig. 4.2. Changes in natural food in the pond and fish yield, in relation to standing crop of cultured organisms and supplemental feeds (modified from De Silva, 1995).

operating costs, whereas a further delay in supplemental feeding will result in a reduction in fish growth and total yield.

Relying on natural food during the early stages of SIC and delaying feeding with commercial formulated feed or supplemental feed sources to later stages have been a common practice for small-scale farmers in many developing countries for decades. Large-scale fish producers usually raise their fish intensively, using commercial, high-protein pellets, with low feeding levels, while small-scale farmers use locally available sources at higher rates. Tilapia are produced mainly for local consumption, and their gate prices in most developing countries are relatively low. Therefore, the use of high-quality commercial feed is not recommended in small-scale farming systems (Yakupitiyage, 1995). Instead, pond fertilization and supplemental feeding are more productive and cost-effective. Diana *et al.* (1994) evaluated the effects of addition of supplemental feed to fertilized Nile tilapia ponds by adding feed alone, feed plus fertilizer or fertilizer alone. The combination of feed and fertilizer was more efficient in growing tilapia than feeding or fertilizing alone. In a similar study, Green *et al.* (2002) found that Nile tilapia raised in ponds fertilized with chicken manure at 1000 kg/ha for the first 60 days, followed by supplemental feed beginning on day 61, produced a better yield (7.4 Mt/ha/year) than feeding only (4.4 Mt/ha/year) and fertilizer only (3.2 Mt/ha/year).

4.4.2. Timing of supplemental feeding

A number of studies have been conducted to investigate the most suitable timing for supplemental feeding in semi-intensive tilapia culture. The most comprehensive study in this regard was conducted by Diana *et al.* (1996). These authors found that the most efficient system is to grow the fish up to 100–150 g with fertilizers alone. Feeding the fish before they reach this size was wasteful. The authors also found that the best annual revenue was achieved when the fish were fed 50% *ad libitum* once they reached 100 g (Fig. 4.3). In another study, Lin *et al.* (1997) studied the effects of different fertilizer and feeding regimes on nutrient budget and growth in Nile tilapia ponds in Thailand, in two experiments, each lasting for 160 days. All-male Nile tilapia (23–24 g) were stocked at 3 fish/m^2 and reared under continuous urea fertilization (60.7 kg/ha/week + 35.7 kg TSP/ha/week), while commercial feed (30% crude protein (cp)) was added beginning on day 80. In the other trial, ponds were fertilized until day 80 and followed by feeding commercial feed only. The final weight and net yield were better in the ponds receiving the fertilization throughout the study and feeding on day 80 than with the other treatment. Fish harvest removed 15.45–20.04% N and 10.02–15.10% P from the total inputs, with no significant difference between the experiments. The major portions of N and P were not accounted for in the estimated losses. These N and P losses were 70.66–78.01% and 81.88–87.25%, respectively.

Brown *et al.* (2000) studied the timing of supplemental feeding of Nile tilapia in fertilized ponds in seven different fish farms in the Philippines. Sex-reversed fish fingerlings (1.1 g) stocked in ponds (at 4 fish/m^2) fertilized with urea and ammonium phosphate at a weekly rate of 28 and 5.6 kg/ha. The fish were fed a farm-made diet beginning on day 45 or day 75. No significant differences in fish growth and net yield were found between the two feeding timings; however, fish fed on day 45 consumed significantly more food. More recently, Abdelghany *et al.* (2002) studied the effects of timing of supplemental feed on growth and production of Nile tilapia in polyculture with common carp and silver carp in fertilized ponds for 19 weeks. A commercial diet (25% cp) was provided for the fish at the beginning of the study, after 6 weeks and after 13 weeks, while the control group received no feeding. The authors found that providing the feed after 6 weeks led to better performance and lower feed consumption than the rest of the diet-fed groups.

It is clear from the foregoing discussion that delaying supplemental feeding in well-fertilized ponds does not reduce fish yield but does reduce feed cost significantly. This means that small-scale tilapia farmers who adopt semi-intensive farming system should not feed their fish immediately after stocking, but should rely on fertilization in so far as natural food supports fish growth during the early stages of fish life.

Feeding rates are also another important factor affecting the economic return of tilapia in semi-intensive culture systems. Since natural food is the main food input in the ponds, excessive supplemental feeding may result in a considerable economic loss, in addition to a severe environmental impact (see Chapter 12, Section 12.7, for details).

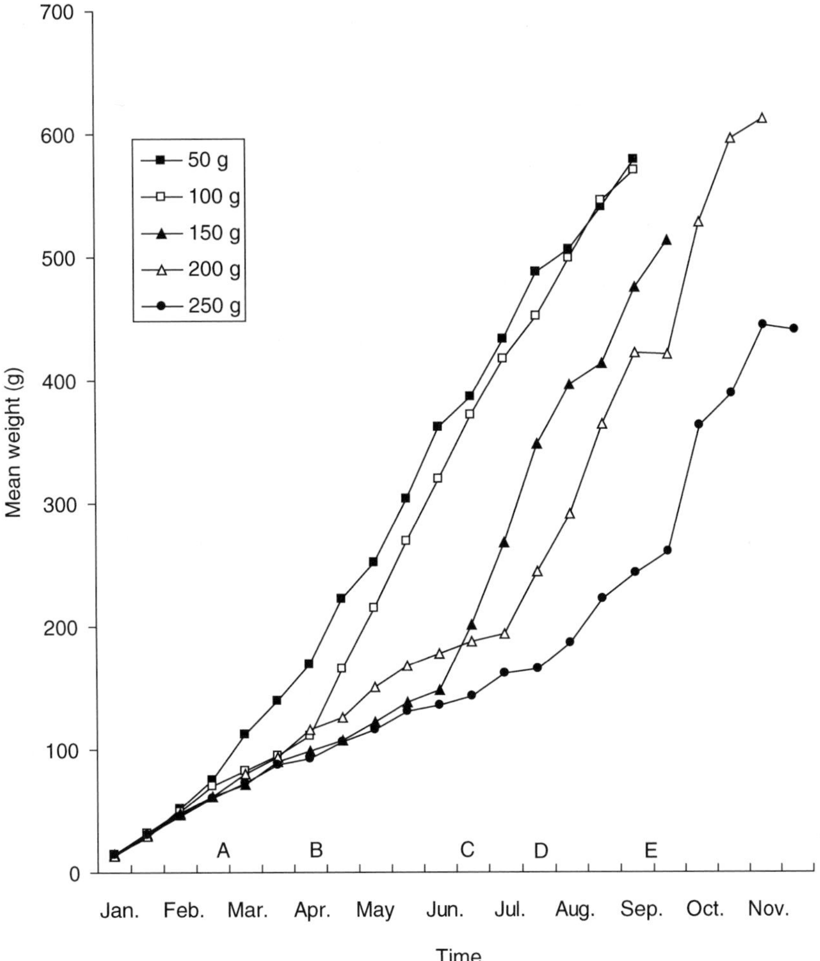

Fig. 4.3. The effects of timing of supplemental feeding on the weight gain of Nile tilapia fed with pelleted feed at 50% satiation level in fertilized ponds. All ponds were fertilized with inorganic fertilizers at the same optimal rate. Treatments were first feeding at: 50 g (■), 100 g (□), 150 g (▲), 200 g (△) and 250 g (●). The letters on the X axis indicate the dates when supplementary feeding was initiated for each treatment. (From Diana *et al.* (1996), with permission from the World Aquaculture Society.)

In support, Lin and Yi (2003) found that Nile tilapia reared in fertilized ponds and fed supplemental diets at 50%, 75% and 100% satiation produced comparable yields. However, considerable reduction in production costs and in nutrient loading was achieved at the 50% level. This means that, at this feeding level, about 50% of the feed can be saved. Similar results were reported with red tilapia fed supplemental feeds in fertilized, brackishwater ponds (Yi *et al.*, 2004a). Abdelghany and Ahmad (2002) also reported that the highest production and net income of Nile tilapia polycultured with common carp and silver carp in fertilized ponds and fed 0, 1%, 3%, 5% biomass and to apparent satiation were achieved at 2.67% of fish biomass/day (equivalent to apparent satiation). These results clearly demonstrate that an optimal feeding regime will reduce both feed costs and nutrient loading in the ponds.

4.4.3. On-farm feed resources

As mentioned in Chapter 5, intensively farmed tilapia depend exclusively on high-quality, commercial, pelleted feeds. These feeds are produced by both small-scale and large-scale feed mills. Therefore, I intentionally do not cover feed formulation and processing in Chapter 5, since farmers buy their feed from feed mills and need not worry about feed formulation and manufacturing. On the contrary, farmers who raise tilapia in semi-intensive systems often adopt fertilization and supplemental feed strategies, as discussed earlier. These farmers can produce the supplemental feeds on farm, using locally available inputs, in order to minimize feed and production costs. It would be appropriate here to provide tilapia farmers with some vital information on farm-made tilapia feeds.

Feed ingredients such as wheat bran, rice bran, cottonseed cake (CSC), etc. (Table 4.2) can be used solely as supplemental feed sources for

Table 4.2. Chemical composition (% dry weight, as-fed basis) of feed ingredients of potential use for on-farm feed production for tilapia in Egypt. (After El-Sayed, 1999.)

Ingredient	Moisture	Crude protein	Crude lipids	Crude fibre	NFE	Ash
Plant sources						
Maize plant	78.72	1.45	0.54	4.41	12.73	2.15
Sweetcorn (first cut)	92.51	0.64	0.15	2.13	3.83	0.75
Sweetcorn (second cut)	76.95	1.87	0.61	5.48	12.65	2.44
Sugarcane leaves	71.13	1.32	0.24	7.66	15.54	4.11
White maize	10.33	9.14	4.82	7.21	67.10	1.40
Yellow corn	7.90	8.30	2.90	2.11	82.50	2.10
Sorghum	11.00	9.16	4.22	2.93	82.71	1.20
Barley	13.20	10.11	1.60	8.20	77.11	3.00
Kidney beans	9.75	27.97	0.95	5.65	50.71	5.51
Agricultural products						
Soybean meal	13.50	45.5	16.92	7.5	24.62	5.51
Cottonseed meal (with hulls)	7.5	27.1	8.5	21.44	27.2	6.9
Cottonseed meal (no hulls)	7.06	44.41	10.06	5.2	25.7	7.53
Sesame seeds (hulls)	9	30.21	14.4	18	20.1	17.4
Lentil seed cake	18.5	13.33	10.73	7.08	35.22	7.49
Rice bran	10.5	13	12.06	10.32	57.2	7.5
Rice middlings	9.22	11	7.81	1.14	75.93	4.2
Coarse wheat bran	12	11.09	3.56	17.83	63.78	4.33
Fine wheat bran	12	17.05	3	9.44	65.51	4.5
Animal sources						
Blood meal	9.3	81.22	1	–	–	5.33
Fishmeal (local)	10.3	65.25	10.5	–	–	16.56
Fishmeal (imported)	9	70	6.11	–	–	11.62
Animal gelatine	11.4	85.69	3.09	–	–	–
Meat and bone meal	4.56	61.82	6	–	–	26
Poultry by-product meal	13	53.85	23	–	–	18.22
Shrimp meal (local)	12.71	51.67	5.55	–	–	26.93

NFE, nitrogen-free extract.

tilapia in semi-intensive systems. This approach can substantially reduce feed cost and improve fish yields, depending on feed source, feeding rate and fertilization effectiveness. Middendorp (1995) found that the net production of Nile tilapia reared in ponds receiving daily cattle manure and CSC at a daily rate of 3 or 6% of their body weight was significantly high (4.8 and 6.5 Mt/ha/year, respectively), while the use of cow manure alone resulted in negative net production. The use of on-farm resources as pond inputs for cultured *Tilapia rendalli* and *Oreochromis shiranus* on small-holder farms in Malawi in monoculture and polyculture systems has also been investigated (Chikafumbwa, 1998). The ponds received daily applications of Napier grass (NP) and maize bran (MB) or isonitrogenous combinations (0.7 kg N/ha/day) of NP, MB, wood ash and urea for 126 and 112 days. The application of NP/MB produced the best performance.

Most farmed tilapia is produced in resource-limited, rural areas in Africa, Asia and Latin America. At the same time, many households produce tilapia for subsistence, using very low inputs. It is necessary, therefore, that unconventional, locally available and cheap feed sources be used as feed/food inputs for the semi-intensive culture of tilapia in earthen ponds. Plant sources and agricultural by-products which are of little or no economic value, such as sugarcane bagasse and water hyacinth, have been successfully incorporated in tilapia feeds (El-Sayed, 1991, 2003). The chemical composition of some feed ingredients commonly used for aquafeed manufacturing in Egypt is given in Table 4.2, while Table 4.3 contains a list of some unconventional feed sources that are of potential use for farm-made tilapia feed in Egypt.

Plant materials and agricultural by-products are generally used without processing (or with minimal processing, such as sun-drying) because it may not be economic to process these sources into the supplemental feed. For example, sugarcane bagasse is essentially carbon, whereas water hyacinth contains very high amounts of water, and therefore processing them may not be cost-effective for the fish farmers. It should also be emphasized that the quality, sustainable availability, composition and costs of these feed inputs must be determined before the farmer decides whether or not to use them as food inputs. Field trials of these sources are also necessary, because a big gap sometimes exists between the results obtained based on laboratory research and those achieved under field practices.

4.4.4. On-farm feed formulation and preparation

In addition to feed ingredients, formulated supplemental feeds are often used to feed tilapia in semi-intensive systems. These feeds must be low in protein but high in energy content, since the fish get some of their protein requirement from natural food available in fish ponds. On-farm feed formulation depends on a number of factors, including the following:

1. The fund resources of the farmers to purchase feed ingredients and feed manufacturing equipment, electricity, fuel, etc.
2. Available feed ingredients, sources, composition and costs. This includes seasonal availability, current usage, proximate composition, quality control, digestibility and assimilation, handling and processing costs prior to mixing or pelleting (Tacon, 1995).
3. Intended processes employed, such as sorting, grinding, mixing and pelleting (steam pelleting, cold pelleting, expansion pelleting, flaking, mash, crumple, paste, ball, moist or dry pellets).
4. Production targets, stocking density of fish and natural food availability in fish ponds.

On-farm feed preparation for tilapia in semi-intensive culture is simple and easy. No preparation or processing is needed if only feed ingredients (such as wheat bran, rice bran, cottonseed cake, etc.) are used as feed inputs. Other ingredients, such as weeds, water hyacinth and plant leaves, may need chopping and cutting into small pieces, particularly when fed to small fish (but the costs of processing should be kept in mind, as mentioned earlier). Farmers should follow the following steps for the preparation of supplemental diets containing mixtures of ingredients:

1. Prepare the equipment needed for feed preparation, such as: weighing scale, sieves, mixer, 10–20 kg capacity meat grinder, corn- or soybean-meal grinder and a steamer (if cooking is intended).
2. Sort the ingredients to ensure that they do not contain any spoiled or rotten parts and hard objects, such as nails, rubble, etc.

Table 4.3. Selected unconventional feedstuffs that can be used as feed inputs for semi-intensive culture of tilapia in Egypt. (After El-Sayed, 1999.)

Feedstuff	Description	Nutritional value/composition (%)	Processing and use
Bakery waste	Includes flour waste, bread rejects, bran	Depends on the source of flour	Fed directly or ground or mixed with other ingredients
Brewers' waste	The residues of sugar and starch extraction from barley. It is relatively rich in fibre and digestible carbohydrates	CP = 15; EE = 8.2; CF = 18.9; ash = 23	Included either wet or dried in the feeds
Poultry manure	Dried poultry excreta collected from poultry farms, usually contain undigested food particles and grains	CP = 7–11; Ca = 4.4–11.7; P = 1.3–3.2	Dried and used directly as a feed ingredient or mineral source at about 5% in the feeds
Buffalo and cow manure	Dried or wet excreta collected from buffalo farms	CP = 5–9; Ca = 4–13; P = 2.3–3.4	Used fresh or dried as feed ingredient for herbivorous fish
Restaurant wastes and rejects	Plant- and animal-based wastes and rejects rich in carbohydrates and ash	Varying according to the source	Sorted and used directly or dried and ground
Fruit and vegetable market wastes and rejects	Leaves, roots, stems and rejects of fruits and vegetables collected from auction and wholesale markets	CP = 8–13; CF = 15–20; ash = 22–28	Used fresh, dried or as silage for cattle, poultry or fish
Rice polishing wastes	Rice grains, middling, bran and hulls	Varying according to the source	Thrown directly into fish ponds, or mixed with the feed
Sugarcane bagasse	Residues of sugarcane pressing	> 40% digestible carbohydrate, > 45% fibre contents	Dried, ground and incorporated into the feed (at low levels)
Duckweed	Grown naturally in freshwater lakes, canals, reservoirs and drainage	CP = 42	Fed fresh to tilapia and carps or incorporated in the feed
Starch and yeast industry wastes	Residues of starch and yeast industry	CF = 37; NFE = 40	Incorporated in animal feed
Aquatic weeds	Freshwater weeds and macrophytes	Vary according to species and location	Used fresh or dried as feed sources for herbivorous fish
Water hyacinth	A macrophyte heavily grown in freshwater canals, reservoirs and drainages	CP = 21.1; EE = 1; CF = 19; ash = 18.2; NFE = 40.7	Cut into small pieces and used fresh or dried and included in the feed (at 10–20%)

CP, crude protein; EE, ether extract; CF, crude fibre; Ca, calcium; P, phosphorus.

3. Dry the ingredients separately (sun-drying is recommended in tropical and subtropical regions, to save energy costs).
4. Mix the ingredients at the predetermined levels.
5. Add water or cooked carbohydrate sources, such as starch, to make a paste.
6. Use a simple meat grinder to extrude the pellets (Fig. 4.4).
7. Dry the pellets (sun-drying is recommended in tropical and subtropical regions).
8. Store the feed in clean, shaded and aerated stores.

In addition to extruded pellets, prepared wet feed paste can also be fed directly to fish in the form of wet balls. However, some of the nutrients are wasted if this approach is adopted, in addition to the need for daily preparation of the feed paste. There is no standard composition of farm-made supplemental feed for tilapia. Most farmers in South-east Asia use fishmeal and a carbohydrate source (such as rice bran) at a ratio of 1 : 3.

4.5. Polyculture

Tilapia can produce high yields in monoculture systems because they can fill several feeding niches, compared to selective feeders such as carp. However, the production of tilapia with other fish or crustacean species in a polyculture system is spreading widely in many parts of the world. It is believed that this system results in a higher net yield than monoculture system, due to the full exploitation of multiple niches. Yet there is a little evidence to support this argument, except for the polyculture of fishes that are characterized by selective feeding habits, such as Chinese carp and common carp.

Fig. 4.4. Extrusion of farm-made tilapia feed in Thailand (photo provided by T. Samrit).

For maximizing tilapia production in polyculture systems, it is essential to fully understand the fish–fish and fish–environment relationships. Understanding such complicated relationships will enable fish farmers and farm managers to select suitable species, sizes, stocking densities, fertilization regimes, nutritional inputs and other management practices. The culture of fish species that have different feeding habits in the same pond will result in more efficient utilization of pond resources. If different feeding niches are used in such a way that the wastes of one species are used as food for another, higher production efficiency is expected to be achieved (Milstein, 1997).

The selection of proper species for polyculture will also minimize antagonistic and maximize synergistic relationships in fish ponds, leading to increasing available food resources for cultured species and improving environmental conditions. Such a fish–fish and fish–environment relationship is exemplified by tilapia and common carp culture in Israel. Monoculture of tilapia and carp usually results in the accumulation and decomposition of organic matter, leading to increasing ammonia and nitrite levels in the pond. Blooming of blue-green algae and other small algae is also developed in monocultured tilapia ponds, where the fish feed on these algae. In a tilapia–carp polyculture system, a synergistic relationship occurs, where carp stir up the mud and facilitate the efficient utilization of organic and inorganic matter. This process significantly reduces the accumulation of ammonia and nitrite and organic loading. Stirring also makes nutrient more available for large-size algae, which can be used as fish food. Higher photosynthesis also leads to increased oxygenation and pH and reduced nitrification.

Tilapia are generally cultured with a number of freshwater or brackish-water species, including carp, mullets, catfish, prawns and shrimps. The culture of tilapia with catfish, anabantids and carp has been a traditional practice in South-east Asia for many years. Tilapia polyculture is also widely expanding in Africa and Latin America. Several studies have considered the ecological and economic efficiency of different combinations of tilapia and other species in polyculture systems. Edirisinghe (1990) stocked common carp, big-head carp and Nile tilapia in fertilized ponds at a density of 18,000 fish/ha at a ratio of 2 : 1 or 1 : 2 big-head carp : tilapia and 2 : 3 : 4 big-head carp : common carp : tilapia, for 154 days. The inclusion of common carp significantly increased the growth rate of big-head carp and tilapia, indicating that the incorporation of a bottom feeder was beneficial. Similarly, Wang *et al.* (1998, 1999) studied the effects of rearing different combinations of aquatic animals in polyculture systems. They found that the polyculture of common sea perch–Chinese shrimp–tilapia hybrids converted nitrogen more efficiently than other fish combinations. In another study, the effects of the structure of tri-species polyculture of Chinese shrimp, Taiwan red tilapia (*O. mossambicus* × *O. niloticus*) and constricted tagelus (*Sinonovacula constricta*), stocked at different combinations, on the profitability and ecological efficiency of the system were investigated (Tian *et al.*, 2001). The COD and chlorophyll a were higher, while transparency was lower in the control (shrimp monoculture) than in polyculture treatments, indicating the efficiency of polyculture systems in N and P conversion. The best stocking density for optimum performance was 7.2 : 0.08 : 14.0 shrimp : tilapia : tagelus/m^2. The total conversion efficiency of N and P in the system was 23.4 and 14.7%.

When tilapia are polycultured with shrimp or prawn, the stocking densities of these crustaceans may or may not affect tilapia performance, depending on the species, stocking size and type of culture water. The growth, survival and yield of tilapia stocked at 1 fish/m^2 with 0, 2, 4 and 6 post-larval freshwater prawns (*Macrobrachium rosenbergii*) were not affected by the stocking density of the prawn (Dos Santos and Valenti, 2002). A stocking density of up to 6 prawn larvae/m^2 did not affect tilapia production and required neither additional feeding nor significant changes in management. This polyculture system allowed increased production with the same amount of feed. On the contrary, the survival, growth and yield of both *O. niloticus* and tiger shrimp (*Penaeus monodon*) reared at different combinations of stocking densities in a polyculture brackish-water system were significantly affected by stocking densities and competition for food (Gonzales-Corre, 1988). The best performance of shrimp was attained at a stocking density of 6000 shrimp and 6000 tilapia/ha, compared to monoculture systems. However, tilapia had low survival and slow growth in all treatments, due to the competition for food between Nile tilapia and tiger shrimp. It is therefore worth mentioning that in polyculture systems there is a degree of competition even between species with completely different feeding niches, especially at high stocking densities.

One of the serious drawbacks associated with tilapia culture in semi-intensive systems is their continuous and excessive breeding in culture ponds, especially at low stocking densities. This phenomenon often results in overpopulation and low production of marketable fish, mainly due to the competition for food and space. Controlling the reproduction and recruitment of tilapia in ponds is therefore essential for successful and profitable tilapia culture. There are a number of methods used for tilapia seed control, including monosex culture, the use of predators and stock manipulation. The production of monosex tilapia is highlighted in Chapter 7. The use of carnivorous fish including walking catfishes, *Clarias gariepinus*, *Clarias macrocephalus* and catfish hybrids, for controlling overpopulation of tilapia has been a common practice in many parts of the world.

Lazard and Oswald (1995) found that the addition of 260, 150-g African catfish controlled the population of Nile tilapia in 10 acre ponds. The authors also found that other carnivorous fish such as banded jewelfish, *Hemichromis fasciatus*, and the catfish *Heterobranchus isopterus*, were successful in controlling tilapia seed. Similarly, the inclusion of catfish fingerlings with blue tilapia significantly controlled their reproduction and improved their growth, in addition to the improvement in catfish growth (Morrison *et al.*, 1995). Nile perch, *Lates niloticus* (Ofori, 1988), and tarpon, *Megalops cyprinoides* (Fortes, 1985), have also been successful in controlling tilapia population in monoculture and polyculture systems, respectively. A prey : predator ratio of 1 : 20 to 1 : 250 for *L. niloticus* with tilapia has been suggested for controlling tilapia seed production (Ofori, 1988). In addition to controlling fish population, the use of these predators improves the growth and yield of cultured fish. The snakehead (*C. striata*) can also be successfully used for controlling tilapia recruitment in Asia, even at very low predator : prey ratios. Yi *et al.* (2004b) reported that snakehead stocked with monocultured or polycultured mixed-sex tilapia completely controlled fish recruitment at a predator : prey ratio of 1 : 80. However, the addition of snakehead into tilapia ponds did not result in any improvement in tilapia growth and significantly reduced tilapia yields.

4.6. Integrated Culture

Integrated aquaculture is the practice of farming fish and shellfish with land crops and domestic animals. In other words, it is a means of diversification of farm outputs through the full utilization of available resources and wastes. This farming system depends mainly on available resources, with minimum use of inputs from outside the farm. Integrated aquaculture has been an ancient practice in China and many other South-east Asian countries (Pillay, 1990). The most popular integrated farming systems of tilapia are their culture in rice fields and their integration with domestic animals. The pen-cum-pond system is also a promising integrated system. The research conducted at the AIT revealed that this production system has a great production potential, in addition to its ability to reduce the environmental impacts resulting from culture effluents (see Chapter 12). Although this system was tested successfully in the early 1990s (Lin, 1990), it has not been applied on a commercial scale. The common integrated systems of tilapia are discussed in the following sections.

4.6.1. Tilapia culture in rice fields

In the 1990s, the world rice area was about 148 million ha (Halwart, 1998), about 90% of which was in Asia. This huge area can be a substantial source of fish for rural households worldwide. Therefore, fish culture in rice fields has received great attention in recent years as a means of sustainable rural development, food security and poverty alleviation in several developing countries, especially in Asia. China is currently the largest fish producer from rice fields, with a production exceeding 377,000 Mt in 1996 from an area of 1.2 million ha, followed by Egypt (Halwart, 1998). Raising fish in rice fields has many advantages, including the following:

1. The recycling of nutrients by the fish through feeding and depositing faeces in the soil, which can be used as a fertilizer for rice, reducing the dependence on commercial fertilizers. In this process, the uptake of nutrients such as phosphorus and nitrogen by the rice is usually increased.
2. The increase in rice yields.
3. The increase in revenue from both rice and fish production.
4. An additional source of protein for rural households, especially in countries facing a shortage of captured fish.
5. An effective method of insect pest control (such as leafhoppers, stem-borers and aphids).
6. An effective method of controlling weeds, which herbivorous fish consume.

Tilapia culture in rice fields is widely practised in many Asian countries, and has also gained considerable momentum in some countries in Africa and Latin America. Tilapia can be raised in rice fields in monoculture systems; however, this approach is not very common. Instead, polyculture of tilapia with carp is widely practised in South-east Asia and Africa (Fig. 4.5). Stocking densities of target species vary depending on geographical location, fish sizes and culture objectives (market size vs fingerling production). The fish can also be grown with or without fertilization or supplemental feed, and therefore the yield varies considerably. Accordingly, the profitability of tilapia culture in rice fields depends on the inputs pumped into the system.

A number of research studies have been conducted on monoculture of tilapia in rice fields, with varying results. Stocking sizes of cultured fishes appear to play an important role in fish growth and rice yield. When tilapia was stocked at 7500 fish/ha, at two sizes, small (2.9–3.5 g) and large (28.2–33 g), the growth rate of small fish (12.8 g/fish) was better than that of large fish (33.76 g/fish) (Haroon and Pittman, 1997). In a 78-day culture period, the net yields were 37.2 and −58.3 kg/ha/crop for small and large tilapia, whereas survival rate was 66%. The low tilapia growth, survival and yield in that particular study have been attributed to: (i) slower swimming of the fish, which were entangled in aquatic macrophytes, leading to predation by frogs, snakes and birds; and (ii) lower nutrient ingestion and digestion. Zooplankton in the ponds made little contribution to the filter-feeding mechanism of tilapia, and ontogenetic shifts in tilapia diets were also minimal or absent. Similar results of poor survival, growth

Fig. 4.5. Raising tilapia and carp in rice field in Egypt (photo provided by the General Authority for Fisheries Resources Development, Egypt).

and yields of tilapia monocultured in rice fields have also been reported by other authors (Mang-Umphan and Arce, 1988; Haroon and Alam, 1992; Sevilleja *et al.*, 1992; Torres *et al.*, 1992).

Because tilapia reach their sexual maturity at an early age (3–6 months), stocking large tilapia in a rice field may result in fish reproduction, which will lead to overpopulation and fish stunting, because of the increased competition for food and space. In rice–fish culture, if large tilapia are to be monocultured, all-male populations are preferable. All-male culture overcomes the problem of early spawning, and is likely to improve fish growth and yield. However, more research is needed along these lines before any concrete conclusion can be drawn.

The polyculture of tilapia in rice fields has become a common practice in South-east Asia. In China, for example, the polyculture of Nile tilapia, common carp, grass carp (*Ctenopharyngodon idella*) and crucian carp (*Carassius auratus*) in rice fields is very common, productive and profitable. This culture system is practised mainly in Guangxi, Sichuan and Hunan provinces (Ming and Yi, 2004). All-male Nile tilapia or hybrid tilapia fingerlings (5 cm) are generally stocked in rice fields, mainly with carp, at a density of 4500–7500 fish/ha, about 10 days after transplanting rice seedlings (Y. Yi, Pathum Thani, Thailand, 2004, personal communication). The fish depend exclusively on natural food available in rice fields, and no supplemental feed is provided. After 100 days of culture, the fish reach about 150 g on average, yielding about 500–1100 kg/ha. The fish produced are generally marketed in local rural markets. Even in countries such as Vietnam, Bangladesh and India, where tilapia culture is not very popular, their polyculture in rice fields has started to gain ground. As a result, polyculture of tilapia with carp in rice fields has become established extensively in some Bheries of West Bengal (India) (Natarajan and Aravindan, 2002).

Many variables, including fish species and sizes, stocking densities, culture inputs (such as inorganic fertilizer, manure and supplemental feed), season, rice vegetative phase and rice reproductive and ripening phase, affect the growth and yield of cultured tilapia. Accordingly, the net production and profitability of tilapia culture in rice fields vary strongly from one geographical region to another and even from one season to another in the same region. In Vietnam, for example, Nile tilapia were successfully raised with silver barb and common carp in rice fields at a total density of 20,000 fish/ha, in different combinations of sizes and stocking densities, for 149 days (Rothuis *et al.*, 1998). Small tilapia (1.8 g) were stocked at 6600, 5400, 3400 and 1400 fish/ha, while silver barb was stocked at 0, 4000, 10,000 and 16,000 fish/ha, whereas common carp was stocked at 13,400, 10,600, 6600 and 2600. Large tilapia (28.8 g), silver barb and common carp were stocked at 3800, 4500 and 0 fish/ha. Tilapia survival was about 64%, and the growth was inversely correlated with initial stocking density. The best growth of tilapia (89 g) was achieved at 1400 tilapia, 16,000 large silver barb (80 g) and 2600 common carp/ha. The good growth of tilapia was attributed to the availability of natural food at the low stocking density and the synergistic effect of cultured species, where the digested plants contained in the excreta of silver barb may have been utilized by tilapia.

The use of fertilizer and/or supplemental feed in rice–fish culture is disputable, and not fully understood. Some researchers found that the growth and yield of Nile tilapia raised in rice fields were low and were not affected by the source of fertilizer (organic or inorganic) (Mang-Umphan and Arce, 1988). Vromant *et al.* (2002) also found that the growth and yield of Nile tilapia raised in a polyculture system with Java barb, *Barbodes gonionotus*, and common carp in rice fields were significantly improved when supplemental fertilizer and artificial feed were provided. However, growth, yield and survival values were still low compared to those of Java barb. The authors concluded that Nile tilapia are not well suited for culture in rice fields.

In contrast, Dela Cruz and Lopez (1980) reported that the culture of Nile tilapia in rice fields supplemented with feed and fertilizer in the Philippines was profitable. About 480 kg of Nile tilapia and 222 kg of common carp/ha were produced from rice fields stocked with 10,000 fish/ha and supplemented with feed and fertilizer. The application of fertilizer increases primary productivity in the fields, and in turn enhances food consumption by the fish. Similar positive effects of rice fertilization on tilapia growth and yield have been reported by Sevilleja *et al.* (1992).

Once again, it is clear from the above studies that the suitability of tilapia culture in rice fields is still controversial, and more research is needed to further investigate the potential of these fish as a candidate for this system. More research is also needed on whether fertilization, supplemental feeding or a combination of both would be more biologically and economically appropriate. More information should also be gathered on fish and crustacean species that are commonly cultured with tilapia in rice fields in order to adopt the most suitable species that occupy different trophic niches from those of tilapia. The growth and yield of tilapia raised in rice fields are summarized in Table 4.4.

One of the crucial factors that the rice–fish farmer should be aware of is the time at which the fish must be stocked in rice fields. The frequent fertilization of the rice fields after transplantation and the low plant density at this early stage, which allows high light penetration of the water, enhances plankton productivity. Progressive growth of the rice is accompanied by progressive shading, low nutrient availability and decreased primary productivity. Accordingly, the average daily food consumption by cultured fish is expected to be higher during the rice vegetative phase (early stage) than during the rest of the cropping season. Therefore, it is essential that the farmer stocks his tilapia as early as possible after rice transplantation.

Along with the additional fish crop obtained from rice fields, rice yield can be increased up to 20%

Table 4.4. The growth and yield of Nile tilapia raised in rice fields.

Culture system	Initial wt (g)	Final wt (g)	Net yield (kg/ha)	Period (days)	Reference
Monoculture	2.2	15	43	65	Torres *et al.* (1992)
Monoculture	3.1	13	37	78	Haroon and Pittman (1997)
Monoculture	30.7	34	–58*	78	Haroon and Pittman (1997)
Monoculture	8.5	36	64	75	Mang-Umphan and Arce (1988)
Polyculture	1.8	56	117	149	Rothuis *et al.* (1998)
Polyculture	28.8	66	62	149	Rothuis *et al.* (1998)
Polyculture	5	61	–	102	Chapman and Fernando (1994)
Polyculture	–	74	45.7	120	Vromant *et al.* (2002)
Polyculture	5 cm	150	500–1500	100	Ming and Yi (2004)

*Net yield is negative because of the high fish mortality.

(Lightfoot *et al.*, 1992). Tilapia culture and rice farming are also complementary systems, since the fish can feed on rice pests and play a significant role in integrated pest management (Halwart, 1998). The fish have also been reported to control case-worms, *Nymphula depunctalis* (Vromant *et al.*, 1998). The saving of pesticide cost and earning from fish sales have been reported to increase the net income from rice–fish farming by 7–65%, compared to rice monoculture (Halwart, 1998).

4.6.2. Animal–tilapia culture

Farming tilapia with domestic animals, including pigs, ducks, chickens and cattle, has become a common practice in many parts of the world, particularly in developing countries (Table 4.5). Ducks and pigs are probably the most successful candidates for farming with fish. In China, which is the largest fish producer from aquaculture, pig raising is widely integrated with Chinese and common carp culture, while tilapia/pig farming is practised in certain provinces in south-west China (e.g. Sichuan) (Y. Yi, Pathum Thani, Thailand, 2004, personal communication). Tilapia/duck farming is very popular and mutually beneficial in some coastal provinces such as Guangdong (south China). The ducks can feed on unwanted organisms in and around fish ponds, while their excreta fertilize these ponds. The following sections throw more light on these integrated animal/tilapia culture systems.

4.6.2.1. Tilapia–duck culture

As mentioned above, tilapia–duck farming (Fig. 4.6) is a promising and profitable integrated culture system, compared to other integrated farming systems (Pillay, 1990). The major fish species integrated with duck raising in China are silver carp, big-head carp and tilapia. The number of ducks to be raised in the fish pond depends on the duck species, the quality and quantity of feeds provided and the method of raising. The stocking density of ducks also depends on the environmental conditions and the stocking ratio and density of fish species polycultured in the pond. Generally, between 2000 and 4000 ducks/ha are raised in fish ponds. Fish performance and yield decrease if the duck number exceeds 4000/ha, due to the higher manure load, low oxygen and high ammonia concentrations and the increased biological oxygen demand (BOD) (Bao-tong and Hua-zhu, 1984).

The effect of the number of ducks on fish production is controversial. Delmendo (1980) reported that increasing the number of ducks to 1000–2000 ducks/ha yielded only 1.5 Mt fish/ha/year, while Hopkins and Cruz (1982) achieved a total yield of 5.48 Mt/ha/year of tilapia and carp (4.7 Mt tilapia) when the number of ducks was only 750/ha, with a manure load of 81 kg dry matter/ha/day, at a stocking density of 20,000 fish/ha. Even a very low number of ducks with tilapia farming can make a big difference in fish growth, yield and system profitability. Similarly, Sin (1980) reported that raising 1000–2000 ducks/ha

Fig. 4.6. Duck–tilapia integrated culture in Thailand (photo provided by P. Edwards).

Table 4.5. Summary of integrated animal/tilapia culture in some countries.

	Fish		Animal		Harvest data			
Country	Species (size)	No./ha	Type	No./ha	FW (g)	Yield (Mt/ha/year)	Remarks	Reference
Egypt	*Mugil cephalus* (6.5) *Cyprinus carpio* (3.6) Silver carp (3.3) Tilapia (3.1)		Ducks	298	146.9 270.6 519.5 274	1.6 (total)	Better growth and yield than non-integrated system	Soliman *et al.* (2000)
Philippines	Nile tilapia Common carp *Channa striata*	8,500 1,000 100	Pigs	20	184 505	1.7 (total)	Tilapia survival 78%; increasing fish density to 20,000 reduced average harvest weight and did not increase total yield	Hopkins and Cruz (1982)
Philippines	Nile tilapia Common carp *Channa striata*	8,500 1,400 100	Pigs	100	118 183 385	2.3 (total)	Tilapia survival > 100% (because of reproduction); increasing fish density to 20,000 reduced average harvest weight and did not increase total yield	Hopkins and Cruz (1982)
Philippines	Nile tilapia Milkfish *Penaeus indicus*	5,000–20,000 2,000 50,000	Broiler chicken	900		0.34–0.67 0.075–0.12 0.19–0.28	Best tilapia yield at a density of 15,000/ha	Pudadera *et al.* (1986)
Nigeria	Nile tilapia (15) Catfish (5)	20,000 10,000	Chicken	1,000	575 1,375	14.9 13.8	Indigenous chickens were better than exotic chickens	Njoku and Ejiogu (1999)

increased fish yield to 5 Mt/ha/year, compared with only 1 Mt/ha/year produced from non-integrated ponds. This controversy may have been related to the type of ducks and the quality of feed and excreta, as stated above. Gopalakrishnan (1988) found that Muscovy ducks were not suitable for integrated farming with tilapia (*Oreochromis andersonii*) in Zambia, because they do not stay in the water long enough to produce sufficient manure. When Peking ducks were used, fish production significantly improved (3.5–4.5 Mt/ha/year).

Soliman *et al.* (2000) studied the effects of raising ducks at low intensity (298 ducks/ha) on the production and profitability of Nile tilapia, blue tilapia, silver carp, common carp and grey mullet. The fishes were stocked at equal numbers in a polyculture system, in earthen ponds fertilized with cow manure and superphosphate and supplemented with a 10.6% cp diet, in Egypt. Integrated ponds resulted in better growth, feed conversion ratio (FCR), yield and profitability than non-integrated ponds. However, only common carp and silver carp showed significantly higher weight gains in the integrated system, while tilapia growth was similar in both systems, presumably due to the sufficient food availability resulted from fertilization and supplemental feeding.

4.6.2.2. Tilapia–chicken culture

As in the case of duck–fish farming, the performance of fish raised with chickens in an integrated system depends on the species and number of chickens, the quality and quantity of their feeds and excreta and water quality. Higher chicken density (i.e. higher manure load) may increase ammonia and BOD and decrease dissolved oxygen and water transparency and finally lead to reduced fish yield (Hopkins and Cruz, 1982; Njoku and Ejiogu, 1999). Therefore, the integration of tilapia with poultry should be carefully implemented, because the adoption of improper poultry type and density could lead to significant economic loss.

A number of studies have been conducted on integrated fish–chicken farming (Fig. 4.7), with varying results. Pudadera *et al.* (1986) evaluated the integrated culture of Nile tilapia, milkfish (*Chanos chanos*) and shrimp (*Penaeus indicus*) with broiler chickens, in brackish-water ponds in the Philippines. Tilapia were stocked at 5000, 10,000, 15,000 and 20,000 fish/ha, while the densities of milkfish, shrimp and chickens were kept constant at 2000, 50,000 and 900, respectively. Tilapia production ranged from 333.8 to 669.7 kg/ha, and the best net production of tilapia was obtained at 15,000 fish/ha. Njoku and Ejiogu (1999) raised 15 g Nile tilapia and 5 g African mud catfish (*Heterobranchus longifilis*) in ponds in Nigeria, at a stocking density of 30,000 fish/ha, with a ratio of 2 : 1. The ponds were integrated with 1000, 1500 and 2000 indigenous chickens/ha or 1500 exogenous broiler chickens/ha (control) for 270 days. The best daily weight gain (2 g/day) and yield (14.9 Mt/ha) were attained at 1000 chickens (3.6 Mt dry manure/ha), compared to 0.5 g/day and 3.09 Mt/ha when exotic broilers were used. This particular study emphasizes the fact that the type of fertilizer adopted for pond fertilization should be carefully considered.

Fig. 4.7. Chicken–tilapia integrated culture in Thailand (photo provided by P. Edwards).

4.6.2.3. Pen-cum-pond system

A pen-cum-pond system is a system in which one or more fish species are reared semi-intensively in ponds and fed naturally on the wastes resulting from other intensively cage (pen)-cultured fish species (Fig. 4.8). This system has been developed and used successfully by a number of researchers (Lin and Diana, 1995; Yi and Lin, 2001; Yi *et al.*, 2003b). However, as mentioned earlier, it is still not applied on a commercial scale. Further details on cage culture of tilapia in fertilized ponds are provided in Chapter 5.

Yi *et al.* (2003b) studied the ability of Nile tilapia reared in a semi-intensive system in ponds to utilize the wastes resulting from intensive culture of hybrid catfish (*C. macrocephalus* × *C. gariepinus*). They found that tilapia yield from the catfish-waste system was comparable to that obtained from fertilized ponds. Tilapia recovered considerable amounts of N and P contained in the wastes of intensive catfish culture. The authors suggested that water circulation between catfish pen and tilapia ponds is cost-effective, as it can sharply reduce nutrient contents in the pond effluent. Similar results have been reported with walking catfish (*C. macrocephalus*) and catfish hybrids (Lin and Diana, 1995), where the wastes from caged catfish increased phytoplankton productivity in the ponds, which supported the growth of Nile tilapia in the same ponds containing the cages. Moreover, tilapia reared in cages, feeding only on phytoplankton, in intensive channel catfish ponds grew reasonably well and improved the water quality in the ponds (Perschbacher, 1995). Yi and Lin (2000) stated the advantages of integrated cage culture in ponds as follows:

1. Wastes derived from high-protein diets of caged fish can be used as a food source for open pond fish.
2. Nutrients in wastes derived from cages are recovered, reducing the concentration of nutrients contained in the effluents, which are usually released directly or indirectly to the surrounding environment, causing increased eutrophication in the waters.
3. It is used in polyculture ponds to confine costly high-protein diets to caged high-value species and thus to achieve higher economic returns.
4. It is used in subtropical or temperate regions, where tropical fish species cannot overwinter, to make full use of growing seasons and make management, such as fish harvest, easy and convenient.
5. This system makes it possible to fatten large fingerlings with high-protein diets in cages and nurse fry with natural food derived from cage wastes in open water of the same pond.

4.7. Economic Efficiency of Integrated Culture

Although integrated fish–animal farming sounds, in most cases, biologically profitable, it may not be economically so, since several factors affect the profitability of this integration system. The economic effectiveness of fish–livestock production systems

Fig. 4.8. Tilapia culture in pen-cum-pond system in Thailand (photo provided by C.K. Lin).

varies sharply among farming communities. For example, conventional fish–livestock farming may be inappropriate in urban areas because of the large areas needed for the construction and integration of fish and animal facilities. Processing of livestock wastes into feed ingredients or fertilizer for aquaculture may also not be viable. Other culture inputs, such as agricultural by-products (rice bran, wheat bran, maize bran, etc.), labour, transportation, etc., may also not be available in urban areas, or available but at a high cost. This means that the cost of fish–animal integration in urban areas can be very expensive and may not be cost-effective.

In rural areas, where livestock is generally produced, the wastes are usually available at very low costs. Agricultural by-products and labour are also available at sustainable levels and low cost. Fish–livestock integration in rural areas is, therefore, very likely to be more sustainable and profitable. The effectiveness of integrated farming systems is thus location-specific. This means that the comparisons between integration practices in different locations are difficult and sometimes unreliable. Another important factor that affects the economic return of feedlot livestock–fish farming is the cost of livestock feeds, which may render the system unprofitable, especially in resource-poor areas.

As mentioned earlier, fertilizer is a key element in integrated fish–livestock farming. Understanding pond dynamics and conducting a broad economic analysis are necessary for successful and cost-effective fertilization regimes. The farmers should be aware of the following economic factors before they decide which fertilization strategy they will adopt (Knud-Hansen, 1998):

1. Market cost of biologically available N, P and C (purchasing versus on-farm production).
2. Transport and labour cost, including the lost opportunity costs and aquaculture integration with other activities.
3. The need for additional capital expenses, such as water aeration, mixing or exchange.
4. Market value of culture organisms (e.g. social acceptance, marketability of manure-raised fish).
5. Ecological impacts, including water quality (e.g. DO, ammonia, pH, turbidity, diseases, etc.) and increased organic loads.

Based on the above criteria, getting the best economic fertilizer is location-specific. Again, comparison between farms in terms of fertilizer requirement, quality and productivity may be inappropriate unless these farms are ecologically, socially and economically similar. Evaluation and comparisons should therefore be conducted on a farm-by-farm basis, taking into account the 'uniqueness' of each farm.

As mentioned earlier, tilapia–livestock integrated farming can be 'biologically' excellent, but an economical 'failure'. For example, when Nile tilapia and African mud catfish were raised in ponds in Nigeria in integration with indigenous chickens or exogenous broiler chickens, economic analyses indicated that the use of indigenous chickens resulted in a substantial economic return, whereas the use of exotic broilers led to a net deficit (Njoku and Ejiogu, 1999). The high profitability with indigenous chickens was attributed to input saving on variable costs, which in turn decreased total production costs, as opposed to the extremely high production costs when exogenous chickens were used. Similarly, Nitithamyong *et al.* (1990b) found that the growth of Nile tilapia reared with snakehead and pig in an integrated system for 7 months was significantly higher than in monoculture or polyculture with snakehead. However, the integrated system resulted in a significantly lower economic return than the other two systems.

It is obvious from the above discussion that, if the farmer/farm operator does not understand the ecological and dynamic processes occurring in the pond, the performance and yield can be very poor. For example, Ayinla *et al.* (1994) found that, when broiler birds were integrated with Nile tilapia and bagrid catfish, *Chrysichthys nigrodigitatus*, at 1500 birds/ha, fish mortality reached 40% in tilapia and 70% in catfish. These high mortality rates have been related to high organic loads, which may have led to deterioration in water quality. The mortality could have been sharply reduced if some management interactions, such as reducing bird number, partial changing of culture water, aeration, etc., had been taken.

4.8. Closing Remarks

1. Organic fertilizers stimulate natural food production through adding detritus, which, in turn, stimulates heterotrophic food chains and produces more phytoplankton and zooplankton. This natural food can be consumed by microphagous tilapia, leading to considerable improvements in fish yield.

Organic manure is better applied in fine particles or a colloidal state.

2. Tilapia should rely completely on natural food produced through fertilization till they reach 'a critical standing crop'; then supplemental feed should be applied. Timing of supplemental feeding ranges from 40 to about 150 days after stocking, depending on tilapia species and stocking size and fertilization regimes.

3. The history of the ponds and water quality, especially alkalinity, ammonia and turbidity, should be fully considered before pond fertilization.

4. Semi-intensive polyculture of tilapia may be preferred to monoculture due to the full utilization of food niches in the ponds. However, understanding the complicated relationships between cultured species and the surrounding environment is essential in order to enable fish farmers to select the most appropriate species, sizes and stocking densities.

5. Carnivorous fish species, such as Nile perch (*L. niloticus*) and catfish, are successfully used for controlling tilapia seed production. The stock ratio varies according to the species and sizes used. A ratio of 1 : 50 Nile perch : tilapia, for example, has been suggested.

6. The suitability of tilapia culture in rice fields is questionable, and extensive research is needed to verify whether or not tilapia should be considered as a candidate for rice–fish culture and whether monoculture or polyculture is preferable.

7. Duck–tilapia culture has been the most successful; however, Muscovy ducks may not be a good candidate. The type and density of ducks and chickens significantly affect fish growth and yield as well as water quality.

8. A pen-cum-pond system could be an excellent system for tilapia culture, since tilapia can use the waste of other fish raised in cages/pens, leading to both economic and environmental benefits.

9. Understanding pond dynamics and conducting a broad economic analysis are necessary for successful and cost-effective fish–livestock farming.

10. For successful and sustainable integrated fish–animal farming, the farmers should consider the economic and social factors, including market cost, variable costs, transport and labour cost, any additional capital expenses, the market value of cultured species and the ecological impacts of the system.

5

Intensive Culture

5.1. Introduction

Intensive aquaculture is simply the employment of high stocking densities of cultured species, in order to maximize the production with the minimal use of water. Such systems depend exclusively on artificial feeding and water reuse and/or exchange. Intensive systems yield high production, ranging from 100 to > 500 Mt/ha/year (Muir *et al.*, 2000). However, they require high capital and operating costs as well as high levels of technology and management tools. Therefore, fish farmers should fully consider the following criteria before adopting intensive farming systems:

1. Assurance of free and permanent access to seed supply.
2. Availability of funds needed for construction, maintenance and operation of the system.
3. Availability of technology, management tools and experienced personnel.
4. Awareness of environmental impacts of intensive farming systems and creating efficient and reliable conditions and mitigation protocols to minimize such impacts.

Tilapia culture has traditionally relied on extensive and semi-intensive systems in earthen ponds. The expansion of tilapia culture across the world, together with the shortage of fresh water and competition for it with agriculture and with urban activities, has gradually shifted tilapia culture from traditional semi-intensive systems to more intensive production systems. These fish are an ideal candidate for intensive culture due to their ability to tolerate high density and a wide range of environmental conditions, in addition to their high resistance to stress, disease and handling (see Chapter 3 for more details). Tilapia of the genus *Oreochromis*, particularly *O. niloticus*, *O. aureus* and *O. mossambicus* and their hybrids, have been reared successfully in intensive systems and have therefore become the preferred cultured species.

The intensive culture of tilapia has been globally expanding, experimentally and/or commercially, in ponds, tanks, raceways, cages, recirculating and aquaponic systems. Water quality (dissolved oxygen (DO), temperature, salinity, ammonia (NH_3), pH, dissolved solid metabolites, etc.), nutrition and feeding, and stocking densities are the most important factors that determine the success or failure of intensive culture of tilapia. Water quality requirements of tilapia were addressed in Chapter 3, while nutrition and feeding are covered in Chapter 6. This chapter discusses the intensive culture systems of tilapia, with emphasis on the advantages, disadvantages and suitability of each system, in addition to the criteria that lead to successful culture practices.

5.2. Stocking Density

Because stocking density is the most important factor affecting intensive fish culture, extensive research has been carried out on the effects of stocking density on tilapia production in different intensive culture systems. Sadly, the available information has revealed controversial results (El-Sayed, 2002). This controversy may have been due to the differences in species, sizes, nutrition and feeding, culture systems and water quality. This means that the relationship between stocking density and tilapia growth and survival is a function of a number of biological and physical factors, including the fish life stages, size, sex, social hierarchies and their tolerance to environmental changes, in addition to the configuration and hydrodynamics of the culture system (Huang and Chiu, 1997; Muir *et al.*, 2000; El-Sayed, 2002).

The relationship between stocking density and tilapia yield is generally positive (i.e. high fish density leads to high yield) (Watanabe *et al.*, 1990; Siddiqui *et al.*, 1997a), while negative correlation usually occurs between stocking density and individual fish growth (Sin and Chiu, 1983). Watanabe *et al.* (1990) and Huang and Chiu (1997) also found that size variations in Florida red tilapia and tilapia hybrids (*O. niloticus* females × *O. mossambicus* males) increased with increasing stocking density. Size variation may also be related to the social hierarchy of the fish. Under crowded conditions, large fish grow bigger and faster and dominate the population, while subordinate individuals grow at much slower rates (A.-F.M. El-Sayed and M. Kawanna, unpublished results).

In tilapia culture, it is the fish supply and market demand that determine the price of the fish produced. The farmer should therefore carefully define his target consumer and determine the appropriate stocking density that will produce the size of fish preferred by that consumer. He should also adopt the stocking density that satisfies maximum production efficiency and the production of fish of a uniform size. In other words, the farmer should determine whether he wants to produce a high yield but small individual fish (e.g. to be sold in rural areas, at low prices), or if he is willing to produce larger individual fish but a lower yield (e.g. for restaurants or export, at high prices).

In addition to the effects of stocking density on tilapia growth and yield, the size at stocking affects tilapia growth and yield. In this regard, McGinty (1986) studied the effects of stocking density on the growth of all male tilapia. He found that the medium-sized group (11–12 cm) gained more weight than the small (7–8 cm) or large (15–16 cm) fish groups. The growth of small individuals was depressed due to competition with large- and medium-sized fish, while large fish expended more energy for nest building and egg guarding. More details on the effects of stocking density and stocking sizes on tilapia growth and yield in various culture systems are provided in the following sections.

5.3. Intensive Culture in Earthen Ponds

Pond culture of tilapia is the oldest production system among all systems used for tilapia culture. Semi-intensive culture of tilapia in earthen ponds has been covered in Chapter 4. Despite the widespread expansion of semi-intensive tilapia culture in earthen ponds, little information is available on intensive culture of tilapia in these systems. This is probably because tilapia are cultured mostly with other fish species, typically in semi-intensive systems, or in integration with other animals and/or plants, especially in Asia, which is the major tilapia producer. In addition, intensive culture of tilapia in earthen ponds is practised on a much smaller scale than other intensive farming systems that are generally adopted.

Intensive tilapia culture in earthen ponds (Fig. 5.1) is affected by a number of factors, including stocking density, species and size, environmental conditions, management strategies and nutrition and feeding. Feeding the fish with complete pelleted feeds and water exchange and/or treatment are also necessary in intensive pond culture. The production of tilapia intensively cultured in earthen ponds in selected countries is summarized in Table 5.1.

5.3.1. Stocking density

As mentioned earlier, tilapia are known to tolerate a high stocking density and can withstand extreme crowding conditions. Tilapia have been cultured intensively in earthen ponds in the tropics and subtropics for decades. The effects of stocking density on tilapia production in ponds have also been investigated by a number of authors. Sin and Chiu (1983) studied the effects of stocking density and aeration on the growth and survival of all-male tilapia hybrids in earthen ponds. The fish were stocked at 30,000, 59,000 and 122,000 fish/ha in aerated or unaerated ponds in Hong Kong and were fed commercial feed (34% crude protein (cp)). When the fish were reared in unaerated ponds, the growth was significantly higher at the intermediate density (59,000 fish/ha) than at the high density (122,000 fish/ha), while the total yields were 8.4 and 11 Mt/ha, respectively. Survival and feed utilization efficiency were also better at the higher than at the intermediate density. When bigger-size fish (98–112 g) were reared

Fig. 5.1. Intensive culture of tilapia in earthen ponds in China (photo provided by K. Fitzsimmons).

at 30,000 and 111,000 fish/ha under continuous aeration for 74 days, average final weights were comparable (192.9 and 214 g/fish at the low and high density). The total yield was much higher at the high density (20.5 Mt/ha) than at the low density (6.3 Mt/ha). Thus aeration was essential for good growth and yield at a high stocking density, since it increased the yield from 67 kg/ha/day to 133 kg/ha/day.

Intensive culture of tilapia in earthen ponds has been practised in Israel for many years. Zohar *et al.* (1984) described intensive tilapia farming in earthen ponds at high densities with continuous aeration and partial flushing of the pond water. Pruginin *et al.* (1988) also evaluated the intensive culture of different tilapia hybrids in brackish-water ponds in Israel. Philippine red tilapia showed the best growth rates and fecundity, while Taiwanese fish were the worst. At a density of 30 fish/m^2 (300,000/ha), the yearly production ranged from 80 to 120 Mt/ha, with a feed conversion ratio (FCR) ranging from 1.8 to 2.1 and a daily weight gain of 1.3–3.2 g. In association with high summer temperatures, both primary production and organic loads increased and more frequent water replacement was needed.

5.3.2. Water exchange

Increased summer temperature and photoperiod generally increase both primary production and organic loads in intensive fish ponds. Under such conditions, the level of faeces and other total solids and dissolved metabolites, such as ammonia and nitrites, in the water will increase. The accumulation of these metabolites has been reported to negatively affect the performance of cultured fish. Therefore, frequent total or partial water exchange and/or aeration will become necessary. The effect of continuous water exchange on the growth of Nile tilapia in earthen ponds (45 m^3) was studied in Brazil (Sipaúba-Tavares *et al.*, 2000). The ponds were subjected to three treatments: no water exchange, no water exchange but provided with aeration at night, or continuous aeration. Water quality (pH, DO, CO_2, nitrite (NO_2), NH_3, total P and chlorophyll a) did not differ among treatments. These results were expected due to the low stocking density (60 fish per pond; 1.3 fish/m^3). Fish growth was highest at continuous water exchange, but this system may not be cost-effective, and water limitation may render it inapplicable in many areas.

Water shortage generally dictates the reduction, or even elimination, of water exchange in intensive pond culture systems. In arid areas, where fresh water is limited, it would be unwise to replace or exchange culture water. Instead, treatment of this water in such a way that would allow its reuse is a must. Several trials have been conducted on the effects of reducing or eliminating water exchange on the performance of tilapia in intensive pond systems. Chamberlain and Hopkins (1994) reported that water exchange and feed protein level had no impact on the performance of

Table 5.1. Intensive culture of tilapia in earthen ponds. Some of the present values were calculated from the original data.

Species (country)	Stock data		Growth data			Harvest data		Culture period (days)	Remarks	Reference
	No./ha	IW (g/fish)	ADG (g/day)	SGR (%)	FCR	FW (g/fish)	S (%)			
O. n ♂ ×	59,000	6.6	1.01	2.09	2.4	161.6	88.1	153	690 m^2 ponds, pelleted feed, 34% cp, 59,000 fish/ha was optimal under unaerated conditions. Aeration improved the growth at high densities	Sin and Chiu (1983)
O. m ♀	122,000	5.0	0.69	2.00	2.1	112.7	80.3	155		
(Hong Kong)	30,000	111.7	1.4	0.89	1.7	214.0	100	73		
	111,000	97.7	1.3	0.93	1.8	192.9	96.4	73		
Phil-red	300,000	10–14		3.22		240		160	Brackish water, earthen ponds, lined with PVC, aeration, solids removal, 28% cp	Pruginin *et al.* (1988)
NOR-1	300,000			2.56		190		220		
NOR-2	300,000			2.50		160		220		
Mos-red (Israel)	300,000			1.43		100		220		
O. n (Brazil)	13,333		1.11			314.3	88.7	180	No water exchange	Sipaúba-Tavares *et al.* (2000)
			1.6			423.0	85.5	180	No water exchange + aeration	
			1.82			474.5	86.7	180	Continuous aeration	
O. n (Costa Rica)	15,000	87	1.59	1.15	1.75	210	87	77	Pelleted feed, 34% cp, feeding stopped for 14 days after 50 days to assess sedimentation and re-suspension rates	Jiménez-Montealegre *et al.* (2002)

O. a, *Oreochromis aureus*; *O. m*, *Oreochromis mossambicus*; *O. n*, *Oreochromis niloticus*; Phil-red, Philippine red; NOR-1 and NOR-2, F1 hybrid of *O. n* × *O. a*; Mos-red, *O. m* red; IW, initial weight; ADG, average daily gain; SGR, specific growth rate; FCR, feed conversion ratio; FW, final weight; PVC, polyvinyl chloride; S, survival.

tilapia reared intensively in earthen ponds if sufficient aeration was provided. The addition of a carbon source (wheat bran, rice bran, etc.) on the water surface, which bacteria can grow and feed on, reduces ammonia accumulation (see Fig. 5.2). Filter-feeding fish such as tilapia and carp can feed on the bacteria produced. In this way, dietary protein can be reduced significantly, from 30% to 20%, and the overall cost can be reduced by 50%.

An active sedimentation and re-suspension technique can also be adopted to reduce water exchange in intensive pond culture. This recent development depends on the re-suspension of solid particles from pond sediments in the water column by continuous aeration, mixing, stirring and turbulence, because settled particle decomposition on the pond bottom can lead to anoxic conditions, while re-suspension favours the aerobic decomposition of the organic matter. The re-suspension of sediments can be an important mechanism of nutrient transfer in the pond, and can also help in water purification by increasing the number of nitrifying bacteria in the water column (Milstein *et al.*, 2000a). This system is very suitable for areas facing a shortage of water sources, since it is operated with much less water than in traditional intensive pond systems. In addition, herbivorous and filter-feeding fish (such as tilapia) feed on suspended particles and bacterial bloom. This process leads to additional saving in operational and feed costs.

Avnimelech *et al.* (1999) described sedimentation and re-suspension systems and the magnitude of sedimentation/re-suspension of particles derived from the water column and those from the pond bottom in ponds stocked with tilapia and common carp. Milstein *et al.* (2001b) compared the water quality and growth of hybrid tilapia (*O. niloticus* × *O. aureus*) and hybrid sea bass (*Morone saxatilis* × *Morone chrysops*) in conventional intensive ponds with 500% water exchange/day with active suspension intensive ponds (with only 8% water exchange/day). They found that bacterial growth and nitrification were higher, while ammonia removal was lower, in active suspension ponds than in classical intensive ponds. Fish growth was similarly good in both systems. This finding confirms the economic advantage of tilapia culture in water-saving active suspension systems, compared to traditional, water-demanding systems.

The rates of sedimentation and suspension can be affected by a number of factors, including

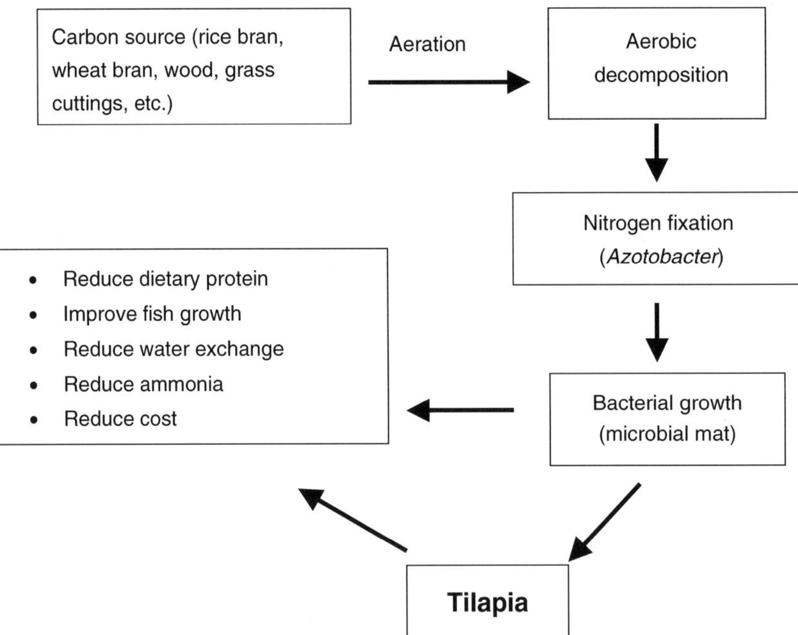

Fig. 5.2. Production of natural food in tilapia fish ponds. Reprinted from *Aquaculture* 179, 149–168, El-Sayed (1999a), with permission from Elsevier.

fish weight, biomass, primary productivity and carbon : nitrogen (C : N) ratio. Jiménez-Montealegre *et al.* (2002) measured the rate of sedimentation and suspension of organic carbon and total nitrogen in earthen tilapia ponds, based on nutrient inputs, water quality and fish size and biomass. They found that the rates of total solids sedimentation and suspension were highly correlated with fish weight and biomass, chlorophyll a, total suspended solids, total feed input and Secchi disc visibility.

The C : N ratio plays a significant role in reducing the accumulation of inorganic nitrogen in intensive fish ponds. Nitrogen accumulation is controlled by feeding bacteria on a carbohydrate source (Fig. 5.2), which results in the uptake of nitrogen from the water (Avnimelech, 1999). Thus, adjusting the C : N ratio in fish feed is essential for controlling the change in pond water quality. Chamberlain and Hopkins (1994) suggested that the optimum C : N ratio for bacterial growth is about 15 : 1. In support, Avnimelech (1999) studied the C : N ratio in tilapia feed, as a control element in aquaculture systems. He found that the addition of a carbonaceous substrate in commercial-scale fish ponds reduced inorganic nitrogen and increased the growth of bacteria which were taken up by the tilapia. When a high-protein (30% cp), low-energy diet with C : N = 11.1 : 1 was fed at 2% body weight (bw)/day or a low-protein (20% cp), high-energy diet (C : N = 16.7 : 1) was fed at 2.6% bw/day, the low-protein, high-carbohydrate diet led to the following:

1. Significant reduction in the accumulation of inorganic nitrogen.
2. Increased utilization efficiency of dietary protein.
3. Significant reduction in feed expenditure, feed requirement and feed cost.
4. Better growth rate (2.0 g/day compared to 1.59 g/day at 30% cp) and FCR.

The author (Avnimelech, 1999) suggested that about 20–25 g of carbonaceous materials are needed to convert 1 g of ammonia nitrogen into microbial protein. One advantage of this process is that it is fast and can reduce ammonia to the desired levels within 1–3 days (Avnimelech, 2003). These findings can be very beneficial to intensive tilapia farmers with earthen ponds. These farmers can use cheap carbon and nitrogen sources to partially replace expensive commercial high-protein feeds. The approach of producing single-cell protein(s) in fish ponds should also be encouraged in developing countries, where tilapia culture is widely practised.

5.4. Cage Culture

Cage culture of fish has received considerable attention in recent years as a means of the intensive exploitation of water systems, with low capital investment compared to other traditional intensive farming systems. This system has several attributes that make it an ideal farming system; these are:

1. Relatively low capital investment compared to other intensive culture systems.
2. Ease of observation and management and early detection of stress and disease.
3. Economic treatment of parasites and diseases.
4. Ease of cage movement and relocation.
5. The use of all the available water resources.
6. Ease of fish harvest and low cost of harvesting.
7. Minimum fish handling and reduction of mortality.
8. Optimum use of artificial feed and close observation of fish feeding response.
9. High stocking density, optimum feed utilization and improved growth rates.
10. Reducing pressure on land.
11. Control of reproduction of some species such as tilapia.
12. Can be used for storage and transport of live fish.
13. Easy to control predators and competitors.

Cage culture also has some limitations, including:

- Risk of theft is high.
- Risk of loss due to cage damage caused by predators or storms.
- High risk of disease outbreaks and difficulty of disease control.
- Low tolerance of fish to poor water quality.
- Complete dependence on high-quality feeds.
- Difficult to apply in rough weather.
- Water exchange is essential to remove fish metabolites and maintain good dissolved oxygen inside the cages.
- Rapid fouling of cage frames and nets, which require continuous cleaning.
- Substantial amounts of feed are lost through cage meshes.
- The accumulation of faeces and metabolites underneath the cage has a negative environmental impact.

Tilapia cage culture has been practised experimentally and commercially since the early 1970s (Bardach *et al.*, 1972). Currently, commercial tilapia culture in cages is expanding at a very fast rate, especially in tropical and subtropical developing countries in Asia, Africa and Latin America (Lin and Kaewpaitoon, 2000; Guerrero, 2001; Watanabe *et al.*, 2002). One of the major advantages of tilapia cage culture is that it can be applied in various water bodies, including rivers, lakes, reservoirs, swamps, ponds, thermal effluents, waste water and seawater (Table 5.2).

5.4.1. Factors affecting cage culture

The success of tilapia cage culture depends on a number of factors, including water quality, water level, tilapia species and strains, stocking density, stocking size, cage size and shape, feed quality and feeding frequency. Some of these factors have already been discussed in the other chapters. Therefore, only cage size and type and stocking density are discussed here.

CAGE SIZE AND TYPE. The sizes of tilapia cages vary considerably according to the farming operations. Breeding cages and fingerling production cages are generally smaller than fattening cages, while experimental cages do not usually exceed a few cubic metres (Coche, 1982). Commercial production cages range from a medium size (6–20 m^3) (Coche, 1982) to a very large size, up to 600 m^3 (Orachunwong *et al.*, 2001). It should be emphasized that the cage size adopted must consider the level of technology available. Large cages appear very suitable for tilapia culture, since they generally result in better growth, reduced feed loss and improved survival even at a low level of dissolved oxygen (Campbell, 1985; McGinty, 1991). Yet tilapia cage culture is generally practised in small- to medium-size cages. Different cage types are also used for tilapia culture in different regions of the world. It is very common for locally available materials to be used in cage construction, especially in small-scale, rural areas.

STOCKING DENSITY. Stocking density is another essential factor that influences tilapia performance in cages. As expected, increasing fish density may increase total yield, but it will reduce individual fish growth. The effects of stocking density on the growth and survival of tilapia in cages have not received enough attention. Only a few studies have been conducted in this regard, and the results have been inconclusive. Watanabe *et al.* (1990) found that Florida red tilapia fingerlings reared in sea cages at densities of 500–1000 fish/m^3 for 30 days had similar growth and feed utilization efficiency. On the other hand, the performance of Nile tilapia reared in floating cages in Thale Noi, Thailand, was significantly affected by stocking density (Chiayvareesajja *et al.*, 1990). The fish were stocked at 30, 100, 300 and 500 individuals/m^3, and were fed a weed-based diet for 3 months. The best production and profit were obtained at 500 fish/m^3, despite the fact that individual growth was better at lower densities. Once again, more research is needed on the relationship between stocking density and fish size on the one hand and tilapia performance in cage culture systems on the other.

5.4.2. Cage culture in Asia

Rearing tilapia in cages in Asia has become very popular, especially in South-east Asian countries, such as the People's Republic of China, the Philippines, Malaysia and Indonesia. Tilapia cage culture in South-east Asia is spreading, mainly due to the favourable weather conditions and the availability of suitable water sources. Tilapia culture in cage systems in South-east Asia has been described by Guerrero (2001). Mozambique tilapia were the main species cultured in cages in Asia for many years. A big switch to Nile tilapia has occurred in recent years, so that Nile tilapia is currently dominating in tilapia cage culture and becoming the most favoured species (see Table 5.2).

Cage culture of tilapia has been carried out in lakes and reservoirs in the Philippines since the early 1970s (Guerrero, 2001). Currently, the Philppines is believed to be the world's largest tilapia producer from cage culture. Tilapia and big-head carp are the most important species raised in freshwater cages in the country (Marte *et al.*, 2000). Both fixed and floating cages are used for tilapia culture in the Philippines. Fixed cages are usually used in shallow water bodies and made of synthetic netting (5 m × 5 m × 3 m to 15 m × 10 m × 3 m) and attached to bamboo poles, which are staked into the bottom. The cages are properly

Table 5.2. Intensive cage culture of tilapia in some countries.

Species (country)	Stock data		Growth data			Harvest data			Culture period (days)	Remarks	Reference
	No./m^3	IW (g/fish)	ADG (g/day)	SGR (%)	FCR	FW (g/fish)	Yield (kg/m^3)	S (%)			
Red tilapia	158	58.3	3.74		1.44	506.5	57.1	71.2	120	20–32% cp feeds, 12 m^3 cages, suspended in a river	Orachunwong *et al.* (2001)
(Chitralada strain)	100	75	4.43		1.50	606.5	59.2	97.7	120		
(Thailand)	133	88.8	4.26		1.44	600.0	59.6	74.4	120		
O. m × *O. h* (Thailand)	27.5	149	4.03		1.52	769	20.1	95	154	20–32% cp feed, cages suspended in pond stocked with tilapia fingerlings (2.8 g)	Orachunwong *et al.* (2001)
	27.2	127	4.26		1.51	736	19.6	98	143		
O. n	30	98.2	0.66		7.80	156.6	3.13	100	90	Bamboo cages, fed pellets containing dry weed, rice bran and FM (4 : 3 : 1), 500 fish/m^3 was best	Chiayvareesajja *et al.* (1990)
(Thailand)	100	86.1	0.77		7.50	145.8	9.77	99.7	90		
	300	88.8	0.52		9.90	129.9	25.25	99.6	90		
	500	76.6	0.60		9.60	117.8	38.6	98.5	90		
O. n (Thailand)	50	103	3.57		1.30	403	19.65	97.6	84	30% cp diet, cages suspended in 330 m^3 ponds. Aeration improved growth compared to non-aerated ponds	Yi and Lin (2001)

Continued

Table 5.2. *Continued.*

Species (country)	Stock data		Growth data			Harvest data			Culture period (days)	Remarks	Reference
	No./m^3	IW (g/fish)	ADG (g/day)	SGR (%)	FCR	FW (g/fish)	Yield (kg/m^3)	S (%)			
O. n (GIFT) (Philippines)	20	73.9	0.8–0.9	0.9–1.0	2.8–3.3	156.6–162.6		96–99	90	6 m^3 cage in ponds, fed either commercial feed, farm-made yeast or compost diets	Fitzsimmons *et al.* (1999)
O. h × *O. m* (Bahamas)	500–1000	1.79	0.4		1.17–1.27	13.4–14.2	6–12.5	84.1–93.5	30	1 m^3 sea cages, 32% cp, no difference between 500 and 1000 fish/m^3	Watanabe *et al.* (1990)
Red tilapia (Mauritius)	61	50.2	1.02	1.13		147.4	9.5	100	95	Seawater cages	Persand and Bhikajee (1997)
O. s (Kuwait)	200	9.65	1.9	1.9		379.6			193	Seawater cages	Cruz and Ridha (1991)
O. n (Lesser Antilles)	300	73	3.8	1.49	1.30	616	18.2	97.7	143	36% cp floating pellets, cages suspended in a 2 ha runoff pond	Rakocy *et al.* (2000a)

O. n, *Oreochromis niloticus*; *O. m*, *O. mossambicus*; *O. h*, *O. hornorum*; *O. s*, *O. spilurus*; S, survival; IW, initial weight; ADG, average daily gain; SGR, specific growth rate; FCR, feed conversion ratio; FW, final weight; GIFT, genetically improved farmed tilapia; FM, fishmeal.

arranged to allow good water circulation. Floating cages are employed in deep lakes, reservoirs, lagoons and bays. They are made of bamboo or steel frames and synthetic netting (5 m × 5 m × 2.5 m to 10 m × 10 m × 4 m). Tilapia fingerlings (0.5–2 g) are stocked in fixed cages at 15–45 fish/m^3 and from 55 to 80 fish/m^3 in floating cages, and are fed with supplemental and/or commercial feeds.

Eguia and Romana-Eguia (2004) reported that large-size net cages (10 m × 20 m × 2.5 m) may be stocked with fingerlings (10 g) at densities of 10–15 fish/m^3. The fingerlings depend on natural food, especially in plankton-rich areas. When higher stocking densities (50–200 fish/m^3) are used, providing the fish with pelleted feeds (25–30% cp) becomes necessary.

Tilapia cage culture is also widely practised in Indonesia, in rivers, irrigation canals, lakes and reservoirs. Floating cages (7 m × 7 m × 2 m) in Jatiluhur Reservoir are stocked with 100–150 kg of Nile/red tilapia to produce 626–1200 kg after 60–110 days (Hardjamulia and Rukyani, 2000). In Malaysia, intensive cage culture of red tilapia is practised mainly in rivers, reservoirs and former mining pools (Guerrero, 2001). Tilapia averaging 700 g are stocked in floating cages measuring 4 m × 3 m × 2 m, at a density of 2000 fish/cage for 2 months; then the number is reduced to 600 fish/cage for another 2 months to reach 1 kg/fish. The fish thus produced are exported to Singapore.

In China, tilapia culture is practised in floating cages (6 m × 4 m × 3 m) set up in lakes, reservoirs and rivers (Qiuming and Yi, 2004). Juvenile tilapia (> 50 g) are stocked at 100–150 fish/m^3 and fed with artificial feed (28–35% cp). The fish are harvested after a culture cycle of 120–150 days, at a final weight of 600–800 g and gross yield of 30–60 kg/m^3. Dey *et al.* (2000a) suggested that the good performance of tilapia reared at high densities in floating cages in the lakes in China is mainly because these lakes are rich in natural food, which the cultured fish can consume.

Despite the fact that freshwater aquaculture production in Thailand exceeded 200,000 Mt/year in recent years, the contribution of cage culture is only 0.3% of this production (Lin and Kaewpaitoon, 2000). In fact, the peak of annual fish production from freshwater cages reached 2700 tons in 1991 and declined after that to a minimum of 600 tons in 1995. However, tilapia cage culture has recently gained considerable momentum in certain parts of Thailand. Generally, cage culture is practised at a small-scale, artisanal level (Fig. 5.3), mainly in flowing waters, with the Chitralada strain being the prime candidate for culture. When 58–89 g fish were stocked in floating cages in Ratchburi River at densities ranging from 100 to 158 fish/m^3 and fed commercial diets (20–32% cp) for 120 days, they attained a final weight of 500–610 g/fish and a net yield of 57–59.5 kg/m^3 (Orachunwong *et al.*, 2001).

Cage culture of tilapia in other Asian countries, including India, Taiwan, Singapore, Bangladesh, Vietnam and Sri Lanka, is also gaining increasing attention. At this time, very little information is available on tilapia culture with this system in these countries.

Fig. 5.3. Tilapia cage culture in Thailand (photo provided by D.H. Giap).

5.4.3. Cage culture in Africa

In Africa, the contribution of tilapia production from cages is low and varies between countries (Jamu, 2001). Small cages constructed from locally available materials are generally used in various parts of Africa. In addition, commercial, large-scale cage culture of tilapia has been adopted in some African countries. Cage culture of tilapia in Africa, especially for small-scale farmers, is affected by the availability and cost of seeds, feed and unpredictable water sources and quality.

The largest and most successful commercial farm for cage culture of tilapia (Lake Harvest) is located on Lake Kariba, Zimbabwe (Fig. 5.4). The production of this farm is some 2000 Mt/year, while the goal is to increase it to 5000 Mt/year. The facility is also provided with a very high-standard processing unit. The tilapia that is produced is exported, either as fresh fillets or frozen, to Europe and the USA (Windmar *et al.*, 2000a).

The culture of tilapia in cages is also spreading in Egypt, particularly in northern delta lakes (Lake Manzala and Lake Borullos) and the Nile River (Fig. 5.5). Cage culture has also attracted the attention of the Egyptian private sector in recent years. Cages measuring 32–600 m^3 are commonly used (A.-F.M. El-Sayed, Alexandria, 2004, personal observation). The fish are generally fed with commercial tilapia feeds (25–30% cp). In some rural areas, crop farmers practise cage culture as an additional crop. Small, floating cages (2–4 m^3) constructed from inexpensive local materials are suspended in drainage canals (Fig. 10.4, Chapter 10). Stocking density is generally low, and agricultural and house wastes are provided as feed inputs. As expected, the production of these cages is low, but they can still be profitable because of the low culture inputs used.

The Egyptian government has also adopted a pioneer policy in recent years for promoting cage culture of tilapia. The Social Development Fund and the Agricultural Development and Credit Bank provide loans for tilapia cage culture enterprises to new university graduates who cannot find jobs within the government or public sector or those who prefer private business. The government runs the feasibility studies (free of charge), assists in constructing the cages and selecting the cage site, and sometimes provides the seeds at very low prices. Each cage (6 m × 6 m × 2 m) produces about 1.5 to 2 Mt per production cycle (5–7 months) (A.-F.M. El-Sayed, Al-Mansoura, Egypt, 1999, personal observation). The project has been successful, despite the fact that it faces some difficulties with respect to fry and fingerling shortages and feed supply in some areas. Other African countries, e.g. Malawi and Kenya, have recently adopted tilapia cage culture operations. Again, very scanty information is currently available on these countries.

Fig. 5.4. Large-scale cage culture of tilapia at Lake Harvest Farm, Lake Kariba, Zimbabwe (photo provided by R. Brummett).

Fig. 5.5. Arrays of tilapia cages in the Nile River (Rosetta branch), Egypt.

5.4.4. Cage culture in Latin America

Tilapia cage culture is expanding greatly in some Latin American countries, such as Brazil (Lovshin, 2000a), Mexico (Fitzsimmons, 2000) and Colombia (Popma and Rodriquez, 2000). In Brazil, Nile tilapia culture in floating cages in reservoirs, lakes and rivers and behind hydroelectric power dams is becoming very common, especially in the north-east region (Lovshin, 2000a). Cage size and stocking densities vary from 4 m^3 stocked at 200–300 fish/m^3 to much larger cages (> 100 m^3) stocked at 25–50 fish/m^3. The yield ranges from 150 kg/m^3/crop in the small cages to 50 kg/m^3/crop in the larger cages. Similarly, Costa *et al.* (2000) reported that the yield of cage-cultured Nile tilapia or red tilapia in Cear State, Brazil, ranges from 40 to 100 kg/m^3. The cage culture of Thai tilapia has resulted in a production of 100–300 kg/m^3. Roubach *et al.* (2003) described a cage culture system consisting of groups of about 15 cages, each of a size of 4–6 m^3, adopted in the north-east region of Brazil, stocked with juvenile tilapia of 25–30 g to reach 500–650 g in 120 days.

In Mexico, net pens (which use staked sides and rest on the bottom) constructed out of local materials are commonly utilized by low-income groups or individuals (Fitzsimmons, 2000). More sophisticated, intensive production floating cages are also in use, mainly in irrigation reservoirs. Floating or slow-sinking pelleted feed is generally used for feeding caged fish. Cage culture in Colombia is practised in large hydroelectric power reservoirs (Popma and Rodriquez, 2000). Cage size ranges from 2.7 to 45 m^3. Sex-reversed tilapia males are commonly stocked at 160–350 fish/m^3, at an initial size of about 30 g, to reach the marketable size of 150–300 g in 6–8 months, using commercial extruded feed (24–34% cp), yielding 67–116 kg/m^3. In Honduras, red tilapia of 5–12 g are cultured in 6 m × 6 m × 2 m cages (about 70–80 m^3) suspended in lake water, and fed pelleted feed (28% cp) (Meyer, 2002). The fish are transferred into cages with increasing netting mesh size according to growth phase until reaching harvest size (450 g).

5.4.5. Cage culture in fertilized ponds and pond effluents

Tilapia can be raised at relatively low densities in fertilized ponds and effluents of fish ponds or industrial sources. The success of this approach depends on fish density, water exchange, fertilization rates and supplemental feeding. As indicated in Chapter 4, tilapia can be raised in fertilized ponds, where dense plankton blooms produce a higher fish yield than moderate blooms, especially when small fish are raised. Farming tilapia in earthen ponds in an integrated system has attracted the attention of many researchers in recent years. In these systems, caged fish are usually fed high-protein feed, which results in the accumulation of considerable amounts of metabolic wastes and uneaten feed in the water masses surrounding the cages, causing water eutrophication and increased organic loads. These wastes can be used as a natural food for filter-feeding fish, such as tilapia and carp, in integrated systems (Lin, 1990). The benefits of this type of integration are listed in Chapter 4.

The cage-cum-pond system has been applied with catfish–tilapia (Lin and Diana, 1995; Yi *et al.*, 2003b) and tilapia–tilapia (McGinty, 1991; Yi *et al.*, 1996; Yi and Lin, 1997, 2001). McGinty (1991) studied the effects of cage size and the number of non-caged (open-pond) fish on tilapia production in cages suspended in fish ponds. Nile tilapia were stocked in the cages at 250 fish/m^3 and fed artificial diets. The ponds were stocked with non-caged Nile tilapia fingerlings at a rate of 3000 fish/ha or swim-up fry at a rate of 50,000 fry/ha. The growth of caged fish was increased at the lower number of non-caged fish, and tended to increase with increasing cage size, presumably due to the decrease in feed loss.

In another study, Yi and Lin (1997) investigated the effects of stocking density of non-caged small Nile tilapia on the growth of both caged and non-caged small and large fish in Thailand. Large tilapia were stocked in 4 m^3 net cages at 400 fish/cage, and the cages were suspended in earthen ponds stocked with small tilapia (15–16 g/fish) at two densities, 1.2 or 2 fish/m^3. The authors found that stocking densities of non-caged tilapia significantly affected the growth and FCR of both caged and open-pond fish. The growth, FCR and yield of both caged and open-pond fish were better at the low density than at the higher density. The net production of the whole integrated system (caged and non-caged) was significantly higher in the low-density treatment (5.5 Mt/ha/crop) than in the high-density treatment (4.5 Mt/ha/crop). In addition, large tilapia grew better in integrated system than in open water in the mixed-pond culture system at the same level of stocking densities of small tilapia.

The effects of stocking density of large caged Nile tilapia on the open-pond fish have also been studied (Yi *et al.*, 1996). Large fish (141 g) were stocked at 30, 40, 50, 60 and 70 fish/m^3 in cages suspended in earthen pond while small Nile tilapia (54 g) were stocked at 2 fish/m^3 to feed on the wastes of the caged fish. The survival of caged tilapia decreased significantly from 91.4% to only 57.2% with increasing stocking density from 30 to 70 fish/m^3. The best yield of both caged and open-pond fish was achieved at 50 fish/m^3.

The biomass of caged tilapia (i.e. the number of cages/pond) and aeration may also affect the performance of tilapia in integrated cage-cum-pond systems. This hypothesis has been tested by Yi and Lin (2001). The authors stocked Nile tilapia (91–103 g) in one, two, three or four cages suspended in earthen ponds at 50 fish/m^3, while fingerlings were stocked in open water at 2 fish/m^3. The survival, growth and FCR of caged and non-caged fish were significantly affected by the biomass of caged fish. The average final weight and survival of caged fish decreased with increasing number of cages. On the other hand, the growth and survival of non-caged fish increased with increasing caged biomass, presumably due to the increased feed inputs and wastes from the cages. Aeration of the water increased the growth of only the four-cage ponds, while the growth of open-pond fish was lower than that in the unaerated ponds, probably due to the increased level of total suspended solids, which resulted in increased aeration-induced mud turbidity and a reduction in primary productivity and natural food supply for the non-caged fish. Water circulation can be an effective solution to water turbidity, and will lead to improved dissolved oxygen at night. It will also stimulate phytoplankton production and increase DO through photosynthesis during the daytime (Sanares *et al.*, 1986; Boyd, 1990).

Walking catfishes are air-breathing fishes that can be grown in ponds at extremely high densities (10 fish/m^3), yielding up to 100 Mt/ha (Areerat, 1987). These scavenger fishes can be fed with trash fish, chicken offal and slaughterhouse wastes, in addition to pelleted feeds. This generally leads to increased waste production, which in turn may cause deterioration in water quality and phytoplankton blooms. However, the rich effluents from catfish ponds can be used for tilapia culture in an integrated cage–pond system. Yi *et al.* (2003b) studied the ability of Nile tilapia reared in a semi-intensive system in ponds to utilize the wastes resulting from intensive culture of hybrid catfish (*Clarias macrocephalus* × *Clarias gariepinus*). They found that tilapia yield from the catfish-waste system was comparable to that obtained from fertilized ponds. Tilapia recovered considerable amounts of N and P contained in the wastes of intensive catfish culture. The authors suggested that water circulation between catfish pens and tilapia ponds is cost-effective, as it can reduce nutrient contents in the pond effluent. Similar results have been reported with walking catfish (*C. macrocephalus*) and catfish hybrids (Lin and Diana, 1995), where the wastes from caged catfish increased phytoplankton productivity in the ponds, which supported the growth of Nile tilapia in the same ponds containing the cages. Moreover, tilapia reared in cages feeding on the phytoplankton in intensive channel catfish ponds grew reasonably well and improved the water quality in the ponds (Perschbacher, 1995).

Preliminary investigations also indicated that climbing perch (*Anabas testudineus*) may also be cultured with tilapia in integrated cage-cum-pond systems (Phuong *et al.*, 2004). Climbing perch (9 g) were stocked in cages at stocking densities of 50–200 fish/m^3 and fed pelleted feed (26–28% cp), while Nile tilapia fingerlings (10 g) were stocked in fertilized earthen ponds at 2 fish/m^3. The survival and growth of caged and non-caged fishes were reasonable, despite the fact that the FCR of the climbing perch was quite high. The best results were achieved at a stocking density of 150 perch/m^3.

5.4.6. Tilapia culture in sea cages

The scarcity of fresh water and the competition for it with other activities, such as irrigation, drinking and other urban activities, mean that its use for tilapia culture is not guaranteed or secure. The challenge facing fish farmers and aquaculture researchers is to use brackish water and seawater, which are available in most of the tropics and subtropics, for tilapia culture. The euryhaline characteristics of tilapia make them an ideal candidate for culture in the saline waters. The tolerance of tilapia for water salinity was discussed in Chapter 3.

Tilapia cage culture in seawater has been tried by a number of researchers, with varying degrees of success, depending on cultured species, size and sex, stocking density, cage size and shape and environmental conditions. Watanabe *et al.* (1990) successfully raised Florida red tilapia in seawater cages in the Bahamas with a survival rate of 84.1–93.5%. However, the results of this study may not be reliable due to the short duration over which it was carried out (30 days). Persand and Bhikajee (1997) reared red tilapia (*Oreochromis* spp.) in cylindrical seawater cages in Mauritius for 95 days. The specific growth rate (SGR) (1.13%) and yield (9 kg/m^3) were considered low as compared to other similar rearing studies, such as that of Cruz and Ridha (1991), who found that the SGR of *Oreochromis spilurus* grown in sea cages in Kuwait was 1.9%. As mentioned earlier, the difference between the two results may have been related to stocking density and species and size, duration of the trials and environmental conditions. Al-Ahmed (2002) described the culture of *O. spilurus* in sea cages in Kuwait. Fry are stocked at 600/m^3, while 200 fish/m^3 are considered optimum for the grow-out phase. In one study, the fish grew from 118 to 323.3 g in 101 days, with a growth rate of 2.03 g/day and a production of 44 kg/m^3. In another study, the fish gained 2.31–3.49 g/day, with an FCR of 1.47–2.13 and a survival of 95–97.7%. The main limitation of tilapia culture in sea cages in Kuwait is the decrease in water temperature during wintertime (mid-November to mid-April).

Commercial farming of *O. spilurus* in open-sea cages (125 m^3) at a salinity of 37–38‰ has also been carried out in Malta (Agius, 2001), with very promising results. The fish (50 g) were stocked in the cages at the end of June and reached 450–500 g at the end of November (after 5 months), with a mortality rate of less than 2%. Once again, as in the case of Kuwait, the main obstacle that limits the expansion of this culture system in Malta is the low water temperature during winter months. It is clear, nevertheless, that rearing certain species of tilapia in seawater cages can be an excellent alternative tilapia culture system if environmental conditions are favourable.

5.5. Tank and Raceway Culture

Tank culture of tilapia is becoming more popular, especially in arid and semi-arid areas, where fresh water or brackish water is limited. Tanks are generally of a smaller size than earthen ponds, and are constructed of concrete, fibreglass, wood, metal or any other suitable materials; concrete and fibreglass tanks are the most popular. Tanks are generally characterized by smooth internal surfaces to protect the fish from injury. The following features have been reported to be important for tank culture of tilapia: (i) tank size and shape; (ii) stocking density and size; (iii) water exchange and flow rate; and (iv) drainage system and sludge removal. The first three factors are discussed in the following sections, while the drainage system and sludge removal are explored in Section 5.7.

5.5.1. Tank size and shape

The size and shape of tilapia culture tanks are variable, depending on the culture objectives. Fry and nursery tanks are generally small (< 1–3 m^3), while production tanks can reach > 30 m^3 (Balarin and Haller, 1983; Martin *et al.*, 2000). In a tilapia farm at Al-Nubaria, near Alexandria, Egypt, the farmer uses 600 m^3 concrete production tanks (20 m × 30 m × 1 m) with a muddy bottom, while fish effluents are used to irrigate agricultural crops (A.-F.M. El-Sayed, Alexandria, 1998, personal observation).

Tank shape may also affect fish growth and production. Circular tanks have been recommended, since they permit higher water velocities than other tank shapes (Wheaton, 1977), leading to better-conditioned fish. Circular tanks also distribute the water better, are more self-cleaning and use a lower water flow and their construction costs are less than those of raceways (Balarin and Haller, 1983). However, they create a greater metabolic demand (i.e. a reduction in feed utilization efficiency), presumably due to the continuous movement of the fish.

Rectangular tanks (mainly concrete) are also widely used in tilapia culture, as they are easy to construct (Fig. 5.6). However, they have several problems compared with circular tanks. Fish may crowd into tank corners, causing oxygen depletion. Small rectangular tanks are also characterized by 'dead spots' and low water circulation, so that oxygen depletion may occur as a result of waste build-up, causing fish stress (Balarin and Haller, 1982). In order to avoid these problems, other tank shapes were adopted. For example, the D-shaped concrete tanks (Fig. 5.7) are successfully used for commercial tilapia production. A large-scale fish farm near Riyadh, Saudi Arabia, uses D-shaped tanks for intensive culture of Nile tilapia (A.-F.M. El-Sayed, Riyadh, Saudi Arabia, 1993, personal observation).

5.5.2. Stocking density and fish size

Stocking density may have a significant effect on final weight and total yield of farmed tilapia raised in tanks and raceways. Maximum density depends on fish size, water flow, aeration and the culture systems adopted. Siddiqui *et al.* (1991a) raised Nile tilapia in concrete tanks using drainage water at a density of 40 4-g fingerlings/m^3. The fish attained an average final weight of 544 g (21.7 kg/m^3) in 415 days. When larger fingerlings (19 g) were used at a density of 64 fish/m^3, the final fish weight was 361 g (23.1 kg/m^3). When much larger fish (40 g) were stocked at a density of 42.6/m^3, they reached 323 g (13.4 kg/m^3) in 164 days (see Table 5.3 for further details). Bailey *et al.* (2000) found that Nile tilapia stocked at 200 fish/m^3 in 2 m^3 tanks provided with biofiltration and aeration and fed a 32% cp diet attained a better daily gain and FCR than fish stocked at 450/m^3. On the contrary, Watanabe *et al.* (1993a, b) reported no difference in the growth of Florida red tilapia reared in 23 m^3 flow-through seawater pools at 15, 25 and 35 fish/m^3.

5.5.3. Water exchange and flow rate

Despite the significant effects that water exchange and flow rate may have on the growth and physiological functions of tilapia, little information is available on the subject. Continuous water exchange generally sustains the good quality of culture water, while low or zero water exchange may reduce the quality of the water. In addition, at a very high water flow rate, the fish spend a substantial amount of dietary energy for continuous swimming, leading to reduced growth and increased mortality (El-Sayed *et al.*, 2005b). At a low water flow rate, uneaten food, faeces and other fish metabolites may accumulate in fish tanks and adversely affect the water quality. This means that proper water exchange and flow rates should be adopted in order to obtain maximum fish performance.

The effects of water exchange on growth and feed utilization of Nile tilapia reared in outdoor tanks in Saudi Arabia have also been studied (Siddiqui *et al.*, 1991b). Best growth and FCR were achieved at a continuous flow rate of 0.5–1.0 l/min/kg. Similarly, Balarin and Haller (1983) reported that 100–200 kg/m^3/year of tilapia could be produced from commercial tank farms in Kenya with a continuous water flow of 0.5–1.0 l/min/kg, using commercial balanced feeds.

One major advantage of tank culture of tilapia is that it can be practised in the backyards of family houses in rural areas, generating

Fig. 5.6. Intensive culture of tilapia in rectangular concrete tanks in Saudi Arabia.

Fig. 5.7. Intensive culture of tilapia in D-shaped concrete tanks in Saudi Arabia.

additional income for rural households. Afolabi *et al.* (2000) evaluated the economic and technical viability of tilapia culture in homestead concrete tanks constructed in the backyards of family houses in suburban Lagos, Nigeria. Tilapia hybrids (48 g) were stocked at 6 fish/m^3 in the tanks, fertilized with poultry droppings (900 kg/ha) in addition to feeding with cottonseed cake (5%/day). The fish reached 183 g in 120 days with a yield of 1.07 kg/m^3. Cost/benefit analysis indicated that this system is profitable.

5.5.4. Raceway culture

A raceway culture system (Fig. 5.8) can be an ideal system for tilapia culture, especially in rural areas where irrigation raceways and canals are available. While rearing tilapia in raceways was reported more than 20 years ago (Balarin and Haller, 1982), production of tilapia in raceways is still limited. Studies on the production of tilapia in raceways are also still limited. D'Silva and Maughan (1995) evaluated the effects of stocking density and water quality on red tilapia (*O. mossambicus* × *Oreochromis urolepis hornorum*) reared in pulsed-flow culture systems in irrigation ditches. The fish were stocked at 10, 20, 30, 50 and 70 fish/m^3. The results indicated that 20 fish/m^3 produced the best marketable yield. Increasing density resulted in increased ammonia concentration and decreased dissolved oxygen levels. Windmar *et al.* (2000b) described a raceway 650 m long, divided into five compartments, with a water flow rate of 1.7 m^3/s, which is used for the production of 480 Mt of Nile tilapia in Zimbabwe.

Table 5.3. Tank culture of tilapia in selected countries.

Species (country)	Stock data		Growth data			Harvest data			Culture period (days)	Remarks	Reference
	No./m^3	IW (g/fish)	SGR (%)	ADG (g/day)	FCR	FW (g/fish)	Yield (kg/m^3)	S (%)			
O. n (Saudi Arabia)	40	4	1.18	1.3	2.18	544	21.7		415	Pelleted feed, 34% cp, drainage water, growth stopped during winter months	Siddiqui *et al.* (1991a)
O. n (Saudi Arabia)	42.6	40	1.27	1.72	1.96	323	13.4		164	Pelleted feed, 34% cp, drainage water, growth stopped during winter months	Siddiqui *et al.* (1991a)
O. n (Saudi Arabia)	64	19	1.7–1.8	0.94–1.06	1.7–2.3	92–104			98	Outdoor concrete tanks, continuous water flow, a rate of 0.5 l/min/kg biomass was recommended	Siddiqui *et al.* (1991b)
O. n (Virgin Islands)	26	71.5	1.17	2.64	1.41	514.4	13.4	99.3	168	Circular outdoor tanks, aeration, solids removal twice a day	Cole *et al.* (1997)
O. n (Virgin Islands)	26	70.4	1.06	2.06	1.51	416.7	10.8	98.9	168	6 m^3 circular outdoor tanks, aeration, no solids removal	Cole *et al.* (1997)
O. n (Virgin Islands)	200	4.1		0.78	1.22	69.3		92.1	84	32% cp floating pellets, tanks supplied with microscreen drum and biological filtration	Bailey *et al.* (2000)
	450	4.3		0.60	1.25	55.1		91.9	84		
Hybrids (Nigeria)	6	48.4	1.11	1.13	2.9	183.4	1.07	89.6	120	Homestead concrete tanks in the backyard of family house, poultry waste and SBC were applied	Afolabi *et al.* (2000)
O. n (Zimbabwe)	46	40	1.11	2.33	1.2–2.0	600	24.0	85	240	30.2% cp commercial feed, 650 m long raceway, water flow 1.7 m^3/s	Windmar *et al.* (2000b)
O. n (Egypt)	60	52		0.7	2.1	163	9.6	97.8	158	30% cp feed, 150 m^3 circular tanks, aeration, periodic solids removal	Sadek *et al.* (1992)
	100	22		0.5	2.3	110	10.3	93.4	175		
	140	15		0.4	2.4	87	11.0	95.0	180		
Red tilapia (Thailand)	50	75		0.77	2.25			87.4	70	22% cp diet, circular concrete tanks, solids removal, biological filtration	Suresh and Lin (1992b)
	100	75		0.65	2.57			85.4	70		
	200	75		0.64	2.61			79.7	70		

O. n, *Oreochromis niloticus*; IW, initial weight; SGR, specific growth rate; ADG, average daily gain; FCR, feed conversion ratio; FW, final weight; S, survival; SBC, soybean cake.

Fig. 5.8. Intensive culture of tilapia in raceways in Mexico (photo provided by K. Fitzsimmons).

5.6. Green-water Tank Culture

Green-water tank culture is an intensive culture system where fish wastes and other metabolites are treated in such a way as to increase natural primary productivity in the culture units. The process involves the oxidation of toxic ammonia (NH_3) and nitrite (NO_2) into relatively non-toxic nitrate (NO_3) through nitrifying bacteria grown on suspended organic matter. The bacteria remove the organic matter from the tanks by using it as food, while the bacteria themselves are used as natural food for filter-feeding fish such as tilapia and carp. However, in green-water systems, solid wastes should be removed continuously and both biofiltration and aeration should be provided. Vigorous aeration is necessary to support the suspension of the microbial community, maximize contact between bacteria and waste products and, in turn, increase phytoplankton productivity.

Green-water tank culture has many advantages, including:

1. High system productivity and profitability.
2. Efficient use of water (i.e. minimum or even zero water exchange where only losses due to evaporation and/or solid removal are replaced).
3. The solid waste (sludge) can be used as organic fertilizer.
4. Easy management.
5. Lower feed cost, due to nutrient recycling and natural food production.

One other important advantage of the green-water system is that it can control the population of pathogenic bacteria in tilapia tanks and reduce their effects. Tendencia and dela Penã (2003) studied the effects of some components of the green-water system on the initial control of luminous bacteria (*Vibrio harveyi*), using tilapia (*Oreochromis hornorum*) and *Chlorella* separately. Tilapia were reared in indoor, circulated tanks at 0, 1 and 3 kg/m^3 and the bacteria were inoculated into the tanks at a density of 103 colony-forming units (cfu)/ml. The bacteria were not detected in fish tanks 4–6 days after injection. The authors attributed the ability of tilapia to prevent the initial growth of *Vibrio* to the inhibitor properties of lactic acid bacteria (*Enterococcus duran*) isolated from fish intestine. When pure *Chlorella* culture was injected with *V. harveyi*, the bacteria were not detected 2–3 days after injection, presumably due to the secretion of antimicrobial compounds from *Chlorella*. More work is necessary to investigate the effects of other microalgae on other pathogenic agents in filter-feeding and herbivorous fish such as tilapia and carp.

Tilapia culture in green-water systems is practised in Israel (Avnimelech, 1998), the USA (Lutz, 1996) and the Virgin Islands (Martin *et al.*, 2000, 2001; Rakocy *et al.*, 2000a, b). Rakocy *et al.* (2000a, b, 2004) described the green-water tank system currently in use in the US Virgin Islands and evaluated its suitability for red tilapia and sex-reversed Nile tilapia culture in a series of studies. Water exchange at 5%/day did not improve fish survival, growth or yield over a zero water exchange. Sludge removal significantly improved fish performance, but not survival. Weekly application of alum (aluminium sulphate) at 51.5 mg/l had resulted in a significant increase of Nile tilapia growth and yield. Martin *et al.* (2000) also reported that Nile tilapia stocked in green-water tanks at 26 fish/m^3 and fed with a 32% cp feed attained biomass of 13.4 kg/m^3, an FCR of 1.41 and a survival rate of 99.3%, with an exchange rate of only 0.23% of the tank water/day. Lutz (1996) reported that tilapia production capacity in greenhouses using green-water systems in Louisiana is about 250 Mt/year.

5.7. Tilapia Culture in Recirculating Systems

Tilapia have traditionally been cultured in inland-based systems, such as earthen ponds and concrete tanks and, to a lesser extent, raceways. These systems have started to encounter a number of problems that make their long-term sustainability unwarranted. These are:

1. The growing environmental concerns about the effect of the effluents from land-based aquaculture facilities.
2. The limitation and shortage of supplies of fresh water, together with the competition for it with agricultural, industrial and other urban requirements, especially in arid areas.
3. Increasing land costs, which will limit land use opportunities.
4. Climate changes, which may limit management and production in these systems.
5. Introduction of tilapia into many countries where the environmental conditions are outside their tolerance limits has made the traditional land-based systems unsuitable for tilapia culture in these countries.

Recirculating culture systems, also known as recycle or closed systems, are an ideal alternative for confronting these difficulties. These systems are characterized by water reuse, minimal effluent discharge and optimal water conservation. Culture water goes through various treatment processes to restore its quality. These include the removal of solid wastes and suspended metabolites (such as carbon dioxide, ammonia and nitrite), sterilization and aeration. The typical closed system consists of settling tanks (for solids removal), biological and/or mechanical filters (for ammonia removal), ultraviolet (UV) light (for sterilization) and aerators (compressors) or a source of oxygen (Fig. 5.9). A full description of this system has been reviewed by a number of authors (Mires and Anjioni, 1997; Losordo *et al.*, 1999; Muir *et al.*, 2000).

Tilapia culture in recirculating systems is spreading widely in different parts of the world, especially in areas that face freshwater shortages and/or harsh climatic conditions. In southern Israel, commercial culture of tilapia in closed systems is carried out year-round in greenhouses, even at low water temperature (19°C) and up to 29°C (Muir *et al.*, 2000). At a stocking density of 15 kg/m^3, the system produces 25 kg/m^3/year. Fingerling red tilapia (*O. niloticus* × *O. aureus*) were also grown in indoor tanks or raceways at a density of about 300 fish (40 g)/m^3 until they reached an average weight of 200 g. The density was then reduced to 150 fish/m^3 during the fattening stage to reach 450 g (Muir *et al.*, 2000).

Mires and Amit (1992) described tilapia culture in a closed system known as the DEKEL system, where water recycles between concrete grow-out ponds and earthen reservoir serving as a biofilter. This system was able to maintain water quality suitable for tilapia culture. The net yield reached 19.5 kg/m^2 in 1990. In a later study, Mires and Anjioni (1997) evaluated the technical and economic performance of red tilapia in the DEKEL system and in another closed system supported by a biofilter and pure oxygen supply and referred to as O2BIO. Production in the O2BIO was higher than in the DEKEL system, but it was less cost-effective. However, after some technical improvements of the DEKEL system, a net present value (NPV) > 0 can be obtained after only 2 years, with an internal rate of return (IRR) = 22% after 4 years. On the contrary, the NPV > 0 of O2BIO can only be reached after 11 years. Only when production reaches 100 kg/m^2 and the initial

Fig. 5.9. A recirculating system used for intensive culture of tilapia at the United Arab Emirates University.

investment is reduced from US\$600/m^2 to US\$500/m^2, is an NPV > 0 in O2BIO system obtained, after 6 years. Meanwhile, both systems were uneconomic due to high production costs and should be improved to produce a positive cash flow.

Tilapia culture in water recycling systems in the USA is also gaining great popularity because their culture in traditional land-based systems is limited due to their tropical nature. Closed systems are currently the most sophisticated, using liquid oxygen, microscreens, fluidized bead biofilters and UV sterilizers. They also have the highest stocking densities of any tilapia culture systems (Rosati *et al.*, 1993; Fitzsimmons, 2003). Rosati *et al.* (1993) determined the carrying capacity and performance of a recirculating system, consisting of an 18.5 m^3 fibreglass raceway and a vertical screen filter, when stocked with Nile tilapia under commercial conditions. The fish (15 g) were stocked at 263.2 fish/m^3. Final survival was 70%. Oxygen and un-ionized ammonia became limiting factors when the feed approached 8 and 12 kg/day. Rosati *et al.* (1997) further evaluated the performance of Nile tilapia (95% males) reared in a prototype commercial-scale recirculating system. The system consisted of six raceways, with a total water capacity of 160 m^3, provided with a rotating drum with a screen biofilter, submerged biofilter media and pure oxygen. The fish grew from 15–20 g to 560 g in 6 months. The total yield was 11.33 Mt/year (78.8 kg/m^3/year).

Tilapia culture in recycle systems has also been reported in other tropical and subtropical regions. Cheong *et al.* (1987) reported a growth of Taiwanese red tilapia raised in a recycle system from 0.78 g to 438 g in 239 days, with a total production of 49.3–50.2 kg/m^3. Watanabe *et al.* (1997) also described a recycle hatchery system for Florida red tilapia in the Bahamas.

It should be emphasized here that the recirculating system is technically sophisticated and very costly, while tilapia is considered a low-priced fish in many countries. Production of these fish with high investment in commercial recycle systems may, therefore, be neither cost-effective nor sustainable in these countries. The message is clear: cost/benefit analyses should be carefully conducted prior to the adoption of tilapia culture in recycle systems in developing countries.

Recirculating systems are characterized by the ability to support extremely high stocking densities and high net production compared with other culture systems. However, stocking density has a significant effect on individual fish growth, survival, total production and water quality. A decrease in final individual fish weight and an increase in total yield with increasing stocking density is very commonly observed. Considerable variations in fish size may also occur with increasing fish density in recycle systems (Rosati *et al.*, 1993). Suresh and Lin (1992b) evaluated the effects of stocking density on the water quality and production of all-male red tilapia juveniles (75 g) stocked in circular tanks at 50, 100 and 200 fish/m^3 for 70 days. Fish growth, feed efficiency and water quality were inversely correlated with stocking density. However, 100 and 200 fish/m^3 showed a similar performance. Feed digestion and body composition were not affected by stocking density. As mentioned earlier, Bailey *et al.* (2000) raised Nile tilapia juveniles (4.3 g) in 2-m^3 circular tanks in a recirculating system, at two densities, 200 and 450 fish/m^3, for 12 weeks. Fish stocked at 200 fish/m^3 grew slightly faster than those stocked at 450 fish/m^3. The desired survival, uniform size and FCR were not achieved, as a result of using large feed pellets.

5.8. Effluent Treatment and Management

In recycle systems, as in other culture systems, considerable amounts of fish metabolites, including ammonia, nitrites, organic phosphates, suspended solids, etc., along with uneaten feed, are released into the culture water. These compounds are generally toxic to fish and have a high biological oxygen demand (BOD). If not removed or treated, they may be subject to oxidation, leading to the removal of oxygen from the water and, in turn, producing anaerobic conditions that will stress and/or kill the fish. The magnitude of this problem is much higher in recirculating aquaculture systems than in flow-through systems, where these compounds can be discharged with effluent water and eliminated through natural processes. Thus, effluent treatment and management become essential for successful aquaculture in recycle systems. Typically, management includes: (i) waste settlement and removal; (ii) removal of ammonia and nitrites via biofiltration, sterilization and aeration; and (iii) partial water replacement.

5.8.1. Waste settlement and removal

Solid wastes can be removed through water exchange or back-flushing of filters (i.e. mechanical or biofiltration with fixed-film media) (Cole *et al.*, 1997). Waste removal from recirculating systems of tilapia has been reported to improve their performance very significantly. Cole *et al.* (1997) studied the effects of solids removal (twice a day) on tilapia production and water quality in continuously aerated tanks. The removal of solids significantly increased final fish weight and net yield, but not fish survival, compared to a no-solids-removal system. It also sharply reduced total solids, volatile solids, suspended solids, organic nitrogen, inorganic nitrogen, total phosphorus, chemical oxygen demand (COD) and BOD. However, DO, temperature, NH_4-N and NO_3-N were not significantly affected by solids removal.

Traditional biofiltration, including water exchange, back-flushing, fixed-film media, etc., may be inconvenient for small-scale tilapia farmers in developing countries, mainly due to their: (i) occasional failure; and (ii) high costs, which make them unaffordable for many farmers. Other cheap, locally available biofiltration methods have been tried for solids removal from tilapia culture tanks with great success. For example, the addition of carbon-rich (C-rich) agricultural by-products to fish tanks can be used for solids removal. These C-rich wastes can be combined with nitrogen-rich (N-rich) organic solids discharged from a recycle systems to produce marketable compost. Along this line, Scott *et al.* (1998) investigated the use of rice hulls as a primary medium for solids removal from a recirculating tilapia culture system. Rice hulls were selected because of their slow degradation, availability at no charge and light weight with high porosity. These hulls removed 94%, 66% and 63% of settleable solids, total suspended solids (TSS) and COD, respectively. In another study, the authors evaluated the effects of the direct addition of rice hulls to the culture water on solids removal, and found that about 52% of TSS can be removed using this method. This type of research is necessary and should be encouraged, as it indicates that simple and, at the same time, efficient technologies can be used in recirculating aquaculture systems in areas where sophisticated technologies are either not available or not affordable.

5.8.2. Removal of ammonia and nitrites

Several methods of ammonia (NH_3) and nitrite (NO_2) removal from culture water are in use, including the biological conversion of ammonia to nitrites and nitrates with nitrifying bacteria in filters or aeration tanks, the adsorption onto ion exchange resins, transfer to the atmosphere by air bubbling through water and chemical reactions with chlorine (Balarin and Haller, 1982). Biological treatment remains the most practical and cost-effective method. Nitrification is based mainly on biofiltration on fixed media, such as biosedimentation or biodiscs, or suspended media, such as activated sludge.

Different biofiltration types have been described for ammonia removal in recirculating tilapia culture systems. Ridha and Cruz (2001) evaluated the efficiency of two types of biofilter media, polypropylene plastic chips and polyethylene blocks, in ammonia removal, water quality and waste accumulation with respect to Nile tilapia reared in a recycle system. No difference in performance efficiency between the two types of media was evident. However, the authors suggested the use of the polypropylene plastic chips because they are cheaper.

Bead filters can be used efficiently for nitrification and clarification of culture water in recirculating culture systems. They can capture solid wastes, stimulate the growth of heterotrophic bacteria and reduce NH_3, NO_2 and O_2 availability for the nitrifying autotrophic bacteria (Lenger and Pfeiffer, 2002). Different types of bead filters are currently in use. Sastry *et al.* (1999) studied the efficiency of a bubble-washed bead filter in solids capture and biological filtration in a recirculating Nile tilapia culture system at feed loading rates of 16, 24 and 32 kg/day/m^3 of beads. They found that more frequent back-flushing improved filter performance and was necessary at higher feeding rates. Five backwashes in 24 h were necessary at the highest feeding load. Similar results have been reported when floating bead filters were used (Lenger and Pfeiffer, 2002). Slow sand filters have also been widely used for the treatment of effluents within recycle aquaculture systems. These filters are simple, cheap, easy to construct and operate and effectively remove suspended solids. Drennan *et al.* (1993) evaluated the use of a slow sand filter for the containment of tilapia eggs and/or fry in a recycle system. The system was very efficient despite the frequent clogging of the beads due to high solids loading.

Aquatic weeds can also remove a considerable portion of solids from recirculating aquaculture system. Through competition for light and nutrients in the water, macrophytes usually prevent or, at least, minimize phytoplankton blooming and keep culture water clear. The macrophytes themselves can be used as food for cultured herbivorous fish such as tilapia and carp. Rakocy and Allison (1981) investigated the efficiency of two floating macrophytes, water hyacinth (*Eichhornia crassipes*) and *Spirodela oligorhiza*, in oxygen uptake from waste water of blue tilapia (*O. aureus*) culture. The fish were stocked in concrete tanks at 16.2, 32.5 and 48.7 fish/m^3. Plant populations removed 15.8, 13.4 and 12% of the waste nitrogen at the three stocking densities, respectively. Marine algae have also been used for biofiltering waste effluents from closed systems of marine fish and shellfish. *Gracilaria chilensis* (*Rhodophyta*) was highly efficient in removing soluble nutrients, but had little effect on particulate nutrients, during separate tank culture of a chelan fish *Isacia conteptionis*, oyster *Crassostrea gigas* and sea urchin *Loxechinus albus* (Chow *et al.*, 2001). The ammonia from the fish culture and nitrate from the oyster culture were completely removed from the water by the weed.

5.8.3. Water discharge

As discussed earlier, recirculating systems are adopted mainly in areas having limited freshwater sources and harsh weather conditions. Economic use of the freshwater sources becomes a necessity, and water discharge may not be appropriate. Water recycling and reuse, through post-treatment and/or treatment within the system, have been tried for several years as a means of reducing water discharge. Minimizing, or even eliminating, water discharge from tilapia cultured in recycle systems has been investigated with varying degrees of success. Shnel *et al.* (2002) evaluated the 'zero' discharge from tilapia recirculating system for 331 days in a greenhouse. Neither water nor organic matter was discharged, and water was added only to compensate for loss and evaporation. Sludge was also biologically digested and nitrate was reduced to nitrogen gas. Total tilapia yield was 81.1 kg/m^3 at a maximum stocking density of 61.8 kg/m^3. Inorganic nitrogen concentrations were within acceptable levels for tilapia culture. Water quality remained within the acceptable levels for tilapia throughout the study. Daily average ammonia removal was 0.16 g N/m^2, while nitrate removal was most profound in the sedimentation/digestion basin. About 70% of the phosphorus added with the feed was recovered, mostly in the anoxic treatment stage of the system.

5.9. Tilapia Production in Aquaponic Systems

Hydroponic systems are generally the production of plant/vegetable crops in areas smaller than those required under field production, or the production of these vegetables without soils. This practice is carried out by providing the nutrients that the plant requires with irrigation water. Aquaponics is the integration of hydroponics with aquaculture in a recirculating system, where the wastes and metabolites produced by cultured fish are removed by nitrification and taken up by the plants. The bacteria living in the gravel and in association with the plant roots play a critical role in nutrient removal. Thus, as the aquaculture effluent flows through the aquaculture/hydroponic system, the plants act as a 'biological filter' by removing fish wastes and improving the quality of the culture water. Aquaponic systems have many advantages, including the following:

1. The wastes of one system are used as inputs or fuel for another biological system.
2. The integration of fish and plants increases diversity and enhances system stability and sustainability.
3. Reducing the volume of waste discharged to the environment and minimizing the negative environmental impacts by water filtration and removing nutrients from water before it leaves the system.
4. Minimizing water exchange and reducing operating costs.
5. The introduction of 'organic products' into the market, which increases farmers' income and supports the local economy.
6. The production of two (or more) food products from one production unit.
7. In arid areas where water is scarce, aquaponics is an appropriate technology that allows food production with reused water.
8. The system is simple, inexpensive and easy to adopt.

Aquaponics is, therefore, a promising, sustainable food production technology, especially in developing/arid countries, where resources are limited, fresh water is scarce and the population is increasing.

Tilapia are among the best candidates for aquaponic systems due to their tolerance to high stocking loads and crowding and their resistance to stress, disease, handling and environmental changes. Several studies have considered the production of tilapia in aquaponic systems, with promising results. The Agricultural Experimental Station at the University of the Virgin Islands (UVI) (Fig. 5.10), USA, has a long history of aquaponic research, especially with tilapia production, which started more than two decades ago (Rakocy and Allison, 1981). The production of tilapia in aquaponic systems has also been reported in other geographical regions.

The use of effluents from intensive tank culture of Florida red tilapia for field production of sweet peppers has been evaluated (Palada

Fig. 5.10. Tilapia culture in aquaponic system at the University of the Virgin Islands (photo provided by J. Rakocy and D. Bailey).

et al., 1999). Water from fish rearing tanks with low (8 fish/m^3) and high (16 fish/m^3) stocking densities and two daily water exchange rates (0% and 5%) were applied to plots grown with sweet peppers two to three times weekly. Pepper production from these treatments was compared with that produced with solids (sludge) removed from fish rearing units, cow manure and commercial N fertilizers. Yield from fertigated plots was significantly higher than with all other treatments. The yields of pepper from plots fertilized with rearing tank water from two exchange rates were not significantly different from yields produced through inorganic N fertilizer. Tank sludge produced the highest marketable yield. Similarly, Anadu and Barho (2002) compared the production of tomatoes and lettuce grown with blue tilapia in an aquaponic system with those grown in conventional soil. The vegetables grew much faster in the aquaponics system, and tomatoes started flowering 2 weeks earlier than soil-grown tomatoes. A prototype tilapia/hydroponic greenhouse recirculating production system for institutional application has been described by Newton *et al.* (1998) in Blacksburg and Petsburg, Virginia, USA. Tilapia was used as the aquaculture species, while vegetables tested included tomatoes and lettuce. The objectives were to produce a tilapia crop within 5–6 months and to reuse the effluent water for irrigation and nutrient inputs in the vegetable crops. Rakocy *et al.* (2001) also described and evaluated outdoor, commercial-scale production of Nile tilapia in an aquaponic system with 13 different types of vegetables.

The production of tilapia and vegetables in aquaponic systems appears to be very profitable. Rakocy and Bailey (2003) conducted an initial economic analysis of tilapia and lettuce in an aquaponics system (Table 5.4). The results showed that the return/risk and management for one production unit was US$30,874.62 (US$185,247.73 for six units). The authors suggested that break-even

Table 5.4. Enterprise budget for aquaponic system producing tilapia and lettuce in Virgin Islands in US$ (from Rakocy and Bailey, 2003).

	Units	Price or cost/unit	Quantity per system	Value or cost per system	Value or cost per six systems
Receipts					
Tilapia	Kg	5.51	5,000	27,562.50	165,375.00
Lettuce	Case	20.00	1,404	28,080.00	168,480.00
Total				55,642.50	333,855.00
Variable costs					
Tilapia				14,082.94	84,497.64
Lettuce				7,071.88	42,431.30
Total				21,154.82	126,928.94
Fixed costs					
Tilapia				1,656.76	9,940.57
Lettuce				1,349.79	8,098.71
Total				3,006.55	18,039.28
Other costs					
Tilapia				397.70	2,386.18
Lettuce				208.81	1,252.88
Total				606.51	3,639.06
Total costs				24,767.88	148,607.27
Returns to risk and management				30,874.62	185,247.73

prices of tilapia and lettuce could be even lower in other places where production costs (construction costs, electricity, water, labour, etc.) are lower. Producing other vegetables such as okra and herbs (such as basil) in aquaponic systems with tilapia was also profitable (Rakocy and Bailey, 2003). Rakocy *et al.* (2004) updated the production of tilapia and vegetables in the UVI hydroponic system, and compared the yield of okra and basil in this system with that of a conventional field system. The production of these crops was much higher in the aquaponic system. These results clearly justify the adoption of aquaponic systems for fish/vegetable production.

5.10. Closing Remarks

1. The widespread aquaculture of tilapia, especially in areas outside their natural geographical existence, together with the shortage of and competition for freshwater sources, has gradually shifted tilapia culture from traditional semi-intentive to intensive systems.

2. There is an inverse relationship between stocking density and individual fish growth. The farmer should carefully identify his target consumer (i.e. he should decide whether to produce a higher yield but smaller fish or a lower yield with larger fish size).

3. Water exchange and/or aeration may be necessary in earthen ponds. The cost-effectiveness of this practice should be carefully investigated, taking into account the availability and cost of water, fuel, electricity, etc.

4. Water-saving suspension systems may have economic advantages over traditional water-demanding systems in earthen ponds. The re-suspension of sediments can be an important mechanism of nutrient transfer in the pond, and can also serve as a water purifier due to the increased number of nitrifying bacteria in the water column.

5. The adjustment of the C : N ratio (mainly in fish feeds) is necessary for reducing the accumulation of inorganic nitrogen in intensive tilapia ponds. The proper C : N ratio can reduce ammonia to the desired levels within 1–3 days. A ratio of about 15 : 1 C : N has been recommended for tilapia. Cheap carbon and nitrogen sources can be used as a partial replacement for expensive, commercial, high-protein feeds. The approach of producing single-cell protein(s) in fish ponds should also be encouraged in developing countries, where tilapia culture is widely practised.

6. In arid areas, where fresh water is limited, water treatment and reuse would be an ideal alternative for intensive tilapia culture.

7. Farming tilapia in a cage-cum-pond system appears to be economic, favourable and successful. However, the farmer must carefully consider the density and size of non-caged and caged species, as well as the number of cages (i.e. total biomass). Aeration may also be necessary.

8. Recirculating culture systems are an ideal alternative for tilapia culture. However, these systems are technically sophisticated and very costly, while tilapia is considered a low-priced fish in many countries. The production of these fish with high investment in commercial recycle systems may, therefore, not be sustainable.

9. The addition of carbon-rich agricultural by-products (such as rice hulls) to fish tanks in closed systems can be used for solids removal. This practice can reduce the running costs of the system.

10. A green-water tank system is an ideal alternative for tilapia culture, especially in arid areas, where freshwater resources are limited.

11. A green-water system can control pathogenic bacteria in culture water. Microalgae, such as *Chlorella*, secrete antimicrobial compounds, which inhibit pathogenic bacterial growth. This approach should be further explored.

12. Tilapia production in aquaponic systems can be an efficient and sustainable way of producing tilapia and plants/vegetables, especially in rural, resource-limited areas.

6
Nutrition and Feeding

6.1. Introduction

Nutrition is the most expensive component in the intensive aquaculture industry, where it represents over 50% of operating costs. Therefore, proper feeding management is a necessary tool for successful tilapia culture practices. A great deal of attention has been paid to aquaculture nutrition in recent years. The challenge that faces tilapia farmers in general, and tilapia nutritionists in developing countries in particular, is the development of commercial, cost-effective tilapia feeds using locally available, cheap and unconventional resources.

Under extensive culture conditions, tilapia depend exclusively on natural food, through fertilization, while both natural foods and supplemental feeds are normally used in semi-intensive farming systems. Tilapia nutrition and feeding in extensive and semi-intensive farming systems are discussed in Chapter 4. The present chapter addresses tilapia nutrition under intensive and super-intensive farming systems. It deals with the requirements of farmed tilapia for the five classes of dietary nutrients, namely proteins, lipids, carbohydrates, vitamins and minerals. Consideration is also given to feeding regimes (feeding levels and frequency) and diet form.

6.2. Protein Requirements

Proteins are large, complex molecules composed typically of carbon, hydrogen, oxygen and nitrogen, as well as small amounts of sulphur and, sometimes, phosphorus. The principal components of proteins are known as amino acids. Proteins are indispensable for the structure and function of all living organisms.

Protein is the most expensive dietary source in intensive aquaculture. It represents about 50% of total feed costs. Moreover, the dietary protein requirements for the maximum performance of fish are generally higher than those of farmed terrestrial animals. Therefore, special attention is paid to the dietary protein requirements of farmed aquatic animals.

Protein requirements of tilapia have been extensively studied using dose–response procedures. In this regard, semi-purified test diets containing casein, casein/gelatin mixtures or casein/amino acid mixtures as protein sources or practical diets in which animal and/or plant ingredients served as dietary protein sources have been widely used. The utility of the results of many studies is questionable, because they were indoor, short-term studies and the outcomes may not be directly applicable in commercial rearing facilities. In addition, casein was traditionally used as a sole dietary protein source in many studies. Casein contains adequate levels of all essential amino acids (EAA) except arginine. When casein is used as the sole dietary protein source, higher levels are required to compensate for the arginine deficiency (Teshima *et al.*, 1985a; El-Sayed, 1989). This means that the protein requirements of tilapia would probably have been lower if proper combinations of casein and arginine sources had been used instead of casein alone. Since gelatin is deficient in all EAA except for arginine, casein/gelatin mixtures are a more suitable protein source in semi-purified tilapia diets. Casein/gelatin-based diets were found to be utilized more efficiently than casein/amino acid diets (El-Sayed, 1989). It is, therefore, no surprise that the results of protein requirements of tilapia vary and are sometimes contradictory, as shown in Table 6.1.

Protein requirements of tilapia depend, among other things, on fish size or age, protein source and the energy content of the diets. Generally speaking, protein requirement decreases with increasing fish size (Table 6.1). During the larval stages, tilapia (especially Nile tilapia) require about 35–45% dietary protein for maximum growth performance (Siddiqui *et al.*, 1988; El-Sayed and Teshima, 1992). Some researchers reported higher values (> 50%) (Winfree and Stickney, 1981; Jauncey and Ross, 1982), but these values appear impractical under commercial conditions. For tilapia juveniles, the protein requirement ranges from 30 to 40%, while adult tilapia require 20–30% dietary protein for

Table 6.1. Protein requirements of cultured tilapia.

Species and life stage	Weight (g)	Protein source	Requirement (%)	Reference
Nile tilapia				
Fry	0.012	FM	45	El-Sayed and Teshima (1992)
	0.51	FM	40	Al Hafedh (1999)
	0.56	Casein/gelatin	35	Teshima *et al.* (1985a)
Fingerlings	1.29	Casein	40	Teshima *et al.* (1982)
	2.40	Casein/gelatin	35	Abdelghany (2000a)
	3.50	Casein	30	Wang *et al.* (1985)
Adults	24	FM/SBM/BM	27.5	Wee and Tuan (1988)
	40	FM	30	Siddiqui *et al.* (1988)
	45–264	FM	30	Al Hafedh (1999)
Broodstock		FM/SBM	40	El-Sayed *et al.* (2003)
		FM	45	Siddiqui *et al.* (1998)
		Casein/gelatin	35–40	Gunasekera *et al.* (1996a, b)
Mozambique tilapia	Fry	FM	50	Jauncey and Ross (1982)
	0.5–1.0	FM	40	Jauncey (1982)
	6–30	FM	30–35	Jauncey and Ross (1982)
Blue tilapia	2.50	Casein/albumin	56	Winfree and Stickney (1981)
	7.50	Casein/albumin	34	Winfree and Stickney (1981)
Tilapia zillii	1.40	Casein/gelatin	35	El-Sayed (1987)
O. n × *O. a*	0.60–1.10	FM	32	Shiau and Peng (1993)
	21	SBM	28	Twibell and Brown (1998)
O. n × *O. h*	1.24	–	32	Luquet (1989)
O. m × *O. h*	8.87	–	28	Watanabe *et al.* (1990)

O. a, *Oreochromis aureus*; *O. h*, *Oreochromis hornorum*; *O. m*, *Oreochromis mossambicus*; *O. n*, *Oreochromis niloticus*; FM, fish meal; SBM, soybean meal; BM, blood meal.

Table 6.2. Essential amino acid requirements of tilapia as % of dietary protein and of total diet (in parentheses).

	Requirement			
Amino acid	*Oreochromis mossambicus*[a]	*O. mossambicus*[b]	*Oreochromis niloticus*[c]	*O. niloticus*[d]
Lysine (Lys)	4.05 (1.62)	3.78 (1.51)	5.12 (1.43)	–
Arginine (Arg)	3.80 (1.52)	2.82 (1.13)	4.20 (1.18)	4.1
Histidine (His)		1.05 (0.42)	1.72 (0.48)	1.5
Threonine (Thr)		2.93 (1.17)	3.75 (1.05)	3.3
Valine (Val)		2.20 (0.88)	2.80 (0.78)	3.0
Leucine (Leu)		3.40 (1.35)	3.39 (0.95)	4.3
Isoleucine (Iso)		2.01 (0.80)	3.11 (0.87)	2.6
Methionine (Met)	1.33 (0.53)	0.99 (0.40)	2.68 (0.75)	1.3
Cysteine (Cys)			0.53	2.1
Phenylalanine (Phe)		2.50 (1.00)	3.75 (1.05)	3.2
Tyrosine (Tyr)			1.79	1.6
Tryptophan (Try)		0.43 (0.17)	1.00 (0.28)	0.6

[a]Jackson and Capper (1982); [b]Jauncey *et al.* (1983); [c]Santiago and Lovell (1988); [d]Fagbenro (2000).

optimum performance. On the other hand, tilapia broodstock require 35–45% dietary protein for optimum reproduction, spawning efficiency and larval growth and survival (Gunasekera *et al.*, 1996a, b; Siddiqui *et al.*, 1998; El-Sayed *et al.*, 2003).

6.3. Amino Acid Requirements

As mentioned earlier, proteins are composed of amino acids, of which about 25 commonly occur in food proteins. These amino acids are divided into two groups:

1. Essential amino acids (EAA), which cannot be synthesized by living organisms, including fish, and must be supplied in the diets. These EAA are: arginine (Arg), lysine (Lys), histidine (His), threonine (Thr), valine (Val), leucine (Leu), isoleucine (Iso), methionine (Met), phenylalanine (Phe) and tryptophan (Try).

2. Non-essential amino acids (NEAA), which can be synthesized by the organism, in the presence of suitable starting materials, including EAA.

Tilapia, like other fish, require the same ten EAA. However, specific EAA requirements of most farmed tilapias have not been determined. Few studies have considered EAA requirements of Nile tilapia (*Oreochromis niloticus*) and Mozambique tilapia (*Oreochromis mossambicus*). These requirements are summarized in Table 6.2. The EAA requirements of tilapia vary by species and amino acid (AA) source (e.g. crystalline AA vs casein/gelatin). It was reported that tilapia fed crystalline EAA-based diets grew at lower rates than those fed casein or casein/gelatin diets (Jackson and Capper, 1982; Teshima *et al.*, 1985a), presumably due to the low pH of the AA-based diets (Wilson *et al.*, 1977). Neutralizing AA-based diets to a pH of 7–7.5 would improve their quality for tilapia (Mazid *et al.*, 1979). More work is needed in this area, since other studies indicated that lowering food pH enhanced the appetite of *Tilapia zillii* for food (Adams *et al.*, 1988).

Tilapia requirements for sulphur-containing AA can be met by methionine or a methionine/cystine mixture. Recent studies have recommended a methionine : cystine ratio of 50 : 50 for the best performance of Nile tilapia (Abdelghany, 2000b). Similarly, the requirement of tilapia for phenylalanine (aromatic amino acid) could partially be met by tyrosine (NRC, 1993).

6.4. Protein Sources

Since purified and semi-purified protein sources are not recommended for tilapia under commercial culture conditions, other conventional and unconventional, locally available dietary protein sources should be sought. Research has evaluated many such sources for different species of tilapia, with varying results. It is appropriate to highlight these protein sources for tilapia, with emphasis on the sources that have economic potential and are locally available, especially in developing countries (see El-Sayed, 1999a, for more details). The suggested inclusion levels of these sources in diets of different tilapia species and sizes are given in Table 6.3.

6.4.1. Animal protein sources

6.4.1.1. Fishmeal

Fishmeal (FM) has been traditionally used as the main protein source in the aquafeed industry, due to its high protein content and balanced EAA profile. FM is also an excellent source of essential fatty acids (EFA), digestible energy, minerals and vitamins. The increased demand for FM, coupled with a significant shortage in global FM production, has created sharp competition for its use by the animal feed industry. As a result, FM has become the most expensive protein commodity in animal and aquaculture feeds in recent years (El-Sayed, 1998, 1999a). Many developing countries have realized that, in the long run, they will be unable to afford FM as a major protein source in aquafeeds. Many attempts have been made to partially or totally replace FM with less expensive, locally available protein sources.

A wide variety of unconventional protein sources, including animal proteins, plant proteins, single-cell proteins and industrial and agricultural wastes, have been evaluated in respect of their utility in farmed tilapia feeds. Some sources proved cost-effective, while others were not. The following evaluation of alternative protein sources will provide farmers and nutritionists with information on the advantages and disadvantages of such feed ingredients as well as their proper inclusion levels in tilapia feeds.

6.4.1.2. Fishery by-products

Despite the fact that large amounts of fishery by-products and by-catch are produced annually in the world, little attention has been paid to the commercial use of these by-products for tilapia. The exception is fish silage (FS) and shrimp meals, several studies having considered their use as an FM replacer in tilapia feeds. The results indicated that between 30 and 75% fish silage can be successfully incorporated in tilapia feed, depending on fish species and size, silage source and diet composition (Fagbenro, 1994; Fagbenro *et al.*, 1994). Wassef *et al.* (2003) also found that grow-out Nile tilapia (28.6 g) responded better to FS than fry or fingerlings. They also reported that 50% FS was most cost-effective, while about 75% was required when FS + soybean meal (SBM) (1 : 1) was used. It is evident that fish silage has a potential as a protein source for tilapia.

On the other hand, the quality of fish silage is affected by the fermentation and/or silaging methods. For example, diets containing formic acid-preserved fish silage reduced the growth performance of tilapia, presumably due to acidity of the diet and a high proportion of free amino acids in the fish silage. It has been suggested that acidity reduces diet acceptance and affects protease activity in fish guts (Hardy *et al.*, 1983), while free amino acids may depress fish appetite (Wilson *et al.*, 1984).

Shrimp meal has also been successfully used as a protein source for tilapia. Blue tilapia (*Oreochromis aureus*) utilized shrimp head meal at up to 60% of the diet without adverse effects on their performance (Nwanna and Daramola, 2000). Moreover, Mansour (1998) and El-Sayed (1998) reported that shrimp meal can replace FM in red tilapia (*O. niloticus* × *Oreochromis hornorum*) and Nile tilapia diets at 50% and 100%, respectively without significant retardation in weight gain and feed efficiency.

More recently, Abdelghany (2003) evaluated gambusia (*Gambusia affinis*) fishmeal (GFM) as a replacement for herring FM protein in diets for red tilapia (*O. niloticus* × *O. mossambicus*). The growth of fish fed GFM at up to 50% substitution level was not different from that of fish fed a herring FM-based diet. Partial or complete substitution of GFM for FM did not affect feed utilization efficiency and digestibility of protein, dry matter and gross energy of the diets. Economic analysis showed that the best cost-effective results were achieved at the 50% substitution level.

6.4.1.3. Terrestrial animal by-products

Terrestrial animal by-products, including poultry by-product meal (PBM), blood meal (BM), hydrolysed

Table 6.3. Tested and recommended levels of different protein sources for tilapia. Levels tested are a substitution of standard dietary protein (mainly FM), SBM or whole diet. Recommended levels are based on biological and/or economic evaluation.

Source (specification)	Levels tested (%)	Requirement (%)	Species (weight, g)	References
Animal sources				
Gambusia meal	0–100	50	*O. niloticus* × *O. mossambicus* (0.42)	Abdelghany (2003)
Shrimp meal	100	100	*O. niloticus* (20)	El-Sayed (1998)
Shrimp meal	100	100	Red tilapia (9)	Mansour (1998)
Shrimp head waste	0–60	60	*O. niloticus* (1.4)	Nwanna and Daramola (2000)
FS + SBM or MBM, PBM or HFM (1 : 1)	0–75	50–75	*O. niloticus* (8)	Fagbenro (1994), Fagbenro *et al.* (1994)
MBM + Met	40–50	50	*O. niloticus* (11 mg)	Tacon *et al.* (1983)
MBM	100	100	*O. niloticus* (20)	El-Sayed (1998)
MBM	100	100	Red tilapia (9)	Mansour (1998)
MBM + BM (2 : 3)	0–100	100	*O. mossambicus* (1)	Davies *et al.* (1989)
BM	0–100	100	*O. mossambicus* (1)	Davies *et al.* (1989)
BM	100	< 100	*O. niloticus* (20)	El-Sayed (1998)
BM	100	< 100	Red tilapia (9)	Mansour (1998)
HFM	0–100	66	*O. niloticus* (0.01)	Bishop *et al.* (1995)
HFM ± EAA	10–50	30	*O. niloticus* (4–5)	Tacon *et al.* (1983)
Animal by-products	0–100	100	*O. niloticus* (0.1)	Rodriguez-Serna *et al.* (1996)
Chicken offal silage	0–20	20	*O. niloticus* (10.8)	Belal *et al.* (1995)
Oilseed plants				
SBM ± Met	75	75	*O. niloticus* (0.8)	Tacon *et al.* (1983)
SBM	0–100	75	*O. mossambicus* (50)	Jackson *et al.* (1982)
SBM ± Met	0–100	67	Hybrids (4.47)	Shiau *et al.* (1989)
Soy protein concentrate	0–100	100	*O. niloticus* (3.2)	Abdelghany (1997)
SBM + EAA + DCP + oil	0–100	100	Hybrids (84)	Viola *et al.* (1988)
SBM + DCP + oil	0–100	100	Hybrids (169)	Viola *et al.* (1988)
SB flour + PMM (75 : 25)	25–75	25	*O. niloticus* (7)	Sadiku and Jauncey (1995)
CSM ± Lys	100	100	*O. niloticus* (20)	El-Sayed (1990)
CSM	0–100	50	*O. mossambicus* (12)	Jackson *et al.* (1982)
CSM	0–100	80	*T. zillii* (1.5)	El-Sayed (1987)
Sesame seed meal	0–75	25	*T. zillii* (2.4)	El-Sayed (1987)
Groundnut cake	0–100	25	*O. mossambicus* (30)	Jackson *et al.* (1982)
Rapeseed meal	0–75	75	*O. mossambicus* (13)	Jackson *et al.* (1982)
Copra meal	0–50	25–50	*O. mossambicus* (31)	Jackson *et al.* (1982)
Defatted cocoa cake	100	100	*Tilapia guineensis* (52)	Fagbenro (1988)

Continued

Table 6.3. *Continued.*

Source (specification)	Levels tested (%)	Requirement (%)	Species (weight, g)	References
Palm kernel cake	0–100	60	*O. niloticus* (2.5)	Omoregie and Ogbemudia (1993)
Palm kernel cake	20–50	50	*O. mossambicus* (8.4)	Lim *et al.* (2001)
Macadamia press cake	0–100	50	*O. niloticus* (7.5–12)	Balogun and Fagbenro (1995)
Macadamia press cake	100	100	*T. guineensis* (NA)	Fagbenro (1993)
Sunflower meal	0–50	20	*T. rendalli* (0.93)	Olvera-Novoa *et al.* (2002)
Aquatic plants				
Spirulina	0–100	40	*O. mossambicus* (0.3)	Olvera-Novoa *et al.* (1998)
Azolla pinnata	8–42	42	*O. niloticus* (0.011)	Santiago *et al.* (1988)
Azolla pinnata	0–100	< 25	*O. niloticus* (4–40)	El-Sayed (1992)
Hydrodictyon	0–100	20	*O. niloticus* (1)	Appler (1985)
Hydrodictyon	0–100	20	*T. zillii* (1)	Appler (1985)
Eleocharis ochrostachys	20–40	20–30	*O. niloticus* (7)	Klinnavee *et al.* (1990)
Potamogeton	25–50	25	*O. niloticus* (14.5)	Essa (1997)
Duckweed	0–100	50–100	*Tilapia* sp. (NA)	Skillicorn *et al.* (1993)
Duckweed (*Lemna*)	0–50	50	*O. niloticus* (14.5)	Essa (1997)
Duckweed (*Spirodela*)	0–100	30	*O. niloticus* (13.9)	Fasakin *et al.* (1999)
Grain legumes				
Leucaena leaf meal	0–50	< 25	*O. mossambicus* (50)	Jackson *et al.* (1982)
Cassava leaf meal	20–100	< 100	*O. niloticus* (14–15)	Ng and Wee (1989)
Jackbean meal (cooked in distilled water)	20–30	20	*O. niloticus* (7.6)	Fagbenro *et al.* (2004)
Jackbean meal (cooked in trona solution)	20–30	30	*O. niloticus* (7.6)	Fagbenro *et al.* (2004)
Lucerne LPC	15–55	35	*O. mossambicus* (0.3)	Olvera-Novoa *et al.* (1990)
Cowpea LPC	0–50	20–30	*O. niloticus* (0.16)	Olvera-Novoa *et al.* (1997)
Maize gluten feed + SBM	100	100	*O. niloticus* (30)	Wu *et al.* (1995)
Maize gluten + SBM	100	100	*O. niloticus* (30)	Wu *et al.* (1995)
Maize co-products	0–100	50	Hybrids (21)	Twibell and Brown (1998)
Coffee pulp	0–39	13–26	*Oreochromis aureus* (4–10)	Ulla Rojas and Weerd (1997)
Toasted lima bean + Met	20–80	40–80	*O. niloticus* (5)	Adeparusi and Olute (2000)

FM, fishmeal; SBM, soybean meal; MBM, meat and bone meal; PBM, poultry by-product meal; HFM, hydrolysed feather meal; BM, blood meal; SB, soybean; CSM, cottonseed meal; PMM, poultry meat meal; DCP, dicalcium phosphate; NA, not available; LPC, leaf protein concentrate.

feather meal (HFM) and meat and bone meal (MBM), have been widely used as protein sources for tilapia, due to their high protein content and good EAA profiles. However, they may be deficient in one or more of the EAA. The most limiting EAA in these by-products are lysine (PBM and HFM), isoleucine (BM) and methionine (MBM, BM and HFM) (NRC, 1983; Tacon and Jackson, 1985). If these by-products are included in feeds at the proper ratios, the EAA deficiencies can be overcome and the quality of such diets is likely to improve (Tacon *et al.*, 1983; Davies *et al.*, 1989). Tacon *et al.* (1983) found that hexane-extracted MBM or MBM : BM (4 : 1) supplemented with methionine successfully replaced up to 50% of FM protein in Nile tilapia fry diets. Furthermore, Davies *et al.* (1989) found that optimum MBM/BM ratios could replace up to 75% of FM in diets fed to *O. mossambicus* fry. They also found that diets containing MBM or high MBM/BM ratios (3 : 1 and 2 : 3) were superior to FM even at a 100% substitution level. Cost–benefit analyses indicated that these sources can be used as single dietary protein sources for Nile tilapia (El-Sayed, 1998). On the contrary, BM and HFM are not efficiently utilized by tilapia due to low digestibility and poor EAA profiles (Viola and Zohar, 1984; Davies *et al.*, 1989; Bishop *et al.*, 1995).

Terrestrial animal by-product silage has been successfully used as a protein source for tilapia. Belal *et al.* (1995) fed *O. niloticus* fingerlings (10.8 g) test diets containing 0–20% chicken offal silage (COS), made from chicken viscera, as a replacement of FM. They found that the growth and body composition of fish fed COS up to the 20% level were similar to those of fish fed an FM-based diet. High inclusion levels of COS should be tested in order to determine the proper inclusion level.

6.4.2. Plant protein sources

6.4.2.1. Oilseed plants

SOYBEAN MEAL. Among plant protein sources, soybean meal (SBM) contains the highest protein content and has the best EAA profile. However, SBM is deficient in sulphur-containing amino acids (Met, Lys, cysteine (Cys)) and contains endogenous anti-nutrients, including protease (trypsin) inhibitor, phytohaemagglutinin and anti-vitamins. Some of the factors can be destroyed or inactivated during thermal processing (El-Sayed *et al.*, 2000).

SBM can be used as a total or partial protein source for farmed tilapia, depending on fish species and size, dietary protein level, SBM source and processing methods, and culture systems employed. For example, it was found that prepressed, solvent-extracted SBM, with or without Met supplementation, successfully replaced up to 75% of FM in the diet of Nile tilapia fry (Tacon *et al.*, 1983) and *O. mossambicus* (Jackson *et al.*, 1982) and 67% in the case of tilapia hybrids (Shiau *et al.*, 1989). In the meantime, supplementing SBM with the deficient EAA did not improve fish growth, and therefore was proved unnecessary (Teshima and Kanazawa, 1988).

The utilization of SBM in tilapia feeds may be limited by dietary minerals (phosphorus in this case), rather than the deficient EAA. It was reported that the non-inclusion of the deficient EAA in SBM-based diet did not result in any growth retardation, while SBM supplemented with 3% dicalcium phosphate (DCP) and oil completely replaced FM without any adverse effects on fish growth (Viola *et al.*, 1986, 1988). The non-necessity of EAA supplementation has also been reported with other oilseed plants (El-Sayed, 1987, 1990).

It should be realized that the quality of SBM (and other plant protein sources) for tilapia depends on the processing methods. For example, Wassef *et al.* (1988) found that the germination and defatting of SBM reduced the activity of protease inhibitors. Heating SBM also destroys the antinutritional factors and helps rupture the cellulose membrane surrounding the cell, releasing the cell contents and making them more available. Boiling full-fat SBM at 100°C for 1 h also improves its quality and reduces trypsin inhibitor activity for Nile tilapia. However, El-Sayed *et al.* (2000) found that full-fat SBM contained traces of protease inhibitors even after thermal treatment (at 200°C for 10 min) or soaking for 3 days, leading to an increase in trypsin secretion (to compensate for the reduced activity) in Nile tilapia.

COTTON SEED MEAL/CAKE. Cottonseed meal (CSM) is one of the most available plant protein sources in tropical and subtropical regions. It is also one of the best protein candidates for tilapia in developing countries, due to its high availability, relatively low price, good protein content (26–54%, depending on processing methods) and amino acid profile. However, it contains relatively low levels of Cys, Lys and Met in addition to a high content of gossypol (a phenolic anti-nutrient), which may limit the use of CSM in tilapia feeds. Results on the use of CSM and cottonseed cake (CSC) indicated that replacement of more traditional protein sources at between 50 and 100% can be effective in tilapia feed, depending on CSM source, processing method and fish species and size (see Table 6.3).

OTHER OILSEED BY-PRODUCTS. Other oilseed by-products, including groundnut, sunflower, rapeseed, sesame seed, copra, macadamia, cocoa cake and palm kernel, may have a good potential as protein sources for tilapia. However, only a few studies have been conducted to evaluate these ingredients, with varying and sometimes contradictory results. For example, Jackson *et al.* (1982) found that rapeseed meal could effectively replace up to 75% of FM protein in *O. mossambicus* diets without significant retardation in fish performance. On the other hand, Davies *et al.* (1990) found that only 15% rapeseed meal could effectively replace FM/SBM in *O. mossambicus* diets, while higher levels resulted in poor growth and poor feed efficiency due to the high content of glucosinolate (anti-nutrient) in rapeseed. Similar results were reported in respect of the use of macadamia press cake (MC) as a protein source for tilapia.

Sesame seeds are deficient in Lys and zinc. The supplementation of either Lys or zinc significantly improved the growth and survival of *T. zillii*, in addition to the disappearance of disease symptoms (El-Sayed, 1987). Therefore, Lys and zinc may meet the requirement for each other, supporting the argument that certain minerals rather than EAA deficiency may be the limiting factor in sesame seeds.

EFFECTS OF PHYTASE SUPPLEMENTATION. Many plant protein sources contain high levels of phytic acid, which binds with divalent minerals such as Ca, P, Zn, Mn, Mg and Fe to form water-insoluble salts, rendering the minerals unavailable. When these plants are used as the primary source of protein in a tilapia feed, higher supplementary mineral levels may be required, particularly if the culture water is deficient in one or more of the required minerals. The inclusion of bacterial phytase in tilapia diets can be effective in reducing phytic acid activity and improving the utilization of plant protein sources. Phytase may also reduce the effect of anti-nutritional factors, protect amino acids from degradation and decrease leaching of water-soluble components (Riche *et al.*, 2001; Heindl *et al.*, 2004).

Many recent studies have indicated that the addition of phytase to tilapia diets has improved growth rates, digestibility and the utilization of dietary protein phosphorus (Riche *et al.*, 2001; Heindl *et al.*, 2004; Phromkunthong *et al.*, 2004; Liebert and Portz, 2005). These studies demonstrated that about 750–1000 phytase units/kg feed were required for optimum performance, depending on the dietary plant protein/animal protein ratios and mineral contents of the diets.

6.4.2.2. Aquatic plants

Several studies have been conducted on the use of aquatic plants in tilapia feeds. Among these plants, the duckweed (family: *Lemnaceae*) is the most promising. Fresh duckweed is an excellent food source for tilapia, as it contains about 35–45% crude protein with good amino acid and mineral profiles. It can be easily cultivated, and it grows at a tremendous rate to yield 10–50 dry Mt/ha/year, depending on cultivation environment (Leng *et al.*, 1995). Duckweed can be used as a single food source for farmed tilapia. A study in Bangladesh (Skillicorn *et al.*, 1993) indicated that, when duckweed was used as a single nutritional input for tilapia in earthen ponds, fish production reached 7.5 Mt/ha/year. Furthermore, with better pond management and stocking density, the annual production/ha can reach 10 Mt or more. Dry duckweed is also a good protein and energy alternative for tilapia. It can replace up to 50% of the commercial feed without adverse effects on fish performance (Essa, 1997). Fasakin *et al.* (1999) also found that replacing FM with up to 30% sun-dried duckweed supported the growth of Nile tilapia and was cost-effective.

The use of *Azolla*, a freshwater fern that has a symbiotic relationship with the nitrogen-fixing cyanobacterium *Anabaena azollae*, as a fresh or dried feed ingredient for farmed tilapia has also been investigated, with controversial results. El-Sayed (1992) evaluated *Azolla pinnata* as an FM replacer for Nile tilapia fingerlings and adults at 0–100% substitution levels. Fish fed with *A. pinnata* showed extremely poor performance even at the lowest inclusion level (25%). Similar results were reported with *O. niloticus* (Almazan *et al.*, 1986) and *Tilapia rendalli* fed *Azolla microphylla* (Micha *et al.*, 1988). It seems from these studies that the acceptability of *Azolla* to tilapia is poor. In contrast, between 30 and 42% of FM-based diets fed to Nile tilapia have been successfully replaced by *Azolla* meal without negatively affecting fish performance (Santiago *et al.*, 1988; Naegel, 1997). More long-term field studies are needed to resolve this controversy.

Other aquatic plants, including *Hydrodictyon reticulatum*, coontail (*Ceratophyllum demersum*), chuut-nuu (*Eleocharis ochrostachys*) and *Potamogeton gramineus*, can be used as a partial replacement of standard protein for different tilapia species. However, these sources should be carefully looked at, since some other aquatic plants, such as *Elodea trifoliate* and *Muyriophyllum spicatum*, have been reported to reduce tilapia performance.

6.4.2.3. Grain legumes

Many leguminous or cereal plants and by-products can be used as partial protein sources for tilapia. Among these, leucaena leaf meal (LLM, 30% crude protein), brewery wastes, maize products (maize gluten, maize distiller's grain, maize co-products, maize gluten feed), cassava leaf meal, green gram legume, lima bean and leaf protein concentrates are of prime importance. However, most leguminous or cereal plants are deficient in certain EAA (e.g. Arg, Thr, Iso, His and Met are deficient in LLM) and may contain anti-nutrients, such as mimosine (a toxic non-protein amino acid), found in LLM (Lim and Dominy, 1991). Therefore, proper processing of these sources (heating, soaking, cooking, etc.) may improve their quality for tilapia. Osman *et al.* (1996) reported that cooked or sun-dried LLM produced better performance of Nile tilapia than did sodium hydroxide-treated or rumen liquor-incubated LLM.

Other legume seeds have been tested as protein sources for tilapia. Ng and Wee (1989) found that the performance of Nile tilapia fed cassava leaf meal (CLM) was reduced with increasing CLM levels in the diets. However, fish growth was significantly improved when CLM was supplemented with 0.1% Met. Similarly, when the green gram legume (*Phaseolus aureus*) was fed to Nile tilapia fry at different dietary protein levels, the best growth rates were observed at 25% substitution level (De Silva and Gunasekera, 1989). In a recent study, Fagbenro *et al.* (2004) found that jack bean (*Canavalia ensiformis*) can be a useful partial substitute for FM in tilapia feed. Up to 20% of SBM protein was provided by jackbean seeds cooked in distilled water, while cooking in trona solution increased the inclusion level to 30%.

As pointed out earlier, most of the above-mentioned plant sources may contain high levels of phytic acid, which binds with divalent minerals, such as Ca, P, Zn, Mn, Mg and Fe, to form water-insoluble salts, thus reducing the bioavailability of these minerals. When these plants are used as a primary source of protein in a tilapia feed, higher supplementary mineral levels may be required, particularly if the culture water is deficient in one or more of the required minerals.

6.4.3. Single-cell proteins

Single-cell proteins (SCP) are a group of microorganisms, including unicellular algae, fungi, bacteria, cyanobacteria and yeast. The use of SCP for tilapia in semi-intensive and intensive farming systems has attracted attention in recent years. SCP production is a simple, cheap and effective way of producing natural fish food. For example, if a carbon source (such as wheat bran, rice bran and cellulose) is sprayed on the surface of pond water with continuous aeration, at the optimum carbon : nitrogen ratio (15 : 1), bacterial growth will increase (Chamberlain and Hopkins, 1994). The bacteria consume the carbon source as energy and reduce the ammonia concentration through nitrification, while the fish feed on the bacteria. In such cases, cheap carbon and nitrogen sources can partially replace expensive commercial protein sources in tilapia feeds. More attention should be given to the production of SCP in fish ponds in developing countries, where tilapia culture is widely practised.

6.5. Economic Evaluation of Protein Sources

Most of the protein sources used in tilapia feeds have been evaluated from the biological and nutritional points of view. Economic evaluation of such feed inputs for tilapia has not been given enough attention. Only a few studies have demonstrated that, despite the fact that many of these feed inputs produced lower biological performance than standard proteins (mainly fishmeal), cost–benefit analyses indicate that they were economically better. For example, economic evaluation of cottonseed meal (El-Sayed, 1990), maize gluten feed and meal (Wu *et al.*, 1995) and animal by-product meal (Rodriguez-Serna *et al.*, 1996; El-Sayed, 1998) as protein sources for Nile tilapia, brewery waste (Oduro-Boateng and Bart-Plange, 1988) for *Tilapia busumana* and cocoa cake (Fagbenro, 1988) for *Tilapia guineensis* indicated that profit indices of these protein sources were better than for FM-based feeds. The authors suggested the use of these sources as total fishmeal replacers for tilapia. More work is needed to economically evaluate the other unconventional protein sources that are commonly used in tilapia feeds.

6.6. Measurement of Dietary Energy

Dietary energy content is generally expressed in three different ways: gross (total) energy (GE), digestible energy (DE) and metabolic energy (ME). Theoretically, ME is preferred to GE and DE, since it accounts for energy loss from protein metabolism and, in turn, provides a more precise estimate of the energy used for growth. But the determination of ME is difficult, because of the problems associated with collecting fish metabolites. It is also suspected that ME offers little advantage over DE since energy losses via gills and urine are small. Therefore, it may be more appropriate to use DE as a measurement of dietary energy because it can be easily determined. The digestibility coefficients of a number of ingredients commonly used in tilapia feeds have been determined (Table 6.4). It is more appropriate to use the values reported for tilapia than ME or DE values reported for other fish species. If DE is not available, GE is likely to be the only useful alternative.

One other major problem associated with the determination of dietary energy requirements of tilapia is the interchangeable and inconsistent use of terminology. Many authors have used ME and DE values interchangeably, and have used varying energy values for the same ingredient under the same terminology. In addition, energy values reported for other fish species have been widely used in the preparation of tilapia diets (see El-Sayed and Teshima, 1991, for details).

6.7. Dietary Lipid Requirements

Tilapia, like other vertebrates, require dietary lipids for the following physiological functions:

1. A source of essential fatty acid (EFA) supply.
2. Energy production and protein sparing.
3. Normal growth and development.
4. Carrying and assisting in the absorption of fat-soluble vitamins.
5. Structure and maintenance of cell membrane integrity and flexibility.
6. Precursors of steroid hormones.
7. Improving the texture and flavour of the diet and fatty acid composition of the fish.

Tilapia are known to utilize dietary lipids very efficiently. Dietary lipids may spare more protein for growth than do carbohydrates (Teshima *et al.*, 1985b; El-Sayed and Garling, 1988). Despite these attributes, lipid requirements of tilapia have not been well studied. The requirements depend on lipid source, dietary protein and energy contents and tilapia species and size. Generally speaking, tilapia require about 10–15% dietary lipids for maximum growth performance (Table 6.5). However, tilapia farmers often use lower lipid values (6–8%).

6.7.1. Essential fatty acid (EFA) requirements

It is known that cold-water fish and marine fish require n-3 polyunsaturated fatty acids (n-3 PUFA), while freshwater fish inhabiting warm-water environments tend to require n-6 PUFA. This may suggest that tilapia would utilize plant oils (rich in n-6 fatty acids) more efficiently than fish oils (rich in n-3 fatty acids). However, information on fatty acid requirements of tilapia has been contradictory. Several studies have indicated that tilapia require n-6 EFA rather than n-3 EFA (Kanazawa *et al.*, 1980a; Stickney *et al.*, 1982; Stickney and McGeachin, 1983). The growth of Nile tilapia fed on a fish oil-containing diet (rich in n-3 EFA) was significantly reduced as compared with one containing soybean oil or maize oil (rich in n-6 EFA) (Takeuchi *et al.*, 1983). Similar results were reported for Nile tilapia broodstock, where fish fed with fish oil-containing diets had a significantly poor reproductive performance compared to those fed soybean oil diets (Santiago and Reyes, 1993).

Table 6.4. Protein and organic matter (in parentheses) digestibility (%) by tilapia of some feed ingredients commonly used in feed manufacturing.

Ingredient	1a (*O. n*)	2a (*O. a* × *O. n*)	3a (*O. n*)	4b (*O. n*)	5a (*O. n*)	6a (*O. n*)	7c (*O. n*)	8c (*O. n* × *O. a*)
Animal sources								
Fish meal	92.2		72 (58)	92	94			90.2 (89.2)
Sardine						87 (80)		
Tuna						82		
Anchovy						94	90 (86)	
Meat and bone meal	92.2							
Poultry offal meal				74 (59)		74 (59)		87.2 (79.2)
Shrimp meal		74		87				
Silkworm pupa meal	91.1							
Plant sources								
Azolla				75				
Brewers' grains		63	62 (42)			63 (30)		
Cooked cassava meal						74		
Casein				97				
Copra meal				56	81	81		
Maize grains (raw)						83 (76)		75.1 (61.4)
Maize grains (cooked)						90		
Maize gluten meal	90.7			97				96.5 (83.4)
Cottonseed meal		90	31(–24)			90		
Cottonseed cake					90			
Groundnut meal			79 (72)					
Palm kernel meal			–26 (–89)					
Rapeseed meal								85 (57.3)
Soybean meal	90.9	91		93	93	91 (57)	93	96.2 (84.5)
Sunflower seed meal								99.2 (65.5)
Sunflower seed cake							86–89 (42)	
Wheat germ meal	95.5							
Wheat middlings						76 (58)		
Wheat bran							75 (30)	83.6 (38.8)

1, Watanabe *et al.* (1996); 2, Degani *et al.* (1997); 3, Luquet (1989); 4, Lorico-Querijero and Chiu (1989); 5, Moreau (1996); 6, Hanley (2000); 7, Maina *et al.* (2002); 8, Sklan *et al.* (2004); a, % protein and (organic matter) digestibility; b, true digestibility; c, apparent digestibility.
O. a, *Oreochromis aureus*; *O. n*, *Oreochromis niloticus*.

Table 6.5. Lipid requirements of tilapia.

Tilapia species	Requirements (% of diet)	Reference
T. zillii	15	El-Sayed and Garling (1988)
O. niloticus	15	Teshima *et al.* (1985b)
O. aureus	10	Stickney and Wurts (1986)
O. aureus × *O. niloticus*	12	Jauncey and Ross (1982)
O. mossambicus × *O. niloticus*	18	De Silva *et al.* (1991)
O. niloticus × *O. aureus*	12	Chou and Shiau (1996)

On the other hand, many other studies have indicated that tilapia may require both n-3 EFA and n-6 EFA. Stickney and McGeachin (1983) found that 10% soybean oil or 10% fish oil produced similar growth of blue tilapia. Furthermore, Chou and Shiau (1999) found that both n-3 and n-6 highly unsaturated fatty acids (HUFA) are required for maximum performance of tilapia hybrids (*O. niloticus* × *O. aureus*). More recently, El-Sayed *et al.* (2005a) found that Nile tilapia broodstock reared in brackish water required n-3 HUFA for optimum spawning performance, while the reproductive performance of fish reared in fresh water was not affected by dietary lipid source. It is evident that the requirements of tilapias for EFA are species-specific, and more work is needed to quantify the EFA requirements of different tilapias under different culture systems and in different water environments.

6.8. Carbohydrate Utilization

Carbohydrates are the cheapest sources of dietary energy for humans, fish and domestic animals. Carbohydrates are also the most available food source in the world. Tilapias are mainly herbivorous and are expected to utilize dietary carbohydrates more efficiently than carnivorous fishes. Tilapia can efficiently utilize as much as 35 to 40% digestible carbohydrate (Anderson *et al.*, 1984; El-Sayed and Garling, 1988). In addition, increasing dietary carbohydrate results in the sparing of dietary protein for growth.

Traditionally, wheat bran, maize and rice bran have been used as the major dietary carbohydrate sources for tilapia. A number of other unconventional sources have been evaluated for different tilapia species. El-Sayed (1991) found that up to 20% sugarcane bagasse could be included in *T. zillii* diets without significant retardation in fish growth and feed digestibility. However, Nile tilapia fed sugarcane bagasse diets exhibited poor performance. Cacao husks successfully replaced wheat bran, wheat flour or rice bran in tilapia feeds at up to 20% inclusion level (Pouomogne *et al.*, 1997). Barley seeds replaced up to 30% of dietary maize in Nile tilapia diets without adverse effects on fish performance (Belal, 1999). In a recent study, El-Sayed (2003) studied the use of fresh and fermented water hyacinth by Nile tilapia fingerlings as a supplement of wheat bran at 20 and 40% levels (10 and 20% of the diet). He found that, at the 10% level, fermented and fresh water hyacinth produced a similar performance to the control diet. At the level of 20% of the diet, fermented water hyacinth was better utilized than fresh water hyacinth.

Carbohydrate utilization by tilapia is affected by a number of factors, including carbohydrate source, other dietary ingredients, fish species and size and feeding frequency. Tilapia have been reported to utilize complex sugars (starch) more efficiently than disaccharides and monosaccharides (Anderson *et al.*, 1984; Shiau and Chen, 1993; Shiau and Lin, 1993). In support, it was found that lipogenic enzyme activities in tilapia liver were higher in fish fed starch diets than in fish fed a glucose diet (Lin and Shiau, 1995), indicating that these activities are adaptive to the dietary carbohydrate source.

The requirement of niacin, which is a precursor of certain coenzymes needed for carbohydrate metabolism, is affected by dietary protein source (Shiau and Suen, 1992). In addition, dietary chromium chloride or chromic oxide was found to improve glucose (but not starch) utilization by tilapia (Shiau and Lin, 1993; Shiau and Liang, 1995).

The quality of dietary carbohydrates for tilapia is influenced by their fibre content. Shiau *et al.* (1988) found that increasing dietary carboxymethylcellulose (CMC) depressed stomach emptying, carbohydrate digestion and absorption, fish growth and feed conversion efficiency. The inclusion of 10% guar gum, carrageenan, CMC and cellulose in tilapia diets resulted in a significant decrease in fish performance. Moreover, as low as 2% dietary chitin or chitosan depressed tilapia growth (Shiau and Yu, 1999).

Carbohydrate utilization by tilapia is also affected by daily feeding frequency. Tung and Shiau (1991) and Shiau and Lei (1999) have demonstrated that increasing feeding frequency (from 2 to 6 times/day, or continuous feeding) improved carbohydrate utilization, growth rates and protein sparing. Fish size also affects carbohydrate utilization. When two sizes of tilapia (0.46 and 4.55 g) were fed glucose or starch, the larger fish utilized glucose more efficiently than the smaller fish, while starch was utilized equally by the two sizes (Tung and Shiau, 1993).

Carbohydrates may contain amylase inhibitors or other anti-nutrients that reduce their utilization by fishes. El-Sayed *et al.* (2000) found that wheat bran contains protease inhibitor, the activity of which may negatively affect food digestibility. Processing of dietary carbohydrates may therefore improve their quality for tilapia. In this regard, El-Sayed (1991) found that cooking sugarcane bagasse slightly improved its utilization by *T. zillii*. As noted above, in a recent study, El-Sayed (2003) found that fermented water hyacinth was better utilized than fresh water hyacinth when incorporated in Nile tilapia diets at 20% levels, while at the 10% inclusion level fermented and fresh hyacinth were utilized equally.

6.9. Vitamin Requirements

Vitamins are organic compounds that are generally required in small quantities in animal diets. They act as cofactors or substrates in normal metabolic reactions. Vitamins are subdivided into water-soluble (macrovitamins, based on the levels required) and fat-soluble groups. Vitamin deficiency may result in nutrition-related diseases, poor growth or increased susceptibility to infections.

Vitamin supplementation is not necessary for tilapia in semi-intensive farming systems, but in intensive systems supplemental dietary vitamins may be necessary. However, it is difficult to accurately determine the vitamin requirements of tilapia in intensive systems because they consume significant amounts of vitamins along with the natural food they consume from the culture water. The vitamin requirements of these fish have not been sufficiently studied. The species that has been studied in relative detail is hybrid tilapia (*O. niloticus* × *O. aureus*), while limited information is available on the vitamin requirements of other tilapia species. The vitamin requirements and deficiency symptoms of tilapia are summarized in Table 6.6.

The following can be observed from Table 6.6:

- The niacin requirement is affected by dietary carbohydrates. Tilapia fed simple sugars (glucose) required lower niacin levels than fish fed polysaccharides (dextrin).
- Because ascorbic acid is unstable and loses its activity during processing and storage, more stable forms of ascorbic acid should be used. L-Ascorbyl-2-sulphate (C2S), L-ascorbyl-2-monophosphate-magnesium (C2MP-Mg) and L-ascorbyl-2-monophosphate-sodium (C2MP-Na) could be used.
- The vitamin E requirement increases with increasing dietary lipid level.
- The requirement of vitamin K for tilapia has not been determined.

It has been reported that tilapia may not require certain vitamins, since the fish are able to synthesize them. Lovell and Limsuwan (1982) and Sugita *et al.* (1990) found that dietary supplementation of vitamin B_{12} is not required by tilapia. They suggested that these fishes can synthesize that vitamin in their digestive tracts by means of intestinal microorganisms.

6.10. Mineral Requirements

Minerals are inorganic elements that are required by animals to maintain many of their metabolic functions. These functions can be summarized as follows (Jauncey, 1998):

- Structure of hard skeletons such as bones and teeth.

Table 6.6. Vitamin requirements and deficiency symptoms of tilapia.

Vitamin	Fish species	Requirement (mg/kg)	Deficiency symptoms
Niacin[a]	*O. n* × *O. a*	26 (fed glucose) 121 (fed dextrin)	Haemorrhage, deformed snout Gill and skin oedema, fin and mouth lesions
Biotin[b]	*O. n* × *O. a*	0.06	
Choline[c]	*O. n* × *O. a*	1000	
Pantothenic Acid[d]	*O. aureus*	10	Low growth, haemorrhage, anaemia, sluggishness, high mortality, hyperplasia of epithelial cells of gill lamellae
Thiamin[e]	*O. m* × *O. n*	2.5 (seawater, 32‰)	Low growth and low feed efficiency, low haematocrit
Riboflavin[f]	*O. aureus*	6	Anorexia, low growth, high mortality, fin erosion, loss of body colour, cataract, dwarfism
Riboflavin[g]	*O. m* × *O. n*	5 (seawater, 32‰)	
Pyridoxine[h]	*O. n* × *O. a*	1.7–9.5 (28% cp) 15–16.5 (36% cp)	Low growth, high mortality, abnormal neurological signs, anorexia, convulsions, caudal fin erosion, mouth lesion
Vitamin C	*O. aureus*	50 (ascorbic acid)[i]	
	O. n × *O. a*	79 (ascorbic acid)[j]	
	O. niloticus	420 (ascorbic acid)[k]	
	O. n × *O. a*	41–48 (C2S)[l]	
	O. n × *O. a*	37–42 (C2MP-Mg)[l]	
	O. n × *O. a*	63.4 (C2MP-Na)[l]	
	O. spilurus	11–200 (C2S)[m]	
D[n]	*O. n* × *O. a*	374.8 IU	
E[o]	*O. aureus*	10 (at 3% lipid) 25 (at 6% lipid)	Low growth and FCR, skin haemorrhage, muscle dystrophy, impaired erythropoiesis, ceroid in liver and spleen, abnormal skin coloration
E[p]	*O. niloticus*	50–100 (5% lipid) 500 (10–15% lipid)	
E[q]	*O. n* × *O. a*	42–44 (5% lipid) 60–66 (12% lipid)	
A[r]	*O. niloticus*	5000 IU	Low growth, restless, abnormal movement, blindness, exophthalmia, haemorrhage, pot-belly syndrome, reduced mucus secretion, high mortality
Inositol[s]	*O. n* × *O. a*	400	

[a]Shiau and Suen (1992); [b]Shiau and Chin (1999); [c]Shiau and Lo (2000); [d]Soliman and Wilson (1992a); [e]Lim and Leamaster (1991); [f]Soliman and Wilson (1992b); [g]Lim *et al.* (1993); [h]Shiau and Hsieh (1997); [i]Stickney *et al.* (1984); [j]Shiau and Jan (1992); [k]Soliman *et al.* (1994); [l]Shiau and Hsu (1995); [m]Al-Amoudi *et al.* (1992); [n]Shiau and Hwang (1993); [o]Roem *et al.* (1990); [p]Satoh *et al.* (1987); [q]Shiau and Shiau (2001); [r]Saleh *et al.* (1995); [s]Shiau and Su (2005).
O. a, *Oreochromis aureus*; *O. m*, *Oreochromis mossambicus*; *O. n*, *Oreochromis niloticus*; cp, crude protein; IU, international units; FCR, feed conversion ratio.

- Osmoregulation.
- Structure of soft tissues.
- Nerve impulse and transmission and muscle contraction.
- Body acid–base equilibrium and regulation of the pH of the blood and other fluids.
- Cofactors in metabolism, catalysts and enzyme activators.
- Minerals serve as components of many enzymes, vitamins, hormones and respiratory pigments.

Little information is available on the mineral requirements of tilapia, and therefore their response to dietary minerals is not fully understood. The requirements of only eight minerals, namely, calcium, phosphorus, magnesium, zinc, manganese, potassium, iron and chromium have been studied for tilapia (Table 6.7). The mineral requirements of tilapia depend on fish size and mineral contents of both culture water and fish feeds. Viola *et al.* (1986) found that about 0.7% dietary phosphorus was sufficient for optimum growth of large male tilapia hybrids, while small fish required about 1% phosphorus. The availability of dietary phosphorus to tilapia also depends on phosphorus source and fish size. It was found that phosphorus in fishmeal was more available than that in plant sources, such as soybean, sorghum and wheat. The best growth and bone mineralization of *O. aureus* fed casein-based diets was obtained at 0.7% dietary calcium and 0.5% phosphorus (Robinson *et al.*, 1987).

Ishak and Dollar (1968) found that *O. mossambicus* reared in water containing 25 ppm manganese and fed diets supplemented with 2.8 mg manganese/kg suffered from poor growth, high mortality, anorexia and loss of equilibrium. Supplementing the diets with up to 35 mg manganese/kg did not improve fish performance. Adding manganese to the water helped the fish overcome some of the deficiency signs, while adding manganese to both the water and the diet resulted in the best performance. On the other hand, Watanabe *et al.* (1988) reported that 12 mg manganese/kg was recommended for Nile tilapia.

The zinc requirement for maximum growth of *O. aureus* was 0.02 g/kg (McClain and Gatlin, 1988), whereas *O. niloticus* require 0.03 g/kg (Eid and Ghonim, 1994). The potassium requirement of tilapia hybrids (*O. niloticus* × *O. aureus*) is 0.02–0.03 g/kg (Shiau and Hsieh, 2001). Watanabe *et al.* (1988) recommended 3–4 mg copper/kg for Nile tilapia.

From 0.59 to 0.77 g magnesium/kg was required for optimum growth of Nile tilapia (Dabrowska *et al.*, 1989), while 0.5 g/kg was needed for *O. aureus* (Reigh *et al.*, 1991). Magnesium and protein requirements of tilapia are significantly affected by one another. Dabrowska *et al.* (1989)

Table 6.7. Mineral requirements of tilapia.

Mineral	Species	Requirement (mg/kg)	Reference
Calcium	*O. aureus*	7000	Robinson *et al.* (1987)
Phosphorus	*O. niloticus*	< 9000	Watanabe *et al.* (1980)
	O. niloticus	4600	Haylor *et al.* (1988)
	O. niloticus × *O. aureus*	7000	Viola *et al.* (1986)
	O. aureus	5000	Robinson *et al.* (1987)
Potassium	*O. niloticus* × *O. aureus*	2100–3300	Shiau and Hsieh (2001)
Magnesium	*O. aureus*	500	Reigh *et al.* (1991)
	O. niloticus	600–800	Dabrowska *et al.* (1989)
Zinc	*O. niloticus*	30	Eid and Ghonim (1994)
	O. niloticus	79	do Carmo e Sá *et al.* (2004)
	O. aureus	20	McClain and Gatlin (1988)
Manganese	*O. niloticus*	12	Watanabe *et al.* (1988)
Iron	*O. niloticus*	60	Kleemann *et al.* (2003)
Chromium	*O. niloticus* × *O. aureus*	139.6	Shiau and Shy (1998)

fed Nile tilapia diets containing two protein levels (24 and 44%) and four magnesium levels (0.07, 0.68, 1.0 and 3.2 g/kg). They found that magnesium-deficient or magnesium-excessive diets resulted in poor growth and poor feed utilization efficiency, while about 0.59 to 0.77 g/kg magnesium was adequate for optimum performance. At the low protein level increasing magnesium to 3.2 g/kg resulted in poor performance, while at the higher protein level increasing dietary magnesium had only a slight negative effect on fish growth. At the high protein level, fish fed a magnesium-deficient diet absorbed magnesium from culture water at a higher rate than those reared on the low-protein diet. In addition, fish fed a low-protein–high-magnesium diet exhibited a decrease in blood haematocrit and haemoglobin contents, accompanied by sluggishness in fish movement.

6.11. Feeding Regimes and Practices

FEEDING LEVELS. Feeding regimes are one of the most disputed areas in tilapia nutrition. Some researchers suggest that daily feed be provided to the fish as a percentage of fish body weight, while others recommend *ad libitum* (or satiation) regimes for best growth and feed utilization. When *ad libitum* feeding is practised, feed should be available all the time (e.g. through automatic or demand feeders), while in the satiation regime the fish are fed as much as they can consume. Regardless of this dispute, the feeding regime of farmed tilapia is not fully understood. Varying results have been reported regarding the feeding levels, amounts and frequency of different tilapia. For example, optimum growth of Nile tilapia was obtained at a frequency of four times per day when the fish were fed *ad libitum* (De Silva *et al.*, 1986). When the fish were fed a restricted ration of 3% of their body weight (bw)/day, the best performance was achieved at one or two feedings per day. These results suggested that restricted feeding regime was better than *ad libitum*, which may result in feed waste. However, Nile tilapia hybrids (*Oreochromis urolepis hornorum* × *O. mossambicus*) reared in marine cages showed maximum growth at feeding rates near satiation (Clark *et al.*, 1990).

FEEDING FREQUENCY. Because tilapia have small stomachs and are characterized by continuous feeding behaviour, more frequent feeding would be appropriate for them. Meanwhile, feeding levels and frequency of tilapia decrease with increasing fish size. During larval stages, the fish require a daily ration of about 20–30% of their body weight, divided into six to eight feedings. Fish fingerlings require 3–4% of body weight, dispensed three to four times daily. It has been indicated that increasing feeding levels above fish requirements may reduce feed digestibility and utilization efficiency (Meyer-Burgdorff *et al.*, 1989).

A feeding trial was conducted to evaluate the appropriate feeding frequency for cage-cultured red tilapia (*O. mossambicus* × *O. hornorum*) in Thailand (Orachunwong *et al.*, 2001). Fish were stocked in floating cages suspended in earthen ponds and fed a commercial tilapia feed (25% crude protein (cp)) *ad libitum* two, three and four times daily. Feeding three to four times a day resulted in better growth and feed conversion ratio (FCR) than twice a day. Dividing the daily ration of tilapia reared in cages into three to four feedings would probably reduce feed loss compared to once or twice a day.

FEEDING METHODS. Feeding methods (hand-feeding, blower feeding, automatic feeders or demand feeders) may affect tilapia performance. Hand-feeding is recommended in small-scale tilapia farms, because it allows the feeder to regulate the amount of feed required, prevent overfeeding and observe fish behaviour and feeding activity. However, automatic feeders, feed blowers and demand feeders are usually used with large commercial farms. Hargreaves *et al.* (1986) found no difference in the performance of *O. aureus* fed by automatic feeders and those fed fixed rations. They recommended that automatic feeders be used, because they reduced 88 to 94% of the labour requirement. Rakocy *et al.* (2000b) also recommended demand feeders for tilapia reared in cages suspended in runoff ponds. Similarly, demand feeders provide excellent growth and feed conversion of tilapia reared in marine cages, in addition to reducing labour. More recently, Endo *et al.* (2002) found that self-fed Nile tilapia had significantly lower blood cortisol, higher phagocytic activity of their macrophages, higher antibody production and a higher number of blood lymphocytes. The authors suggested that self-feeding reduced fish stress and improved their immune response.

DIET FORM. Farmed tilapia are known to accept a wide range of diet forms. The acceptance is affected by fish size, stocking density, culture system and natural food availability. For example,

Nile tilapia fingerlings and juveniles reared in floating cages responded better to mashed than to pelleted feed, while large fish grew better on pelleted diet (Guerrero, 1980). In addition, when the fish were fed moist or dry pellets in cages, moist feed was utilized slightly better than dry feed; however, the leaching rate was higher in moist than in dry feeds. Meanwhile, the growth of Nile tilapia fry fed pellet crumbles and unpelleted diets was not different (Santiago *et al.*, 1987). Tilapia can also consume both floating and sinking pellets very efficiently. Allison *et al.* (1979) found that sinking pellets were better utilized than unpelleted feed by blue tilapia.

FEED COLOUR. Feed colour may affect tilapia performance, particularly in intensive farming systems. In a recent, preliminary study, El-Sayed (2004) evaluated the effects of feed colour on the growth and survival of Nile tilapia fingerlings reared in indoor, recirculating system. The fish were fed four test diets with different colours, red, light green, dark blue and yellow, in addition to a commercial, light-brown diet. Fish fed darker-coloured diets (red and dark blue) had better performance than those fed lighter-coloured diets (green and yellow). The red diet produced the best growth rate while the yellow diet resulted in the poorest performance. Despite the evident effects of feed colour on tilapia performance in this preliminary study, more long-term studies should be conducted using various feed colours before a final conclusion can be drawn.

6.12. Closing Remarks

1. Most of the values reported on dietary protein requirements of tilapia are questionable because of the use of casein as a single protein source (instead of casein/arginine or casein/gelatin mixtures).

2. With the exception of Nile tilapia and Mozambique tilapia, the EAA requirements of farmed tilapias are not known.

3. A wide variety of unconventional dietary protein sources have been tested for different tilapias. Most of these sources could partially or totally replace standard protein sources (mainly fishmeal) in tilapia feeds.

4. The digestibility by tilapia of a number of feed ingredients that are commonly used in tilapia feeds has been determined. DE could be used to calculate the energy contents of practical tilapia feeds if these ingredients are used. If DE is not available, GE should be used instead of using DE or ME values reported for other fish species.

5. Extensive work should be conducted on the utilization of unconventional carbohydrate sources by tilapia, especially in developing countries, where these sources are available at low prices.

6. Despite the fact that tilapia require n-6 EFA, n-3 EFA may also be required.

7. The requirements of tilapia for most water- and fat-soluble vitamins are not known. Extensive, long-term studies should be conducted in this area.

8. Determination of the dietary mineral requirements of tilapia should consider the relationship between dietary protein and energy and essential minerals.

9. The feeding regime (level, amount, frequency) of cultured tilapias is influenced by several factors, including fish species, size and age, diet form (pellets, crumbles, mash, dry, moist, floating, sinking) and culture system.

10. Feed colour may affect the growth and feed utilization of tilapia. It appears that darker-coloured diets are better utilized than lighter-coloured diets. More work is needed to support this argument.

7
Reproduction and Seed Production

7.1. Introduction

As mentioned in Chapter 1, tilapia culture is expanding at an extremely high rate. The more tilapia culture is expanded, the more it becomes vital to produce sufficient amounts of seed to meet the increasing demand of tilapia farmers. The current status of tilapia culture in many countries suggests that there is a gap between seed supplies and farmers' demand. Tilapia seed producers are also usually faced with a number of constraints that limit the management of mass seed production (Little *et al.*, 1993; Bhujel, 2000). These include: early maturation of tilapia, which may lead to overcrowding and fish stunting; low fecundity; successive reproductive cycles; and asynchronous spawning. Therefore, optimizing hatchery management becomes necessary for sustainability of seed production and maximization of tilapia culture output.

This chapter addresses the major issues of tilapia reproduction, including modes of reproduction, broodstock management, hatchery management, environmental factors and larval rearing. It is hoped that this discussion will clarify the many problems associated with tilapia seed production and improve our understanding of tilapia reproduction.

7.2. Modes of Reproduction

Tilapia are currently divided into two main generic groups according to their mode of reproduction:

1. 'Substrate spawners' (guarders), which include the fish belonging to the genus *Tilapia* (e.g. *T. zillii*, *T. rendalli*) (Coward and Bromage, 2000). For substrate spawners, both males and females build and defend the nest (Fig. 7.1) and protect the fertilized eggs and hatched larvae until they become independent of their parents. The courtship in these fish lasts for several days (Coward and Bromage, 2000).

2. 'Mouthbrooders', where fertilized eggs are incubated in the parents' buccal cavities (Fig. 7.2). This group is further divided into two genera:

(a) *Oreochromis* (maternal mouthbrooders), including *O. niloticus*, *O. aureus* and *O. mossambicus*, where only females brood the eggs in their mouths.

(b) *Sarotherodon*: biparental mouthbrooders (e.g. *S. galilaeus*), where both males and females brood and defend the fertilized eggs and hatched fry, and paternal mouthbrooders (e.g. *S. melanotheron*), where only males brood the eggs (Turner and Robinson, 2000). Courtship in these fish lasts for several hours to several days (Rana, 1990a; Coward and Bromage, 2000).

The majority of cultured tilapia belong to the maternal mouthbrooding *Oreochromis* genus, Nile tilapia alone representing 80.8% of global production of farmed tilapia in 2002 (FAO, 2004). The reproduction mode of this species is quite interesting. Breeding starts with the building and defending of the spawning territory by the males, followed by courtship between the resident male and a visiting female, which lasts for only a few hours (Rana, 1990a), after which egg release and fertilization occurs. The Nile tilapia female repeatedly releases a string of about 20–50 eggs (Rothbard, 1979); then the male passes right over

Fig. 7.1. Nests of *Tilapia zillii* (photo provided by A. Al-Harbi).

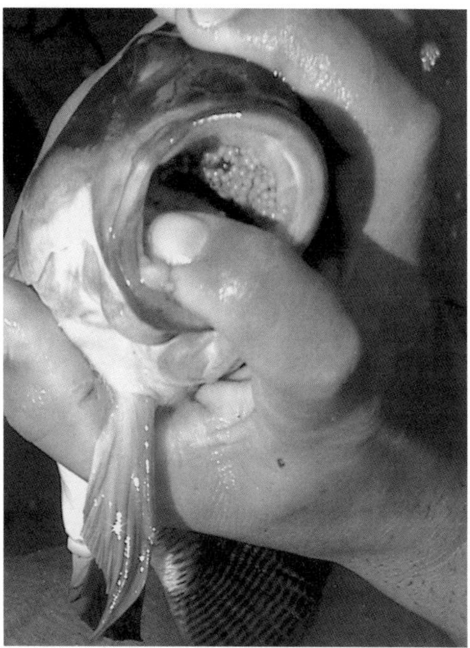

Fig. 7.2. Eggs are being released from mouth-brooding Nile tilapia female (photo provided by I. Radwan).

the eggs, with his genital papilla pressing against the bottom of the nest or tank bottom, and releases a cloud of milky milt. The female returns immediately and takes the fertilized eggs into her mouth.

Studies have shown that Nile tilapia egg fertilization also occurs inside the female's mouth (not outside). Mansour (2001) observed that the Nile tilapia female releases the eggs onto the bottom of the aquarium, and immediately collects them into her mouth before fertilization. Then, the male comes closer with his genital papilla near the female's mouth and starts releasing his milky milt, whereupon the female starts sucking it into the mouth. This process usually lasts for about 4–6 min, after which the female leaves the spawning area. Similar observations have been reported by Myers and Hershberger (1991) and Goncalves-de-Freitas and Nishida (1998). This phenomenon has also been recorded in another tilapia, *Oreochromis macrochir* (Fryer and Iles, 1972b).

Another strange and interesting phenomenon of Nile tilapia reproduction is the 'sneaking' behaviour of subordinate males. Goncalves-de-Freitas and Nishida (1998) found that the dominant male establishes a territory, constructs nests and attracts the female. However, in a number of cases the authors observed that the subordinate male sneaks between a spawning couple after egg delivery and starts releasing milt near the female's mouth and close to the released eggs. The female then sucks the milt and takes up the eggs into her mouth. The authors suggested that these two mating tactics (territorial and sneaking) could be part of an evolutionarily stable strategy. Sneaking behaviour has also been reported in another mouthbrooding cichlid species, *Pseudosimochromis curvifrons* (Kuwamura, 1987).

From the foregoing discussion, it is clear that fertilization of the eggs of some tilapia species (e.g. *O. niloticus* and *O. macrochir*) occurs inside and/or outside the female's mouth. It is not known when or why these fish prefer either method. Also, the question concerning why only some subordinate males display sneaking behaviour remains to be answered.

7.3. Age and Size at First Maturity

The age and size at first maturity of tilapia vary considerably between species, and even within strains of same species (Macintosh and Little, 1995; Bhujel, 2000). Under favourable natural conditions, Nile tilapia reach their sexual maturity at a size of 20–30 cm (150–250 g) (Lowe-McConnell, 1958; Gwahaba, 1973; Trewavas, 1983). However, tilapia may mature at small sizes in many parts of the world, a phenomenon indicating that these fish are under stressful environmental conditions and/or over-exploitation (Gwahaba, 1973; Duponchelle and Panfili, 1998; Duponchelle and Legendre, 2000).

In a natural habitat, age and size at first maturity of tilapia are affected by environmental conditions (Lowe-McConnell, 1982; Stewart, 1988), geographical barriers (Faunce, 2000), overfishing (Gwahaba, 1973), food supply (Santiago *et al.*, 1985; Siddiqui *et al.*, 1997b; El-Sayed *et al.*, 2003) and water body area (Duponchelle and Panfili, 1998). Duponchelle and Panfili (1998) found that age and size at maturity of Nile tilapia collected from large and small water bodies were positively correlated with the reservoir area. They also found that the range of variation in age at first maturity between populations was much higher than that of size at maturity. The authors suggested that these variations may have been related to differences in environmental variables (phenotypic plasticity) rather than genetic differences.

Under aquaculture conditions, tilapia generally mature at smaller sizes than in natural environments. De Graaf *et al.* (1999) and El-Sayed *et al.* (2003) found that farmed Nile tilapia reach their first maturation at about 30 g. Generally speaking, the size at first maturation of Nile tilapia females under aquaculture conditions ranges from 30 to 50 g, as has been reported by Siraj *et al.* (1983), Lester *et al.* (1988), De Silva and Radampola (1990) and Mansour (2001). This size is considerably small compared with the size of 150–250 g (20–30 cm) at which these fish mature under favourable natural conditions (Lowe-McConnell, 1958; Gwahaba, 1973).

Feed quantity and quality and feeding regimes also affect size at first maturity and productivity of tilapia broodfish. Many authors found that the mean weight of Nile tilapia at first spawning tended to increase with increasing dietary protein level (Wee and Tuan, 1988; De Silva and Radampola, 1990; El-Sayed *et al.*, 2003).

7.4. Fecundity and Egg Characteristics

Fecundity can be defined in various ways, including the following:

1. Potential fecundity: the number of mature eggs in the female's ovary immediately before spawning (Payne and Collinson, 1983). This definition assumes that all ripening oocytes are ovulated and only a few eggs are retained in the ovary after spawning (Coward and Bromage, 2000). This definition, however, may not fit with multiple-spawning fish such as tilapia, as it underestimates their fecundities (Wootton, 1979).

2. Fecundity can also be defined as the number of fry produced by the tilapia female throughout its lifetime (Lowe-McConnell, 1955). This definition also may not be appropriate for tilapia under aquaculture conditions, because broodstocks are used only for a relatively short time during their optimum reproduction period (Rana, 1990a). It was suggested, therefore, that the definition of fecundity should be restricted to 'the number of fry produced per year' (Mires, 1982). This modified definition takes into account: (i) the effects of environmental factors; (ii) the underestimation of reproductive potential because the number of fry is often less than the number of eggs released; and (iii) the fact that the correlation between the number of eggs and number of fry released may be weak.

Tilapia are characterized by low fecundity and relatively large egg size (Lowe-McConnell, 1959; Peters, 1983; Duponchelle *et al.*, 1997). For example, Nile tilapia produce from < 100 to > 3000 eggs per spawn, with egg size ranging from < 2 to 7.9 mm (De Graaf *et al.*, 1999). On the other hand, substrate spawners produce much higher numbers of smaller eggs (1–1.5 mm) than mouthbrooding tilapia (Wohlfarth and Hulata, 1983; Coward and Bromage, 1999). The fecundity and egg sizes of various tilapias reared under different environmental and culture conditions are summarized in Table 7.1.

In most tilapias, fecundity varies considerably among fish of the same species, and even among females of similar sizes, especially in large fish classes (Coward and Bromage, 1999). In addition, fecundity and egg size are directly related to fish size, length or age, but with high variability (Rana, 1986; Little, 1989; Ridha and Cruz, 1989). Little (1989) found that small Nile tilapia broodfish produced many more eggs, with shorter spawning intervals, than large fish, over a period of 105–116 days. However, larger females usually produce more eggs per clutch than smaller females. It was also found that large eggs contained more yolk and

Table 7.1. Optimum fecundity of different tilapias at different sex ratios and stocking rates.

Species	Culture system	Fish weight (g)		♂ : ♀ ratio	Stocking density/m^3	Eggs/female (average)	Remarks	Reference
		Male	Female					
O. niloticus	Glass aquaria		42–75	1 : 3	28	305–753 (507)/spawn	Fed 40% cp in freshwater	El-Sayed *et al.* (2003)
	Closed	162–211	112–177	1 : 3	4	3165/kg/month (678)	Fed 45% cp	Ridha and Cruz (1999)
	Closed	117–177	92–160	1 : 3	20	50.9/kg/day	Photoperiod of 2500 lux, for 18 h	Ridha and Cruz (2000)
	Concrete tanks	439	206	1 : 3	1.67	39.1/g (1328)	Fed 45% cp	Siddiqui *et al.* (1998)
O. niloticus × *O. aureus*	Concrete, outdoor tanks	226	189–294	1 : 2	2	18.8/kg/day	Fed 34% cp	Siddiqui and Harbi (1997)
	Concrete, outdoor tanks	226	294–437	1 : 3	2	25.3/kg/day	Fed 34% cp	Siddiqui and Harbi (1997)
T. zillii	Glass aquaria		29–419	1 : 2		461–11,640 (4.9–84.3 g)		Coward and Bromage (1999)

cp, crude protein.

led to larger fry with better growth and higher resistance to starvation and severe environmental conditions (Rana and Macintosh, 1988). A direct correlation between fish size and fecundity was not found in *T. zillii* (Coward and Bromage, 1999), *Tilapia tholloni* and *S. melanotheron* (Peters, 1983).

One may ask whether it is better to use small broodfish, which could collectively yield more eggs per culture unit, or larger fish, with higher individual egg production but at longer intervals. The decision should be taken by hatchery managers, based on the preset objectives and targeted outputs.

7.5. Broodstock Management

The achievement of maximum sustainable seed production of tilapia depends, to a large extent, on broodstock management. Information on broodstock selection, nutrition, fecundity, stocking density, sex ratio and environmental conditions is crucial for maximizing hatchery production efficiency. It is of paramount importance to highlight these parameters.

7.5.1. Broodstock selection

Maintaining pure, high-quality broodstock is probably the core of successful seed production. The selection of suitable broodfish will certainly have subsequent effects on the quantity and quality of the offspring produced. The following measures/characteristics should be considered in broodstock selection:

1. Broodfish should be genetically pure, and fish with unknown or questionable origins must be avoided.
2. Broodfish should not be too small, because small fish are less fecund and less efficient at egg incubation and fry protection (Rana, 1986).
3. Broodfish should not be too old and must not have spawned several times, because fecundity decreases with maternal age and successive spawnings (Rana, 1986).
4. Broodfish should be in good shape and free of deformities and injuries.
5. Inferior and unwanted tilapia must be prevented from entering broodstock ponds; therefore, incoming water should be filtered continuously.

7.5.2. Stocking density and sex ratio

Maintaining the correct stocking density and sex ratio of farmed tilapia will certainly improve hatchery efficiency and mass production of tilapia seed. Suboptimal broodstock density will result in low seed production (Mires, 1982). Also, high broodfish density often reduces seed production (Bevis, 1994), presumably due to aggression and fighting between males leading to reduction in courtship, egg fertilization and incubation (Behrends *et al.*, 1993; Ridha and Cruz, 1999). Guerrero (1982) also reported that tilapia mucus contains a substance that may cause autoallergic responses and inhibit reproduction at high broodstock densities. Similarly, the percentage of spawning females of *O. niloticus* × *Oreochromis hornorum* hybrids was inversely correlated with broodfish stocking density, due to chemical or behavioural factors (Lovshin, 1982).

Several stocking densities have been suggested for optimum seed production of tilapia, depending on tilapia species, broodstock size and condition, culture systems and water quality. Therefore, it is no surprise that varying, and sometimes controversial, results have been reported on the relationship between stocking density of broodstock and seed production. Hughes and Behrends (1983) suggested that 5 fish/m^2 be adopted for optimum seed production of Nile tilapia in suspended net enclosures. On the other hand, only 29% of Nile tilapia females reared in hapas in ponds in Thailand spawned at 10 fish/m^2, compared to 39% and 42% at 5 and 2.5/m^2 (Bevis, 1994). However, Little (1992) found that 6 fish/m^2 was optimum for Nile tilapia reared in hapas suspended in fertilized ponds, and Ridha and Cruz (1999) found that 4 fish/m^2 had better seed production and spawning synchrony than 8 and 12 fish/m^2. Table 7.1 summarizes the sex ratios and stocking densities suggested for optimum seed production of different tilapias.

Extensive work has been carried out on sex ratios of tilapia broodstock because a suboptimal ratio leads to reduction in seed production, waste of resources and an unnecessary increase in running costs. Therefore, different sex ratios have been adopted for different tilapias, with varying results. The male : female ratio ranges from 1 : 5 in ponds to 1 : 10 in tanks and hapas (Little and Hulata, 2000). A ratio of 1 : $<$ 3 appears optimum, and is suggested especially when synchronized spawning is required. Hughes and Behrends (1983) found that, in Nile tilapia, a male : female ratio of 1 : 2 produced more seed than a 1 : 3 ratio. Conversely, Mires (1982) reported that 3 : 1 and 2 : 1 sex ratios in tilapia hybrids were more productive than 1 : 1 and 1 : 2 ratios, presumably due to the effects of 'male pressure' or the increase in spawning frequency of individual females. Nevertheless, it is uncommon to stock more males than females in spawning units, as this may lead to male aggression and fighting and cause severe harm to the females.

It is clear that no single sex ratio can be adopted for the optimum reproduction efficiency of tilapia. It is the hatchery operator's job to decide the most appropriate sex ratio, taking into account the tilapia species, size and age, culture system and hatchery conditions. The optimum sex ratio may also be affected by the broodstock density. Broussard *et al.* (1983) found that increasing broodstock density, at a fixed male : female sex ratio of 1 : 3, had a negative effect on fry production of Nile tilapia reared in ponds. The authors attributed that effect to increasing competition between territorial males and/or constraints imposed by feed availability.

7.5.3. Spawning intervals

As stated earlier, one of the main characteristics of tilapia is their early maturity and asynchronous and successive spawning. They can spawn several times a year if suitable ambient environmental and culture conditions are attained. In these multiple-spawning fishes, females' ovaries contain oocytes of different developmental stages. Coward and Bromage (1998) found that up to 20% of the immediately post-spawned *T. zillii* ovary was occupied by unspawned mature eggs. Residual eggs that progress into atresia have also been reported in post-spawned *O. niloticus* ovaries (Avarindan and Padmanabhan, 1972; Peters, 1983). Immediately after spawning, the tilapia ovary regenerates very rapidly and previtellogenic stages are recruited into vitellogenic and late vitellogenic stages in as little as 1 week (Coward and Bromage, 1998, 2000).

As a result of different gonadal developmental stages in tilapia, spawning intervals are expected to vary considerably among species, and even within the same species. These intervals are influenced by fish size, stocking density, sex ratio, nutrition status, culture conditions and environmental factors. For example, Siraj *et al.* (1983) found that first- and second-year classes of Nile tilapia spawned at short intervals (7–12 days), while third-year class fish spawned at longer intervals (10–20 days).

Many studies have also indicated that the removal of eggs and fry from females' mouths accelerates vitellogenesis and shortens the intervals between successive spawnings (Lee, 1979; Verdegem and McGinty, 1987; Lovshin and Ibrahim, 1988; Tacon *et al.*, 1996; Baroiller *et al.*, 1997). Mair *et al.* (1993) found that the inter-spawning interval of Nile tilapia was 37.5% shorter when eggs were removed from females' mouths at 4-day intervals, as compared to females allowed to incubate their eggs naturally. Similarly, Baroiller *et al.* (1997) found that the inter-spawning interval of incubating Nile tilapia females was 27 days compared to only 15 days in non-incubating groups. In addition, the removal of Nile tilapia eggs and fry from the mouths of females increases seed production (Little *et al.*, 1993; Macintosh and Little, 1995). Macintosh and Little (1995) reported that removing eggs and fry of Nile tilapia at 5- and 10-day intervals resulted in a significant increase in seed production/kg female/day, compared to seeds produced from natural incubation. A 5-day cycle of seed removal from females' mouths is a common practice in commercial tilapia hatcheries in Thailand (Bhujel, 2000). On the other hand, parental care may suppress the reproductive potential of tilapia (Lee, 1979). Therefore, longer inter-spawning intervals were reported in females naturally incubating their eggs. This has been attributed mainly to feed deprivation during brooding time, which usually leads to the delay of subsequent oogenesis (Lee, 1979; Tacon *et al.*, 1996).

7.5.4. Broodstock exchange

Exchange of tilapia male and female broodfish, after a period of conditioning, could be an effective tool for improving seed production, spawning synchrony and spawning frequency. Conditioning is the separation of males and females in different conditioning units, at high densities, for a period of rest between spawnings (Bhujel, 2000). During this period, the fish must be provided with good-quality feed and appropriate feeding regimes (Bhujel, 2000). Conditioning of tilapia broodstock has been a common practice in some Asian countries for many years.

Partial or total exchange of Nile tilapia females with resting females at 5–10-day intervals significantly improved seed and fry production (Macintosh and Little, 1995; Little *et al.*, 2000). Costa-Pierce and Hadikusumah (1995) found that the best seed production was achieved at a rotation of 10–16 days. Seed production was negatively correlated with initial broodstock density and initial female body weight, but positively correlated with dietary protein level. Meanwhile, Lovshin and Ibrahim (1988) found that exchanging Nile tilapia male and female broodfish after 21 days of conditioning resulted in a 16% increase in egg and fry production over a 105-day spawning experiment, compared to non-exchange or female-only exchange. Similar results have been reported by Little *et al.* (1993), where conditioned males and females produced more eggs and fry and spawned more frequently than non-conditioning or employing only conditioned females.

It should be emphasized that more frequent disturbance and exchange of broodfish or conditioning for long periods may lead to negative effects on reproduction efficiency. Little *et al.* (1993) found that conditioning of Nile tilapia females for 20 days caused an 18% reduction in seed production, possibly due to the reabsorption of ripe eggs in the ovary, which occurs after about a week if the females do not spawn, as reported by Peters (1983). The quality of fish sperm may also decline with resting for a long time, due to the continuous production and release of spermatozoa into the lumen of the testes, leading to negative effects on the sperm quality (Purdom, 1995). Based on these results, it is suggested that relatively short periods of resting (about 5–15 days) appear optimal for tilapia broodfish.

The positive effects of conditioning on the reproduction efficiency of tilapia have been related to the higher activity of rested males in courtship, compared to non-rested males (Bhujel, 2000). The fertilization capacity of tilapia sperm may also be affected by resting conditions. Rana (1986) found that the fertilizing capacity of Nile tilapia sperm decreased sharply from about 90% to only 20% with increasing mating frequency, presumably due to the production of immature spermatozoa. Replacing old breeders with yearling breeders may also improve spawning efficiency. Smith *et al.* (1991) found that first-year class Florida red tilapia breeders produced more eggs and fry than year class two. The authors suggested that old breeders be replaced by younger yearling breeders.

7.6. Broodstock Nutrition

Spawning performance and seed production of tilapia are directly related to broodstock nutrition (Watanabe, 1985; Finn, 1994; Bhujel *et al.*, 2001b). Despite the fact that eggs can absorb some nutrients directly from water, egg yolk remains the major source of nutrition for embryonic development in fish. The exogenous nutrition of broodfish provides the essential nutrients required for the gonadal development of females and the performance of the seed produced (Santiago *et al.*, 1983; Gunasekera *et al.*, 1997). Therefore, an inadequate food supply for fish broodstock will lead to poor reproductive performance and seed production (Gunasekera *et al.*, 1996a, b; Gunasekera and Lam, 1997; Bhujel, 2000).

7.6.1. Protein requirements

Protein requirements of tilapia broodstock depend on species and size, dietary protein and energy sources and rearing systems. Several studies have investigated the protein requirements of Nile tilapia broodstock, while little attention has been paid to other tilapia species. Most of these studies indicated that the size at first spawning, fecundity, spawning frequency and egg hatchability increased with increasing dietary protein level. Generally speaking, these fish require about 30–45% dietary protein for optimum reproductive performance and seed production (Santiago *et al.*, 1985; Wee and Tuan, 1988; De Silva and Radampola, 1990; Gunasekera *et al.*, 1995, 1996a; Siddiqui *et al.*, 1998; Bhujel *et al.*, 2001a; El-Sayed *et al.*, 2003). On the other hand, Cissé (1988) found that fecundity and spawning frequencies of *S. melanotheron* were not significantly affected by dietary protein levels. The protein requirements of tilapia broodstock are summarized in Table 7.2. More studies are needed to investigate the effects of dietary protein on the reproductive output of other tilapia broodstocks. The relationship between dietary protein and energy for tilapia broodstock is also not well studied.

Table 7.2. Protein requirements (%) of Nile tilapia broodstock.

Range tested (%)	Requirement (%)	Culture system	Remarks	Reference
20–50	35	Closed	Isocaloric diets	Wee and Tuan (1988)
20–35	30	Closed		De Silva and Radampola (1990)
20 and 40	40	Outdoor tanks	Only two protein levels tested, natural food may have affected the results	Santiago *et al.* (1985)
10–35	35	Tanks	Isocaloric diets	Gunasekera *et al.* (1996a)
25–45	45			Siddiqui *et al.* (1998)
15.5, 25 and 30	30	Hapas in fertilized ponds	15.5%, 25% and 30% cp refer to herbivorous diet, large and small catfish pellets, respectively	Bhujel *et al.* (2001a)
25–40	40	Glass aquaria	Isocaloric diets	El-Sayed *et al.* (2003)
25–45	40	Closed	At 400 kcal GE/100 g	A.-F.M. El-Sayed and M. Kawanna (unpublished results)

cp, crude protein; GE, gross energy.

7.6.2. Lipid requirements

The quality and quantity of dietary lipids have been shown to significantly affect the reproduction of several fish species (Watanabe *et al.*, 1984; Harel *et al.*, 1994; Watanabe and Kiron, 1995; Zohar *et al.*, 1995). However, very little information is available on dietary lipid requirements of tilapia broodstock, even though the requirements of tilapia fingerlings and juveniles have been well studied (see Chapter 6). As far as I know, no studies have considered the effects of the dietary lipid source on the reproductive performance and tissue lipids of tilapia, except two studies conducted on Nile tilapia, with controversial results. Santiago and Reyes (1993) fed Nile tilapia broodstock diets containing different oil sources (cod liver oil, maize oil, soybean oil, coconut oil or a mixture of cod liver oil and maize oil (1 : 1)) for 24 weeks. They found that fish fed the cod liver oil (rich in n-3 highly unsaturated fatty acids (HUFA)) had the best somatic growth, the highest level of lipid in the liver and muscle, but the poorest reproductive performance, while soybean oil (rich in n-6 fatty acids (FA)) resulted in the highest overall reproductive performance.

However, the results of Santiago and Reyes (1993) may be questionable, due to a number of reasons. First, the initial stocking sizes of males and females were very large (ranging from 95 to 128 g for males and from 90 to 124 g for females). These fish may have spawned a number of times before stocking, which in turn may have affected their fecundity and spawning efficiency. Secondly, the energy content of the diets was very low (2.8 kcal/g, using the digestible energy values of channel catfish). These energy values may have not been sufficient to meet both somatic and reproductive performances of the fish. Thirdly, the male : female ratio adopted (1 : 1) may have led to male aggression and fighting, causing severe harm to the females, whereas the recommended sex ratio under aquarium conditions is 1 : 3 (or 1 : > 3) (Little and Hulata, 2000).

Contradictory results have been reported by El-Sayed *et al.* (2005a), who studied the effect of dietary lipid source and water salinity on the spawning performance of Nile tilapia broodstock and the growth of their larvae. The fish were fed diets with three different soybean oil : cod liver oil (SBO : CLO) ratios (5 : 0, 2 : 3 and 0 : 5) at three water salinities (0, 7 and 14‰) for 165 days.

At 0‰, the size at first maturation, spawning intervals, egg hatchability, days to hatch, yolk-sac absorption time and larval weight were not significantly affected by dietary oil source, while the inclusion of fish oil in the diets increased the absolute fecundity and number of eggs per spawn at each salinity. At 7‰ and 14‰, fish fed CLO or the CLO/SBO mix spawned at shorter intervals and more frequently than those fed SBO. In addition, eggs produced from broodstock fed SBO at 7‰ and 14‰ needed more time for hatching and yolk-sac absorption and resulted in poorer larval weight than those fed CLO or the CLO/SBO mix. This result indicates that either n-6 or n-3 fatty acids could meet the requirement of Nile tilapia broodfish reared at 0‰. At that salinity, fish fed n-3 essential fatty acids (EFA) may also have sufficient ability to elongate and desaturate 18 : 2 n-6 into long-chain FA, as reported by Kanazawa *et al.* (1980b) and El-Sayed and Garling (1988). This result also reveals that Nile tilapia broodfish reared in brackish water (7 and 14‰) require a source of dietary n-3 HUFA for optimum spawning performance, while plant oil (SBO) may meet the requirements of broodfish reared in fresh water (0‰). Again, more work is needed to verify the relationship between dietary lipid levels and fatty acids and spawning performance of different tilapia species reared under different culture conditions.

7.6.3. Vitamin and mineral requirements

Very little attention has been paid to vitamin requirements of tilapia broodstock, despite their positive effects on the reproductive performance of farmed fish (Soliman *et al.*, 1986; Eskelinen, 1989). Soliman *et al.* (1986) found that supplementation of ascorbic acid at 1250 mg/kg of dry diet of *O. mossambicus* significantly improved egg hatchability and the condition and survival rate of their fry. It has also been found that vitamin E deficiency in Nile tilapia feed causes lack of sexual coloration (light skin colour) and reduces reproductive activity (Schimittou, 1993).

Despite the essentiality of minerals for fish growth and general metabolism, no studies have been conducted on the effects of dietary mineral supplementation on the reproductive performance of tilapia. It is evident, therefore, that extensive work is urgently needed on quantitative lipid requirements, energy requirements, protein-sparing effects by dietary lipids or carbohydrates and vitamin and mineral requirements of tilapia broodstock under different culture systems.

7.6.4. Feeding management

7.6.4.1. Feeding levels

Feeding management of tilapia broodstock can significantly affect reproductive performance, hatchery efficiency and profitability. Since mouth-brooding tilapia incubate the eggs and fry in their mouths and do not eat during incubation periods, it is unwise to feed them at abundant and constant rates (Miranova, 1977; Bhujel, 2000). Instead, a higher feeding rate immediately after seed harvesting, followed by a lower feeding rate and/or no feeding (if broodfish are raised in a green-water system), might be more economic and more efficient in terms of reproductive performance (Bhujel, 2000). In support, Miranova (1977) found that the egg production of *O. mossambicus* was higher with restricted food (25–50% of the satiation ration) than with abundant food, despite the fact that abundant feeding resulted in better growth of females than restricted feeding. This result suggested that the reproduction of tilapia is stimulated by a decrease in the food ration. Similarly, Coward and Bromage (1999) studied the effects of high and low rations on the growth and reproductive performance of *T. zillii* for 550 days. They found that fish maintained upon the higher ration were significantly larger than fish fed a reduced ration. Also, high rations produced a higher mean total fecundity than low ration, though the mean egg diameter and gonadosomatic index remained unchanged. When adjusting spawning data to a common maternal fish size, using a one-factor analysis of covariance (ANCOVA), the authors did not detect significant differences between the two ration levels in terms of mean total fecundity and egg diameter. Furthermore, the ovaries of fish fed the low ration possessed high ratios of early and late perinucleolar oocytes and fewer late-vitellogenic/maturing oocytes than the ovaries of fish fed the higher ration. The authors suggested that, under prolonged food restriction, the female *T. zillii* sacrifices somatic growth for maintaining reproductive efficiency.

In addition to the above findings, Siddiqui *et al.* (1997b) found that the fecundity of hybrid tilapia (*O. niloticus* × *O. aureus*) fed at 0.5, 1, 2 and 3% body weight/day increased with increasing feeding rate. However, fecundity/g female was higher at lower feeding levels (0.5 and 1%) than at higher feeding levels (2 and 3%). Egg size, hatchability, fry length and weight were not affected by feeding levels. It is thus clear that, while food restriction leads to growth retardation and reduction in total fecundity, it tends to increase spawning frequency, the total number of eggs over a discrete period and the amount of energy allocated to egg production.

7.6.4.2. Feeding frequency

Feeding frequency is another important factor affecting fish reproduction. A number of studies have investigated the effects of feeding frequency and schedules on spawning performance and seed output of tilapia. Ungsethaphan (1995) found that *O. niloticus* fed twice a day at low feeding rates (0.5 and 1.0% of body weight) produced more seed compared to those fed once a day. When the feeding level was increased to 1.5% of body mass/day, seed production was decreased. As tilapia are continuous grazers in nature, feeding them at shorter intervals may lead to better seed production (Bhujel, 2000). However, feeding more than twice a day may not be cost-effective, due to labour intensity. Feeding tilapia broodstock with high-energy diets at low feeding rates might also be better than feeding larger amounts of a low-energy diet.

The effect of the scheduled use of high-protein and low-protein diets on growth and fry production of Nile tilapia broodstock has been investigated (Santiago and Laron, 2002). The fish were fed a high-protein diet (HP, 40% crude protein (cp)) or a low-protein diet (LP, 25% cp) or HP for 1–3 days followed by LP for 1–4 days for 54 weeks. Fry production was not significantly affected by the feeding schedule for broodstock. However, cost–benefit analyses indicated that feeding broodstock on 1HP–1LP and 3HP–2LP feeding schedules gave the highest spawning and economic performance. This finding indicates that feed quality, feeding schedules and feed costs should be considered for achieving optimum reproductive performance. It also gives fish farmers and managers an option in the management of feeding their tilapia broodstocks.

7.7. Environmental Factors

Environmental factors play the main role in releasing the genetic potential (Duponchelle *et al.*, 1997) and seed production in tilapia (Little *et al.*, 1994; Bhujel *et al.*, 2001b). Therefore, good water quality and other environmental conditions should be maintained for successful and sustainable reproductive activity and seed production of these fish. In this regard, the following environmental parameters are of prime importance.

7.7.1. Dissolved oxygen

As indicated in Chapter 3, tilapias can survive at low levels of dissolved oxygen (DO) due to their ability to utilize atmospheric oxygen (Chervinski, 1982; Popma and Lovshin, 1996). However, the effects of DO on reproductive and spawning activities of tilapia have not been well studied. The available information indicates that low DO (< 0.5 mg/l), generally occurring in the early morning in outdoor fish ponds, has negative impacts on the ovarian growth, courtship, seed production and quality and the ability of tilapia females to mouthbrood eggs and fry (Bevis, 1994; Little and Hulata, 2000). Low DO also causes a reduction in feed intake and behavioural and morphological changes (melanin pigments in the skin) (Bhujel, 2000). These stressful conditions have been reported to cause disease incidence, inhibit reproduction and spawning activities, cause oocyte atresia and decrease fecundity and hatchability (Schimittou, 1993).

7.7.2. Temperature

Tilapias are eurythermal fishes that can tolerate a wide range of temperatures, ranging from 8 to > 40°C (Philippart and Ruwet, 1982). Most tilapia species were reported to reproduce successfully at about 22°C (Chervinski, 1982; Popma and Lovshin, 1996; El-Naggar *et al.*, 2000). This was not the case in *O. mossambicus*, where no spawning occurred at 22°C, while increasing water temperature from 25 to 31°C significantly increased their seed production (Miranova, 1977). Tilapia reproduction generally slows at 21–24°C, while the optimum temperature for reproduction ranges from 25 to 30°C (Rothbard, 1979; Popma and Lovshin, 1996). At temperatures higher than 35°C, reproductive performance has been found to be very poor (Bevis, 1994; Little *et al.*, 1997). Spawning units should therefore be designed and constructed in such a way as to meet severe climatic conditions. Increasing temperature through passive daytime heating and conservation of warmer water overnight can improve hatchery efficiency (Little and Hulata, 2000). Concrete tanks built in greenhouses (Fig. 7.3) can be an ideal design for this system. Greenhouse tanks enable tilapia to spawn all year round, especially in temperate zones where the temperature drops sharply during winter months (I. Radwan, Kafr El-Shaikh, Egypt, 2005, personal communication). This will also help tilapia farmers to nurse the fry during wintertime and to stock them in fattening ponds at much bigger sizes than the size they generally use under natural conditions. Windbreaks and deep hapas-in-ponds have been successfully used to reduce the effects of cold weather in Vietnam and Thailand (Cao, 1998; Dan and Little, 2000b). Reducing the effects of hot weather through shading and deepening fish ponds has also been successful (El-Sayed *et al.*, 1996; Bhujel, 1999).

An interaction between endocrine control of reproduction and temperature has been implied. Cornish *et al.* (1997) found that the secretion of luteinizing-like hormone (LH-like H) and follicle-stimulating-like hormone (FSH-like H) was related to increasing water temperature, leading to increasing spawning efficiency of *O. mossambicus*.

7.7.3. Salinity

Salinity is another important parameter for tilapia reproduction, since these fish are salinity-tolerant. A great deal of attention has been paid to the culture of tilapia in brackish water and seawater, due to the shortage of and competition for fresh water in many parts of world. This means that there is a great scope for tilapia culture in coastal areas, either using salinity-acclimatized seed produced from freshwater hatcheries or establishing the hatcheries in brackish water by diluting the seawater with fresh water (Suresh and Lin, 1992a).

The effects of salinity on tilapia reproduction have been studied by a number of authors, with varying and sometimes controversial results, depending on tilapia species and strains. For example, Wardoyo (1991) studied the effects of different salinity levels (0, 10, 15, 20 and 25‰) and acclimatization regimes (direct and gradual exposures) on survival, weight gains and reproductive performance of three strains of Nile tilapia (Egypt, Ghana and Ivory Coast) and a red tilapia hybrid (red tilapia). He found that spawning success and egg hatchability of Egypt and Ghana strains and red tilapia were reduced with increasing water salinity, but were not affected by strains or

Fig. 7.3. Tilapia spawning concrete tanks in a greenhouse at Hamool, Kafr El-Shaikh, Egypt.

acclimatization regimes. On the other hand, fecundity (eggs/g/female) was not affected by salinity levels or strains. Meanwhile, hatchability rates were higher for late eggs (> 3 days) than for early eggs (< 3 days). Hatchability of early eggs was also higher at 10 and 15‰ than at 0 and 20‰.

Corre (1981) also reported that Nile tilapia spawned at 14‰, while no spawning occurred at 28‰. Similar observations were reported by Watanabe (1985) and Fineman-Kalio (1988), who found that these fish reproduce successfully at salinities ranging from 13.5 to 29‰ and cease reproduction at salinities higher than 30‰, despite the fact that 95% of the females were gravid. However, Watanabe and Kuo (1985) reported that the same species can reproduce at salinities ranging from fresh water to 32‰, but extremely poor egg hatchability was obtained in full-strength seawater. Hatchability was similar for eggs spawned in fresh water, 10‰ and 15‰, and was considerably lower than at 5‰. On the other hand, Mozambique tilapia (*O. mossambicus*) is more salinity-tolerant than other tilapias, and can reproduce at a water salinity ranging from 10 to 49‰ (Chervinski, 1982). Furthermore, their reproduction efficiency is better in brackish water than in fresh water. Uchida and King (1962) found that seed production of *O. mossambicus* was approximately three times higher in brackish water (8.9–15.2‰) than in fresh water.

The effect of salinity on the reproduction of blue tilapia (*O. aureus*) was also evident. Available information indicated that nest building did not occur in fish reared in seawater and, in turn, the gonadosomatic index dropped, causing non-occurrence of reproduction (Chervinski and Yashouv, 1971; Chervinski and Zorn, 1974). Al-Ahmad *et al.* (1988) found that fecundity, egg hatching and fry survival of *Oreochromis spilurus* were twice as successful in groundwater (4‰) as in seawater (40‰). Ernst *et al.* (1991) found that seed production of Florida red tilapia breeders was significantly greater at low salinity (5‰) than in brackish water (18‰) and the proportion of brooding females and number of seed per clutch were greater at 5‰ than at 18‰. In another study, Liao and Chang (1983) noted that the breeding behaviour of Taiwanese red tilapia was inhibited in brackish water and seawater compared to fresh water.

It is obvious that salinity-tolerant tilapia species, strains and hybrids tend to reproduce more efficiently at low and moderate salinities than at high salinities. The failure of tilapia to reproduce at high salinities could be attributed to the resorption of the eggs as a result of the osmotic stress (Chervinski and Yashouv, 1971). High salinities were also found to suppress the aggression of dominant tilapia males (Watanabe *et al.*, 1988b). To overcome this aggression, male broodfish may be kept in saline water during times of low seed demand.

7.7.4. Photoperiod and light intensity

Photoperiod plays a significant role in fish growth, metabolism and reproduction through melatonin secretion (key hormone engaged in regulating endogenous rhythms) (Boeuf and Le Bail, 1999; Bromage *et al.*, 2001). Fish are also characterized by photoreceptor cells in both the eye and the pineal organ (Boeuf and Le Bail, 1999). The receptivity of fish to light depends on fish species and developmental stage (Barlow *et al.*, 1995). The effects of light intensity and photoperiod on the growth, body pigmentation, locomotor activity rhythms, sexual maturation and reproduction of several fish species have been studied (Silva-Garcia, 1996; Boeuf and Le Bail, 1999; Purchase *et al.*, 2000; Trippel and Neil, 2002).

In nature, tilapia show some degree of sensitivity to light intensity and photoperiod (Lowe-McConnell, 1959). The reproductive activity of some cichlids living in clear waters was found to be modulated by the lunar cycle. Okorie (1973) found that advanced gonadal maturation of Nile tilapia in Lake Victoria was higher during the full moon than at the new moon. Similar effects of lunar periodicity on the breeding activity of *Tilapia mariae* have been reported (Schwank, 1987). It was also assumed that moonlight enhances the effectiveness of parental care (Schwank, 1987). Baroiller *et al.* (1997) found that the peak of spawning activity of Nile tilapia reared in outdoor tanks for 20 months occurred when water temperature and photoperiod were at their peaks (about 27°C and 16 h, respectively).

Various tilapia species spawn at certain times of the day and at a favourable temperature and photoperiod. This behaviour suggests nychthemeral cyclicity (daytime spawning and nighttime rest) in these fish. The effect of nychthemeral periodicity on tilapia breeding is evident. About 99%

of Nile tilapia reared in aquaria at a 12 h light : 12 h dark (12L : 12D) photoperiod cycle spawned during the afternoon of light periods, with 84% spawning between 14.00 and 16.00 h (Baroiller *et al.*, 1997).

Despite the effects that light intensity and photoperiod may have on the reproductive performance of tilapia, very few studies have considered these effects under culture conditions. Most of these studies have suggested that about a 12–14.5 h photoperiod be adopted for optimum spawning activity. A.-F.M. El-Sayed and M. Kawanna (unpublished results) found that the fecundity, spawning frequency and inter-spawning intervals were significantly better in Nile tilapia reared at 12L : 12D than at 18L : 6D and 6L : 18D cycles. Rothbard and Pruginin (1975) also recommended a 12–14 h light phase, at a temperature of 25–29°C, for the induction of natural interspecific spawning between *O. niloticus* and *O. aureus*. Similarly, Ridha *et al.* (1998) found that *O. spilurus* breeders reared under longer photoperiods exhibited higher fecundity. However, that result is questionable since only two photoperiods were used (13 and 14 h). Furthermore, Ridha and Cruz (2000) found that the best seed production and spawning performance of Nile tilapia were achieved at a light intensity of 2500 lux and a photoperiod of 18 h/day. Increasing photoperiod during winter using artificial light has led to the production of red tilapia fry all year round (Galman *et al.*, 1988). On the other hand, the use of accelerated artificial light (600 W) for 14.5 h/day at 27–30°C did not affect seed production of Florida red tilapia (Smith *et al.*, 1991). It appears, therefore, that the response of tilapia to photoperiods is species-specific.

It was also found that Nile tilapia broodstock exposed to long photoperiods (18L : 6D and 24L : 0D) had the lowest spawning performance, but exhibited the highest biomass and best growth rates, compared to fish subjected to an intermediate photoperiod (12L : 12D) (A.-F.M. El-Sayed and M. Kawanna, unpublished results). Fish that spawned more frequently (at 12L : 12D) exhibited lower average weights, presumably due to the use of a substantial amount of dietary energy for gonad development and spawning activity, rather than for somatic growth. These fish depend on their energy reserve, as they fast during egg and fry brooding.

On the contrary, Campos-Mendoza *et al.* (2004) found that Nile tilapia reared under long day length (18L : 6D) exhibited higher fecundity and lower inter-spawning intervals compared with fish reared under shorter light hours. It is clear, therefore, that more work is needed to resolve this dispute and to determine the photoperiod required for optimum spawning performance of farmed tilapia.

Very recently, Biswas *et al.* (2005) evaluated photoperiod manipulation to control reproduction and, in turn, to overcome the problem of overcrowding in Nile tilapia. Fish broodstock were exposed to 6L : 6D, 12L : 12D and 14L : 10D. Reproductive parameters, including fecundity, gamete quality and offspring viability, were evaluated. Steroid levels (oestradiol-17β, E2; testosterone, T) were analysed on days 1 and 3 post-spawning and at 3-day intervals afterwards. The authors found that females exposed to 12L : 12D and 14L : 10D spawned successfully throughout the study, while the spawning of fish exposed to 6L : 6D was arrested after three to four spawning cycles, with a significant decrease in plasma levels of E2. These findings suggest that photoperiod manipulation can be used as a useful and safe tool for controlling excessive reproduction of farmed tilapia.

7.7.5. Water level and exchange

Water level has been reported to affect reproduction efficiency of tilapia in nature (Noakes and Balon, 1982; Philippart and Ruwet, 1982). Noakes and Balon (1982) found that a high water level increased the reproductive activity of tilapia females, resulting in smaller males being recruited into the breeding population. Partial water change at frequent intervals may also improve seed output and spawning synchrony (Bhujel, 2000). Guerrero (1982) suggested that tilapia mucus contains a hormone-like substance that inhibits reproduction, especially at high densities. Regular exchange of pond water may remove this substance and improve reproduction efficiency. Water exchange also improves dissolved oxygen and flushes out harmful substances, such as undigested food, faeces and other metabolites, such as ammonia, nitrite and nitrates (Mires, 1982). It is no surprise therefore, that seed output from ponds filled with new water is higher than from ponds with water used for long periods (Bhujel, 2000). This suggests that pond water should be frequently changed, partially or completely; however, the exchange rate and frequency depend on water availability and cost, as well as the type of culture system.

7.8. Production of Monosex Tilapia

Mixed-sex culture of tilapia has been a common practice in many countries for decades. However, extensive attention has been given to monsex culture of tilapia during the past two decades. Monosex tilapia may have many attributes that make them an excellent candidate for aquaculture. Among these are the following:

- High growth rates and feed utilization efficiency.
- High tolerance to severe environmental conditions, including temperature, salinity, low dissolved oxygen, etc.
- Higher energy conservation.
- Reduced aggressiveness.
- Greater uniformity of size at harvest.
- Better flesh quality and appearance.
- High resistance to stress and diseases.
- Role in controlling over-reproduction.

The following methods have been used for the production of monosex tilapia: (i) manual sorting; (ii) hormonal sex reversal; (iii) interspecific hybridization; (iv) androgenesis and gynogenesis; (v) triploidy; and (vi) transgenesis. The first three methods are discussed here, while androgenesis and gynogenesis, triploidy and transgenesis are addressed in Chapter 11.

7.8.1. Manual sexing

Manual sexing of tilapia fingerlings was widely used in the past (Hickling, 1963; Shell, 1967). It is an easy technique, but it is extremely laborious, is stressful for the fish and often leads to inaccurate results due to the presence of females as a result of human error (Penman and McAndrew, 2000). Therefore, this method is rarely used. The technique of manual sorting is based on assessing the number of openings in the urinogenital papillae: the male has a single urinogenital opening, while the female has two separate openings. The accuracy of this method ranges from 80 to 90% (Penman and McAndrew, 2000). Accuracy also increases with increasing fish size.

7.8.2. Hormonal sex reversal

Hormonal sex reversal has been extensively used for sex determination and producing monosex fish for aquaculture purposes. Steroid hormones or hormone analogues as well as non-steroid compounds (Table 7.3) are commonly used for producing monosex tilapia. The hormones are generally incorporated into larval feeds and administered to undifferentiated larvae at very early larval stages (preferably at first feeding) for sufficient time to enable sex reversal. The use of hormones has been under increasing public criticism due to their possible health and environmental impacts. As a result, the use of hormones for sex reversal of tilapia is either licensed (in USA) or banned (in Europe) (Penman and McAndrew, 2000).

7.8.2.1. All-male production

ORAL ADMINISTRATION. Many steroid hormones, steroid analogues or non-steroid compounds have been widely used in male sex reversal of various tilapia species. The percentage of sex-reversed males depends mainly on the type of hormones,

Table 7.3. Steroid hormones, hormone analogues and non-steroid compounds used for producing monosex tilapia.

Hormone (compound)	Abbreviation
17α-methyldihydrotestosterone	MDHT
17α-methyltestosterone	MT
17α-ethynyltestosterone	ET
17α-methyl-5-androsten-3-17β-diol	
Androstenedione	AN
17α-ethymyloestradiol	EE
Oestradiol-17β	E2
Fadrozole	F
Trenbolone acetate	TBA
11β-hydroxyandrostenedione	11β-OHA4
Aromatase inhibitor	AI
Diethylstilboestrol	DES
Tamoxifen	
Acriflavine	
Mibolerone	MI

dose, timing and duration of administration, fish species and size/age of larvae, as indicated in Table 7.4. Available data indicate that 17α-methyltestosterone (MT) hormone is the most common and successful hormone used for tilapia sex reversal (Guerrero and Guerrero, 1988; Ridha and Lone, 1990; Lone and Ridha, 1993; Penman and McAndrew, 2000; Beardmore *et al.*, 2001). The doses of steroid hormones tested for sex reversal of tilapia ranged from < 30 to 100 mg/kg food, provided for the fish for < 15–> 60 days. However, the best results have been reported at a dose of 30–60 mg/kg administered for about 25–30 days (Guerrero and Guerrero, 1988; Vera-Cruz and Mair, 1994; Green *et al.*, 1997). Guerrero (1975) found that the administration of 30 mg 17α-ethynyltestosterone (ET)/kg feed for 18 days produced 98% males in *O. mossambicus*, while 60 mg/kg produced 100% males. MT produced 98 and 85% males at 30 and 60 mg/kg. Excessive doses of hormones may therefore reduce the percentage of sex-reversed fish, increase fish mortality and cause sterility or paradoxical sex reversal (Green *et al.*, 1997; Beardmore *et al.*, 2001).

The use of hormones for sex reversal has been under increasing public criticism due to their potential health and environmental impacts. Therefore, recent studies have considered the use of non-steroidal compounds to manipulate sex inversion. Non-steroidal compounds may exhibit antagonistic or agonistic effects to sex steroids that are generally involved in sexual differentiation in fish. When tilapia hybrids (*O. niloticus* × *O. aureus*) were fed tamoxifen-treated diets, 100% males were produced at a dose of 100 mg/kg feed (Hines and Watts, 1995). When acriflavine was used, 89% and 85% males were produced at 15 and 50 mg/kg diet.

IMMERSION TECHNIQUES. Oral administration of hormones for sex reversal of tilapia is generally safe and successful; however, hormone traces from uneaten food and metabolites are often a major environmental concern. Immersing fish fry in hormone solution for short periods of time has attracted attention as a successful alternative tool to overcome this problem. One of the major advantages of the immersion technique is the substantial decrease in treatment period and the reduction of possible effects of the hormones on the workers (Gale *et al.*, 1999). This technique is also characterized by minimal intervention, which is usually reduced to the time when the fish is most sensitive to the treatments. In spite of these attributes, the use of the immersion technique for masculinization of tilapia has not been developed for practical and commercial use.

The percentage of sex-reversed males produced by the immersion technique ranges from < 60% to 100%, depending on fish species, type and dose of hormone and immersion period (Table 7.4). Varadaraj and Pandian (1987) found that the immersion of *O. mossambicus* in 5 or 10 μg/l of 17α-methyl-5-androsten-3-17β-diol for 10 days post-hatch (dph) caused 100% masculinization. Similar results were reported with *O. spilurus* immersed in MT (2.5 mg/l) for 4 days in brackish water, followed by oral administration of the hormone at 50 mg/kg for 35 days (Lone and Ridha, 1993). The exposure of Nile tilapia fry to 500 μg/l 17α-methyldihydrotestosterone (MDHT) for 3 h on days 10 and 13 post-fertilization (dpf) also resulted in 83–100% males (Gale *et al.*, 1999), while low concentration of MDHT and MT (100 μg/l) caused a lower male percentage. Afonso and Leboute (2003) also investigated the masculinizing potency of ET, MT and MDHT applied to Nile tilapia larvae at 14 dph, in two consecutive trials. In the first trial, the fish were subjected to a single 4 h immersion dose at concentrations ranging from 200 to 1800 μg/l. In the second trial, the fish were immersed in these androgens at 1800 μg/l for 4 h either as a single immersion at 14 days or as double immersions at 10 and 14 dph. The authors found that, when different concentrations were applied, the best male proportion was achieved at 1800 μg/l, but none of the treatments produced 100% male (86–90%). The results of the second trial confirmed those of the first trial. However, two immersions slightly improved the male proportion, especially in the MT-treated group, where the male proportion increased from 92 to 98%, but two immersions decreased the male percentage in the MDHT group to 94% compared to 100% at one immersion.

The exposure of fish larvae to ultrasound is believed to increase the transport of hormone from the water into the fish body, leading to a higher masculinization rate (Bart, 2002). Bart (2002) studied the effects of ultrasound exposure time (1 and 2 h) on sex reversal of Nile tilapia, using four different hormones (androstenedione (AN), MDHT,

Table 7.4. Production of male tilapia using natural and synthetic steroids (full names of the hormones are given in Table 7.3.) Oral doses are in mg/kg feed.

Species	Hormone	Route	Optimum dose	Timing	Duration	% Males	Remarks	Reference
O. n	F	Food	200–500	7 dph	30 days	92.5–96	Genetically females	Kwon *et al.* (2000)
	F	Food	75–100	Fry	30 days	100		Afonso *et al.* (2001)
	ET	Food	60	Fry	25–28 days	91–99.4		Vera-Cruz and Mair (1994)
	MT	Food	30	Fry	21 days	99		Guerrero and Guerrero (1988)
	MT	Food	60	10.4 mm	14–28 days	82–92		Smith and Phelps (1997)
	TBA	Immersion	250 µg/l	10 dpf	2 h	98–100	Exposed to ultrasound for 2 h	Bart (2002)
	AN	Immersion	100–250 µg/l	10 dpf	2 h	92	Exposed to ultrasound for 2 h	Bart (2002)
	MT	Immersion	50 µg/l	10 dpf	2 h	98	Exposed to ultrasound for 2 h	Bart (2002)
	MDHT	Immersion	250 µg/l	10 dpf	2 h	98–100	Exposed to ultrasound for 2 h	Bart (2002)
	MT	Food	60	10 dpf	10–30 dpf	92		Gale *et al.* (1999)
	MT	Immersion	500 µg/l	10 and 13 dpf	3 h	87		Gale *et al.* (1999)
	MDHT	Immersion	500 µg/l	10 and 13 dpf	3 h	83–100		Gale *et al.* (1999)
	MDHT	Immersion	1800 µg/l	14 dph	4 h	100	Single immersion	Wassermann and Afonso (2003)
	MT	Immersion	1800 µg/l	10 and 14 dph	4 h	98.3	Two immersions	Wassermann and Afonso (2003)
	ET	Immersion	1800 µg/l	14 dph	4 h	86.7	Two immersions	Wassermann and Afonso (2003)
O. m	MDHT	Immersion	5–10 µg/l	10 dph	10 days	100		Varadaraj and Pandian (1987)
	MT	Food	30		18 days	98		Guerrero (1975)
	ET	Food	60	9–11 mm	18 days	100		Guerrero (1975)

Continued

Table 7.4. *Continued.*

Species	Hormone	Route	Optimum dose	Timing	Duration	% Males	Remarks	Reference
O. a	MI	Immersion	0.6 ppm	9–11 mm	35 days	82.		Torrans *et al.* (1988)
	MT	Food	50	9 mm	42 days	100		Hines and Watts (1995)
	ET	Food	60	Fry	25–28 days	83–97		Mélard and Ducarme (1993), Mélard *et al.* (1995)
	TBA	Food	25	9.1 mm	28 days	98.3		Galvez *et al.* (1996)
O. s	MT	Food	70	Fry	38 days	90		Ridha and Lone (1990)
	MT	Immersion	2.5 ppm	Yolk-sac fry	4 days	100	In BW, later fed MT at 50 mg/kg feed	Lone and Ridha (1993)
Red tilapia	11β-OHA4	Food	50	10 dpf	28 days	99.1		Desprez *et al.* (2003)
O. n × *O. a*	Tamoxifen	Food	100	9 mm	42 days	100		Hines and Watts (1995)
O. n × *O. a*	ET	Food	60	Fry	25–28 days	96–100		Rothbard *et al.* (1983)

O. a, *O. aureus*; *O. m*, *O. mossambicus*; *O. n*, *O. niloticus*; *O. s*, *O. spilurus*; dph, days post-hatch; dpf, days post-fertilization; BW, brackish water.

MT and trenbolone acetate (TBA)). He found that subjecting the fry (10 dpf) to ultrasound for 2 h resulted in a significant increase in the male percentage, compared to 1 h exposure. In addition, TBA (250 μg/l) produced 98–100% males and was not significantly different from 100 μg/l. A lower percentage of males was produced with the MDHT treatment (90.5%).

The immersion of tilapia eggs, instead of fry, has recently emerged as an effective tool for inducing sex reversal. When fertilized eggs are immersed in the hormone solution, the hormone is absorbed through passive diffusion across the lipid membrane of the egg. One major advantage of this technique is the substantial reduction in the amount of hormone used, since small amounts of water are generally used for egg incubation compared to the amounts used in larval holding units. As yet, little information is available on the sex reversal of tilapia using the immersion technique. Only two studies have been reported in this regard. The first study was reported in Thailand, where the immersion of 2-day-old Nile tilapia eggs in MT at 500 μg/l, for 24 h resulted in 88% males (Anon., 2002, cited in Cagauan *et al.*, 2004).

More recently, Cagauan *et al.* (2004) evaluated the sex reversal of Nile tilapia in the Philippines, by immersing fertilized eggs (3–4 days old) in different concentrations of MT (0, 200, 400, 600 and 800 μg/l) for 24, 48, 72 and 96 h. The highest male proportion of 91% was obtained at 800 μg/l for 96 h. However, the results of this study are questionable. At 72 and 96 h, it is very likely that some of the eggs (which were already 3–4 days old) had hatched into sac fry. It is possible, therefore, that, at long immersion times (72 and 96 h), the authors may have immersed sac fry rather than eggs. Further investigations are therefore urgently needed to improve the immersion technique in order to increase the percentage of sex-reversed tilapia to 100% or close to 100%.

Oestrogen biosynthesis is mediated by the steroidogenic enzyme cytochrome P450 aromatase, which converts androgens to oestrogens (Afonso *et al.*, 2001). Non-steroidal aromatase inhibitors (AI) have been shown to alter the sex of several animals, including tilapia. Afonso *et al.* (2001) fed Nile tilapia larvae with diets containing 0, 50, 75 and 100 mg/kg of the aromatase inhibitor Fadrozole (F) for 15 and 30 days, starting 9 days after hatching. Regardless of the feeding period, the proportion of males was significantly higher in the treated groups, and 100% males were produced at the high doses (75 and 100 mg/kg) for 30 days. This result means that 100% Nile tilapia males can be produced by suppressing aromatase activity.

7.8.2.2. All-female production

In many fish species, females grow at higher rates and attain larger sizes than males. Males may also mature before females reach marketable size, leading to dispersion of sizes and reduction in production. The production of all-female fish is an effective way of solving these problems and controlling reproduction of cultured tilapia. Therefore, the interest in producing an all-female population for aquaculture purposes has increased. As for males, several natural or synthetic steroids, mainly oestrogens or other chemicals with oestrogenic capacity, have been used to feminize tilapia. Both oral administration and immersion protocols have been used with varying degrees of success (Table 7.5). Synthetic chemicals have been used for tilapia feminization at larger scales than natural steroids, with 17α-ethynyloestradiol (EE) and diethylstilboestrol (DES) being the most common and the most effective.

The relative effectiveness of steroids on feminization of tilapia depends on the type of hormone, fish species, larval stage, dose and treatment timing and duration. For example, Rosenstein and Hulata (1994) found that DES was more effective in feminizing blue tilapia (*O. aureus*) than EE, while the opposite was found in the case of Nile tilapia (Gilling *et al.*, 1996). Moreover, Rosenstein and Hulata (1992) failed to feminize *O. mossambicus* and *O. mossambicus* × *Oreochromis urolepis hornorum* hybrids by immersing eggs and fry in oestradiol-17β (E2), progesterone, flutamide and progesterone + flutamide at different solutions and durations. However, these authors reported a 100% feminization when oestrogens (DES and EE) were orally administered to these fish. On the other hand, EE, DES and E2 successfully feminized Nile tilapia fry using the immersion technique (Gilling *et al.*, 1996). It appears from these results that more work is needed to verify whether oral administration or the immersion technique is more effective and would produce higher all-female populations.

Table 7.5. Production of female tilapia using natural and synthetic steroids (full names of the hormones are given in Table 7.3).

Species	Hormone	Route	Optimum dose (mg/kg)	Fish age/size	Duration (days)	% Females	Reference
O. niloticus	EE	Immersion	170–200 µg/l	Fry	18	100	Gilling *et al.* (1996)
	DES	Food	400	8.7 mm	28	80	Potts and Phelps (1995)
	EE	Food	100	7–12 mm	40	91	Mohamed *et al.* (2004)
O. mossambicus	DES	Food	100	10 days old	11	100	Varadaraj (1989)
	EE	Food	100	Yolk-sac fry	10	100	Rosenstein and Hulata (1994)
O. aureus	EE	Food	100	Yolk-sac fry	10	100	Rosenstein and Hulata (1994)
	EE + Methallib	Food	100	Fry	42	95	Mair *et al.* (1987)
	EE	Food	100–200	Fry	40	93–98	Mélard (1995)
O. spilurus	EE	Food	100	9 mm	42	92	Ridha and Lone (1995)

7.8.3. Hybridization

Hybridization of two different species has been extensively and successfully applied to produce monosex populations. There are several conditions affecting the success of hybridization of related fish taxa, including: (i) external fertilization; (ii) unequal abundance of two parental species; (iii) weak behavioural isolating mechanisms; (iv) competition for limited spawning habitat; and (v) decreasing habitat complexity (Scribner *et al.*, 2001). These factors are very likely to fit with most tilapias; therefore, tilapia have received a good deal of attention, and have become the best-known example of hybrid production in aquaculture, since they are characterized with short life cycles and frequent spawning.

A substantial amount of information is now available on all-male and all-female tilapia hybrids (Tables 7.6 and 7.7). Most of the crosses carried out for the production of monosex tilapia have been between maternal mouthbrooders (Wohlfarth and Hulata, 1983; Beardmore *et al.*, 2001). Most of crosses between pure mouthbrooding tilapias result in all-male or nearly all-male hybrids (Lovshin, 1982; McAndrew, 1993; Penman and McAndrew, 2000) (Table 7.6). All-male progenies have been produced from the following crosses: male *O. hornorum* × female *O. niloticus* (Lee, 1979), male *O. aureus* × female *O. niloticus* (Wohlfarth, 1994), male *O. mossambicus* × female *O. aureus* (Pierce, 1980), male *O. hornorum* × female *O. mossambicus* (Hickling, 1960) and male *O. mossambicus* × female *O. spilurus niger* (Pruginin, 1967). Other crosses resulted in 50 to 98% males (Table 7.6). The failure to produce all-male progenies, in many cases, has been attributed to poor segregation of broodfish by sex and species, and also to the introduction of hybrids into broodstock ponds (Beardmore *et al.*, 2001). Tilapia strains may also influence the results of tilapia hybridization, as reported by Marengoni *et al.* (1998). These authors found that, when Nile

Table 7.6. Hybridization of different tilapia species to produce all-male populations.

Species hybridized			
Male	Female	% Males	Reference
O. niloticus	*O. aureus*	75–95	Pruginin (1967)
	O. spilurus niger	93–98	Pruginin (1967)
	O. hornorum	75	Pruginin (1967)
O. aureus	*O. niloticus*	50–100	Lee (1979), Wohlfarth (1994)
	O. niloticus (Stirling strain)	100	Marengoni *et al.* (1998)
	O. niloticus (Uganda strain)	96–100	Pruginin *et al.* (1975)
	O. mossambicus	100	Beardmore *et al.* (2001)
	Tilapia vulcani	93–98	Pruginin *et al.* (1975)
O. mossambicus	*O. aureus*	89	Pierce (1980)
O. hornorum	*O. niloticus*	100	Lee (1979), Wohlfarth *et al.* (1990)
	O. mossambicus	100	Hickling (1960)
	O. aureus	90–100	Pruginin (1967), Lee (1979)
	O. spilurus	100	Pruginin (1967)
O. macrochir	*O. mossambicus*	100	Majumdar *et al.* (1983)
	O. spilurus	97.9	Majumdar *et al.* (1983)
	O. niloticus	100	Pruginin (1967)

Table 7.7. Hybridization of different tilapia species to produce all-female populations.

Species hybridized			
Male	Female	% Females	Reference
O. mossambicus	*S. melanotheron*	100	Peters (1963), Bauer (1968)
	T. tholloni	100	Peters (1963), Bauer (1968)
O. niloticus	*S. melanotheron*	100	Peters (1963), Bauer (1968)
	T. tholloni	100	Peters (1963), Bauer (1968)

tilapia females were bred with blue tilapia males, only the Stirling strain of Nile tilapia produced 100% males, while a Japan strain produced only 91% males.

7.9. Seed Production

Despite the great potential of tilapia culture, shortage of fry production to meet the increased global demands remains one of the main obstacles limiting the expansion of intensive culture of these fish. Information about egg composition, hatchability, yolk-sac absorption, and larval rearing is limited, inconsistent and sometimes controversial. Therefore, extensive efforts should be expended to face these challenges, and better understand the optimum requirements for larval rearing of tilapia.

7.9.1. Hatching systems

7.9.1.1. Earthen ponds

Spawning in earthen ponds is the oldest method used for seed production of tilapia, and is still widely used in different regions of the world, especially in developing countries (Hepher and Pruginin, 1981; Macintosh and Little, 1995; Little and Hulata, 2000). Tilapia can spawn easily in ponds, regardless of pond size and depth, once the environmental requirements (temperature, salinity, etc.) and biological criteria (stocking density, sex ratio, etc.) are met. However, pond size, shape and depth affect harvest efficiency and seed production. Earthen ponds can be used for producing mixed-sex pure tilapia (Little *et al.*, 1994), monosex hybrids (Hepher and Pruginin, 1981) and first-feeding fry for sex-reversal (Verdegem and McGinty, 1987). Eggs and fry are usually harvested, either partially or completely, at varying intervals ranging from 6 to 60 days (Fig. 7.4). It has been reported, however, that shorter harvesting intervals lead to better results. Sorting of the seeds and removal of larger fry also reduce cannibalism and result in higher production of similar-sized fry (Little, 1989). Partial harvest is carried out by dip-netting, seining or trapping, without draining the ponds, while complete harvest requires checking of broodfish for seed incubation, along with draining of the pond and harvesting all seed at once (Little and Hulata, 2000).

Fig. 7.4. Harvesting tilapia fry from earthen ponds in the Philippines.

The main limitations of breeding ponds are the lack of sustainability, predation of small fry by other fish, cannibalism by older fry, asynchronous spawning and reduction of spawning frequency with overcrowding. Most of these limitations can be overcome through sound management, such as increasing water fertility, decreasing sediment concentration, preventing the entry of wild predatory fishes and regular seed harvesting (Little, 1989; Macintosh and Little, 1995).

7.9.1.2. Concrete tanks

Concrete tanks are currently among the most widely used spawning units for production of tilapia seeds (Fig. 7.3), especially for intensive culture in areas where fresh water is limited. Tanks have many advantages over earthen ponds, including high yield per unit area, easy harvest and better water management through exchange and/or filtration, while their main disadvantage is the higher cost. Spawning efficiency can be affected by tank size, dimensions, shape, colour, depth and construction material. Reproductive performance of Nile tilapia was found to be better in deep tanks (1–2 m) than in shallow (0.5 m) tanks (El-Sayed *et al.*, 1996). Courtship, spawning intensity and seed production of tilapia can also be related to artificial spawning shelters. Baroiller *et al.* (1997) found that the seed production of Nile tilapia was five times higher in raceways (12 m^2) containing artificial shelters (three breeze blocks) than in unsheltered raceways. The authors suggested that artificial shelters stimulated and synchronized reproductive activity of the fish. On the other hand, Duponchelle and Legendre (1997) reported that absolute fecundity of Nile tilapia females reared in tanks provided with artificial reefs was lower than in naked tanks. However, this result is questionable, because the study was conducted for only 2 months, whereas a much longer period is required for concrete results.

7.9.1.3. Hapas

The use of hapas (Fig. 7.5) as a hatching system for tilapia has been tried in Asia (the Philippines, Thailand, Vietnam, etc.), the Americas (the USA, Brazil, Puerto Rico) and Africa (Malawi) (Little and Hulata, 2000), with varying degrees of success. Hapas have many attributes that make them an excellent hatchery system for tilapia, especially in developing countries. These include easy construction, easy management, easy seed harvest and low cost. Hapas can also be suspended in fertilized earthen ponds, deep water bodies and concrete tanks supplied with clear water. However, they need continuous maintenance and cleaning of fouling. Fouling leads to blocking the mesh netting of the hapas and deteriorating water quality inside them (Little and Hulata, 2000).

Fig. 7.5. Breeding Nile tilapia in hapas fixed in earthen ponds in the Philippines.

The spawning of tilapia in hapas suspended in ponds depends on broodstock density and sex ratio, broodstock exchange, wind, water turbidity and varying water level. The use of the double hapa system, where broodfish are stocked in an inner net having a larger mesh size, which allows free access of the swim-up fry produced to the outer fine-mesh hapa or directly to the surrounding water tanks/ponds has been studied. This system was found effective in reducing the disturbance of brooding fish and fry and decreasing cannibalism, but the performance was not very encouraging (Guerrero and Garcia, 1983; Costa-Pierce and Hadikusumah, 1995).

7.9.1.4. Artificial incubation

The removal of eggs and sac fry from the mouths of females and incubating them artificially is an effective method of tilapia seed production. Artificial incubation is preferable to natural incubation due to: (i) elimination of cannibalism; (ii) high production of even-sized fry; (iii) increased spawning synchrony; (iv) shortened inter-spawning intervals; (v) reduction of hatching time; and (vi) facilitation of research on tilapia genetics and reproduction.

The incubation units for tilapia eggs can range from simple, inexpensive and easy-to-make units, such as soft-drink bottles and round-bottomed containers, such as carboys and plastic containers (Fig. 7.6.) to commercial, more advanced, conical, upwelling jars (Fig. 7.7). The efficiency of an incubator depends on its type, size and shape, the

Fig. 7.6. A GenoMar tilapia hatchery in the Philippines (photo provided by K. Fitzsimmons).

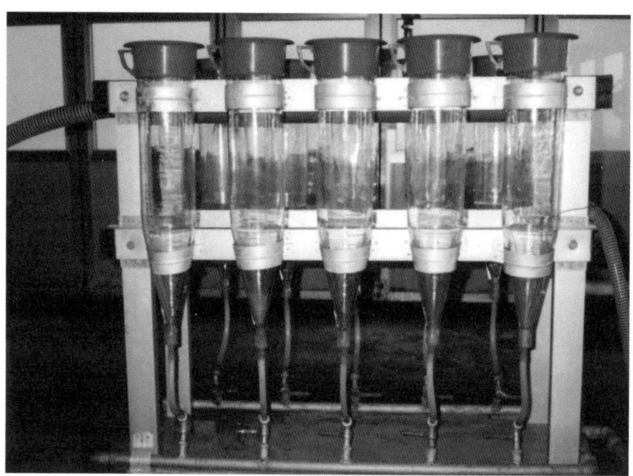

Fig. 7.7. Commercial tilapia hatching jars in Egypt.

developmental stages of the eggs and the water quality and flow. Rana (1986) found that round-bottom incubators resulted in higher survival than conical containers (85% compared to 60%), while the time to hatch was shorter in conical vessels than in round-bottom vessels (Macintosh and Little, 1995). Furthermore, Rana and Suliman (1993) found that hatching time and survival rates of tilapia fry were significantly better in downwelling round-bottomed incubators than in upwelling conical chambers.

7.9.2. Egg hatching and yolk-sac absorption

The development time that the fertilized eggs of tilapia take to hatch ranges from < 3 to > 6 days. Hatching time is affected by a number of factors, including water temperature (Rana, 1990b, c), salinity (El-Sayed *et al.*, 2005a), water flow (Rab, 1989; El-Sayed, *et al.*, 2005b) and broodstock nutrition (Gunasekera *et al.*, 1996a, b; Siddiqui *et al.*, 1997b; El-Sayed *et al.*, 2003). The optimum temperature for best hatching and survival rates ranges from 25 to 32°C. The decrease in water temperature below 22°C in subtropical areas can lead to a delay or decrease in seed production, as has been reported in Nile tilapia in Vietnam (Green *et al.*, 1997) and Egypt (El-Naggar *et al.*, 2000). Improving water temperature through heating, increasing pond depth, shading, deep hapas, etc. improves reproduction efficiency and seed production. On the other hand, high ambient water temperature (33–35°C) was found to reduce spawning efficiency, egg quality and hatchability of Nile tilapia in central Thailand during hot seasons (Little *et al.*, 1997).

Water flow was also found to affect the spawning efficiency and larval growth of tilapia. The time taken from yolk-sac absorption of Nile tilapia eggs to reach swim-up fry was reduced with increasing water flow (Rab, 1989; El-Sayed *et al.*, 2005b). A flow rate of about 8 l/min in a 20 l hatching unit was found optimal for yolk-sac absorption of Nile tilapia eggs (El-Sayed *et al.*, 2005b).

A number of studies have considered the effects of water salinity on hatchability and larval rearing of tilapia. Hatching success of *O. niloticus* females was comparable at 5‰ to that in fresh water, while increasing salinity to 10 and 15‰ lowered hatching success, and no hatching occurred in full seawater (Watanabe and Kuo, 1985). Al-Ahmad *et al.* (1988) also found that egg hatching success and fry survival of *O. spilurus* was twice as high in groundwater (4‰) as in seawater (40‰). On the contrary, Uchida and King (1962) found that fry production of *O. mossambicus* was three times higher in brackish water (8.9–15.2‰) than in fresh water.

It has also been reported that broodstock nutrition significantly affects spawning efficiency, egg hatchability and larval growth. El-Sayed *et al.* (2003) found that the hatchability of Nile tilapia egg hatched at different salinities (0–14‰) was linearly increased with increasing dietary protein levels from 25 to 40%. Eggs produced from broodstock fed 25% protein at 7 and 14‰ needed more time for hatching and yolk-sac absorption and resulted in poorer larval weight than those reared in fresh water. Furthermore, at high dietary protein levels (35–40%), *O. niloticus* eggs took about 3–4 days to hatch, while at a lower protein level (25%) hatching occurred after 4–6 days. Similar results have been reported on *O. niloticus* fed 20 and 35% protein at 26°C (Gunasekera *et al.*, 1996a).

7.10. Larval Rearing and Growth

The growth of tilapia fry depends on many factors, including stocking density, food and feeding regimes, photoperiods and water flow and replacement. The following sections briefly discuss these factors.

7.10.1. Stocking density

The effects of stocking density on larval performance are not well documented. Varying, and sometimes controversial, results have been reported concerning the relationship between stocking density and larval growth and survival. Dambo and Rana (1992) found that the growth rates of Nile tilapia fry were negatively correlated with stocking density ranging from 2 to 20 fry/l. They suggested that 5–10 fry/l be adopted as the optimum stocking density. Rana (1981) also suggested 8 fry/l as the optimal stocking density for *O. mossambicus*. More recently, El-Sayed (2002) found that the growth of Nile tilapia fry decreased with increasing stocking density from 3 to 20 fry/l, but no significant difference was found between 3 and 5 fry/l. Therefore, 5 fry/l was suggested as optimal for hatchery-reared Nile tilapia fry. This result is in agreement with that of Silva *et al.* (2000), who reported that the growth of tetra-hybrid red tilapia decreased with increasing stocking density. On the contrary, Gall and Bakar (1999) reported that body size of tilapia fry was not affected by stocking densities ranging from 10 to 200 fry/l when water flow was uniform. Furthermore, Macintosh and De Silva (1984) found that the relationship between stocking density and survival of *O. mossambicus* and *O. niloticus* female × *O. aureus* male fry was not consistent. It is evident that more work is needed along this line.

The negative correlation between growth rates and stocking density of tilapia fry has been investigated by a number of authors. Macintosh and De Silva (1984) reported that increasing stocking density of *O. mossambicus* fry might have led to diminishing social dominance. This usually leads to ‘social stress’ and causes a ‘chronic stress response’, which, in turn, impairs fish growth, possibly due to the mobilization of dietary energy by the physiological alterations provoked by the stress response (Dambo and Rana, 1992; Kebus *et al.*, 1992). Increasing stocking density may also cause deterioration in water quality, leading to stressful conditions (Barton and Iwama, 1991; Pankhurst and Van der Kraak, 1997).

The high stocking density of tilapia fry may also cause excessive cannibalism, leading to high mortality. This phenomenon has been reported with Nile tilapia (Dambo and Rana 1992; El-Sayed, 2002), *O. mossambicus* (Uchida and King, 1962; Macintosh and De Silva, 1984), and tilapia hybrids (Macintosh and De Silva, 1984). These studies demonstrated that cannibalism can be a main cause of tilapia fry mortality. Continuous grading of the fry in nursing units to separate large fry and using broodstock of approximately similar ages and sizes to produce eggs and fry of similar sizes are common practices to reduce cannibalism among tilapia fry.

7.10.2. Food and feeding regimes

The protein and energy requirements of tilapia fry have been discussed in Chapter 6. About 35–45% crude protein is required for optimum larval performance, depending on protein and energy sources, fish species and culture systems. The effects of feeding levels on performance of tilapia fry are also well documented, although the results are controversial. El-Sayed (2002) found that Nile tilapia fry fed a larval test diet at 10–35% body weight (BW)/day attained their optimum performance at the 35% level, while Santiago *et al.* (1987) found that the growth of Nile tilapia fry increased with increasing feeding levels up to 65%. However, no significant differences in growth rates and survival were found between the 30 and 65% feeding levels. These authors recommended a feeding rate of 30–45% BW/day as optimal for Nile tilapia fry. The growth rates of red tilapia fingerlings (0.56 g) were increased, while feed conversion ratios were impaired, with increasing feeding levels from 4 to 16% (Sirol *et al.*, 2000).

The effects of food colour on growth and feed efficiency of tilapia fry is not well understood. Arumugam (1997) found that tilapia fry showed no preference for diet colour. In contrast, more recently, El-Sayed (2004) found that feed colour significantly affected the growth and feed utilization efficiency of Nile tilapia fry (Fig. 7.8). The fish were fed larval diets with different colours (red, green, dark blue, yellow and brown). Darker-coloured diets (red and blue) resulted in better performance than lighter-coloured diets. The yellow diet produced the poorest growth and feed efficiency. It appears that feed colour could be a key factor for feed acceptance in Nile tilapia.

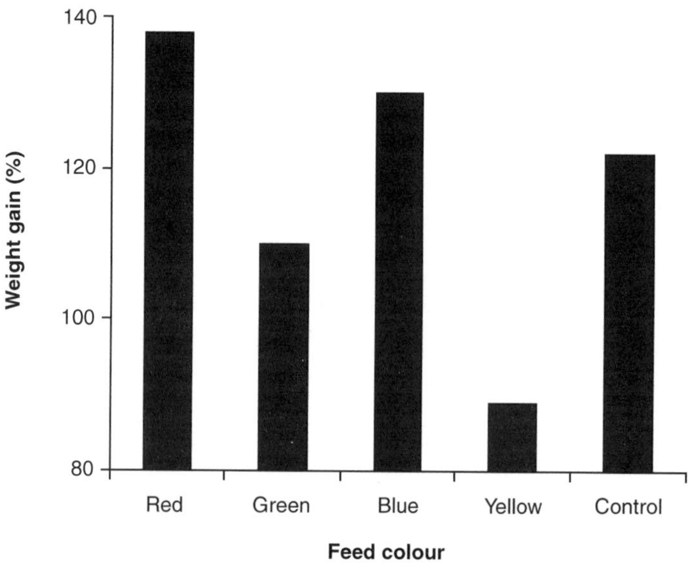

Fig. 7.8. Effect of feed colour on percent weight gain of Nile tilapia fingerlings.

But more studies are needed before a concrete conclusion can be drawn.

7.10.3. Photoperiod

It has been suggested that freshwater fish species show more of a response to photoperiod than marine and diadromous species (Imsland *et al.*, 1995). However, the response of marine species to photoperiods has been well investigated, while little attention has been paid to freshwater species. The available information indicates that the response of tilapia to photoperiod cycles depends on fish developmental stage and sex. El-Sayed and Kawanna (2004) found that the larval stage of Nile tilapia was more sensitive to photoperiod than fingerling and juvenile stages. Fish fry subjected to long photoperiods (24 and 18 h) had significantly better performance than those exposed to intermediate or short photoperiods (12 or 6 h). Similar results have been reported for several marine fish larvae, where the growth rates were improved with increasing photoperiods (Tandler and Helps, 1985; Barlow *et al.*, 1995; Hart *et al.*, 1996).

One possible explanation of the decreased growth of tilapia fry during shorter photoperiods is that during these short light phases insufficient time was available for the establishment of a 'robust rhythmicity', as has been reported by Biswas and Takeuchi (2002) and Biswas *et al.* (2002). The effect of photoperiod on synchronizing an endogenous rhythm with the external environment may also require more energy in the shorter photoperiods, leading to a reduction of somatic fish growth (Biswas and Takeuchi, 2002; Biswas *et al.*, 2002).

7.10.4. Water flow and replacement

Partial or total replacement of culture water may significantly affect the performance of tilapia fry. Absalom and Omenaihe (2000) reared Nile tilapia fry (1.14 g) using the following water replacement treatments: no water replacement, daily replacement, replacement every 3 days and every 6 days. They found that growth rates were not significantly affected by water replacement, and suggested that less frequent water replacement may be better for larval growth and survival. The growth and survival of swim-up fry reared in 20 l fibreglass tanks in closed system, under different water flows (1, 2, 3, 4 and 8 l/min) were not affected by water flow of up to 4 l/min (El-Sayed *et al.*, 2005b). Further increase in the water flow to 8 l/min caused extremely poor growth and high mortality (Fig. 7.9). However, the effects of water flow should be considered in relation to tank size. In other words, the water exchange rate has the major effect rather than water flow per se.

7.11. Closing Remarks

1. Removal of eggs and fry from females' mouths accelerates vitellogenesis and shortens the intervals between successive spawnings. A 5-day cycle of seed removal from females' mouths is a common practice in commercial tilapia hatcheries in Thailand.
2. Exchange of tilapia male and female broodfish, after a period of conditioning, could be an effective tool for improving seed production, spawning synchrony and spawning frequency. A 5–15-day resting period may be appropriate.

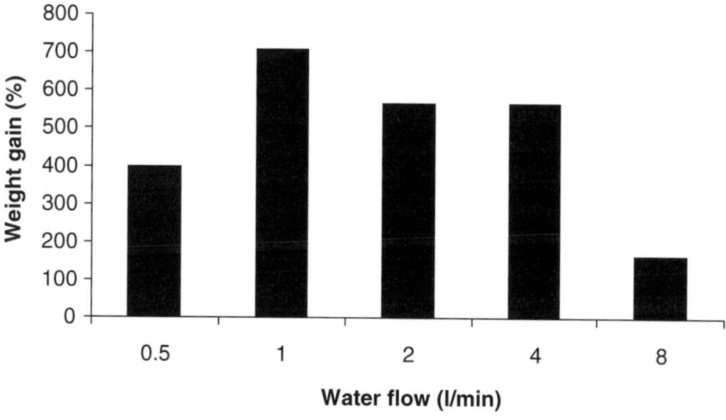

Fig. 7.9. Effect of water flow on percent weight gain of Nile tilapia fry.

3. Extensive work is urgently needed on quantitative lipid requirements, energy requirements, protein-sparing effects by dietary lipids or carbohydrates and vitamin and mineral requirements of tilapia broodstock under different culture systems.
4. Partial or complete water change of spawning units, at frequent intervals, is suggested. This process may improve seed output and spawning synchrony of tilapia.
5. Immersing fish fry in hormone solutions for short periods of time could be a safe alternative for oral administration, but timing, duration and number of immersions should be carefully determined.
6. One hundred percent Nile tilapia males can be produced by suppressing aromatase activity, using aromatase inhibitors such as Fadrozole.
7. The effects of stocking density of tilapia fry on larval performance are not well documented. The stocking density for optimum performance has been controversial, and more work is needed to settle this dispute.
8. Feed colour could be a key factor for feed acceptance in Nile tilapia. These fish appear to prefer dark-coloured diets to light-coloured diets. However, further long-term studies are needed before a concrete conclusion can be drawn.
9. Nile tilapia larvae may be more sensitive to photoperiod than fingerlings and juveniles, and they appear to perform better under long photoperiods (24 and 18 h). However, more work is needed to support this assumption.

8
Stress and Diseases

8.1. Introduction

Tilapia, like other aquatic animals, can be subjected to various infectious and non-infectious diseases. Tilapia diseases have attracted attention in recent years, for the following reasons:

1. The wide expansion of tilapia culture, which may increase the risk of diseases.
2. The extensive introduction of tilapia into many countries, which increases the chances of disease occurrence.
3. Increasing public awareness about the role of fish culture in the spread of human diseases.
4. Increasing public concern about environmental protection.
5. Increasing global exportation and importation of tilapia, with high quality standards.

It is no surprise, therefore, that fish health and disease control are now reviewed in integration with environmental protection, pollution control, human health, aquaculture technologies, sanitation of culture facilities, diagnosis and treatment of diseases of cultured species, formulation and implementation of regulatory measures to control the spread of diseases and development of disease-resistant aquatic strains (Pillay, 1990). This chapter addresses the major fish stressors and common diseases that infect wild and cultured tilapia, with emphasis on disease agents, symptoms and methods of treatment.

8.2. Stress

Stress in the fish is generally defined as the disturbance of the internal equilibrium (homoeostasis). It is the first step towards disease occurrence, since it reduces the resistance of the fish and makes it more susceptible to diseases (Everitt and Leung, 1999). Therefore, the extent of stress and the ability of fish to resist and maintain homoeostasis are most important for survival and growth. Stress can be caused by different factors, including:

- Nutritional differences (e.g. vitamin imbalances).
- Environmental quality and culture conditions.
- Well-being of cultured animal.
- Physical, chemical and biological interference (crowding, handling, transportation, pollution, organic enrichment, etc.).

Tilapia are well-adapted to prevailing environmental conditions and can tolerate a wide range of environmental factors, such as water temperature, salinity, dissolved oxygen, ammonia, etc. (see Chapter 3). However, stressful conditions adversely affect tilapia and make them more susceptible to different diseases, presumably due to immunosuppression. Foo and Lam (1993) found that handling stress caused a significant increase in serum cortisol levels of Mozambique tilapia (*Oreochromis mossambicus*), and the cortisol level depended on the severity of the stressor, duration of exposure and fish health status. The removal of the stressor resulted in a rapid return to the normal cortisol level. Barcellos *et al.* (1999a) studied the effects of acute stress on Nile tilapia (*Oreochromis niloticus*) previously exposed to chronic stress (handling stress). They found that fish exposed to chronic stress for 59 days had lower growth rates than non-stressed fish. Plasma cortisol in fish previously exposed to chronic stress (169 ng/ml) was significantly lower than that obtained from fish exposed only to acute stress (9267 ng/ml). The authors concluded that Nile tilapia can adapt to chronic stress, and this adaptation reduces, but does not eliminate, their response to additional acute stress.

An essential cause of chronic stress in tilapia is social interactions and hierarchies. A negative correlation between growth rates and stocking density of fish fry has been reported in tilapia fry (Macintosh and De Silva, 1984; Dambo and Rana, 1992; El-Sayed, 2002). Increasing fish density leads to what is known as 'social stress' causing a chronic stress response. This leads to impaired fish growth, presumably due to the mobilization of dietary energy by the physiological alterations provoked by the stress response (Kebus *et al.*, 1992). As a result, resting plasma cortisol concentrations of Nile tilapia fingerlings were reported to have increased with increasing stocking density, indicating a chronic stress response attributable to social stress (Barcellos *et al.*, 1999b). Tilapia may also be exposed to secondary stress after medical (chemical) treatments. The response of Nile tilapia to secondary stress after treatment with a mixture of formalin, malachite green and methylene blue (FMC) has been investigated (Yavuzcan and Pulatsü, 1999). Treating the fish with FMC increased plasma glucose and haematocrit and reduced plasma phosphorus and calcium, while magnesium was not affected. These changes in blood chemistry were attributed to secondary stress response to FMC treatment. This indicates that therapeutic agents (such as FMC) are an under-recognized stress source.

Reducing stress is, therefore, an important means of disease control and improvement of immunity in cultured fish. Endo *et al.* (2002) studied the stress and immune response of Nile tilapia under self-feeding and scheduled-feeding regimes. They found that self-fed fish had lower levels of plasma cortisol, higher phagocytic activity of their macrophages, higher antibody production and a higher number of lymphocytes than scheduled-fed fish. The authors concluded that a self-feeding regime is less stressful than scheduled feeding. Thus, it is evident that the best way of preventing, or at least reducing, tilapia diseases is the adoption of best management practices. These practices include improving and controlling the culture environment, adopting proper handling and transportation techniques, reasonable stocking densities, using proper amounts of high-quality feed with the adoption of the right feeding regimes, using legal drugs judiciously and using the most effective vaccines (Plumb, 1999).

8.3. Major Diseases

Some tilapia diseases such as bacterial and parasitic diseases have received considerable attention in recent years. Many articles describing the symptoms, diagnosis, treatment and prevention of these diseases appeared in specialized journals. On the other hand, much less work has been done on other disease groups, particularly fungal, viral and non-infectious diseases. For this reason, the reader may realize that this chapter covers parasitic and bacterial diseases in much more detail than other disease groups. The following sections address the major diseases that infect both wild and farmed tilapia.

8.4. Parasitic Diseases

8.4.1. Protozoan diseases

Several external and internal protozoan parasites are known to live on (or inside) tilapia, causing illness, mortality and economic loss (Table 8.1). The severity of protozoan diseases depends on the number of parasites infesting the fish, culture systems and fish species, sex, size and health status. Protozoan parasites have been reported to cause serious mortality in wild and farmed tilapia, in both hatcheries and rearing facilities. The most important parasitic protozoan diseases of tilapia are summarized below.

ICHTHYOPHTHIRIUS MULTIFILIIS (ICH). *Ichthyophthirius multifiliis* (Ich) is a ciliated protozoan with a round or ovoid body, small mouth and longitudinal rows of cilia on the body surface, converging at the anterior end (Pillay, 1990). It has a large, horseshoe-shaped macronucleus, a small micronucleus and many contractile vacuoles. The infective stage of this parasite is known as a theront or tomite. This disease is known as white spot, because an infected fish is characterized by the appearance of white spots where the parasites live

Table 8.1. A list of the common disease agents, symptoms and treatment protocols of tilapia.

Agent	Species (location)	Disease symptoms	Treatment protocols	References
Parasites				
Ichthyophthirius multifiliis	*O. m, O. a, T. z, O. m* × *O. h* (USA)	Low growth, white spots on fins and epidermis		Lightner *et al.* (1988)
	O. m fry (UK)	High mortality		Subasinghe and Sommerville (1986)
	O. a (USA)		0.5 mg/l $KMnO_4$, single dose	Straus and Griffin (2001)
Trichodina	*O. n* (Cameroon)	Loss of escape reaction, scraping against tank walls, rapid opercular movement, jumping out of water	(1) Formalin bath, 250 ppm for 35–49 min (2) $KMnO_4$, 5 ppm for 10–15 min	Nguenga (1988)
	O. m, O. a, T. z, O. m × *O. h* (USA)	Gill hyperplasia, eroded fins, epidermal ulcers, high mortality		Lightner *et al.* (1988)
	O. n (Vietnam)	Massive mortality in nursery stage		Lua *et al.* (1999)
Myxobolus ovariae	*O. n, S. g* (Nigeria)	Inflammatory reaction, reabsorption of ovary tissue, replacing gonad tissues with the spores, gonadal atrophy and low GSI	Culling infected fish, water filtration to eliminate the spores	Okaeme *et al.* (1989)
Myxobolus sp.	*Oreochromis* sp. (Mexico)	White nodules in the gills, dark skin, exophthalmia, loss of appetite, increased abdominal volume, congestion and thickening of intestines		de Ocampo and Camberos (1998)
Cryptobia branchialis (Costia)	*O. m* (USA)	Thick mucus on gill surface, swelling of gill filaments, reduction of respiratory lamellae and hypertrophy of respiratory epithelium		Kuperman *et al.* (2002)
Dactylogyrus sp.	*O. n* (Cameroon)	Rapid opercular movements, opercula held open, thickened edges of gills and destruction of branchial epithelium	(1) A single dose of formalin, 250 ppm for 35–40 min (2) Two repeated doses of $KMnO_4$, 5 ppm for 10–15 min	Nguenga (1988)
Alitropus typus	*O. n* (Thailand)	50–100% mortality		Chinabut (2002)

Continued

Table 8.1. *Continued.*

Agent	Species (location)	Disease symptoms	Treatment protocols	References
Bacteria				
Aeromonas hydrophila	*O. n* (China)	Erosion of skin and dorsal fin, body surface filled with blood, high mortality		Wang and Xu (1985)
	O. n (China)	Slow movement, caudal fin rot, swimming near water surface, poor appetite, mortality		Liu *et al.* (1993)
	O. n (Philippines)	Skin lesion, ulceration, fin rot, body discoloration, mouth sore, eye opacity, exophthalmia, dislodged eyeball, sluggishness		Yambot (1998)
Pseudomonas sp.	*O. n* (Japan)	Fine white nodules in the spleen, exophthalmia, dark body, nodular lesions and focal necrosis in the liver, spleen, kidney and gills, inflamed swim bladder, abscesses in the eyes, spleen and swim bladder		Miyashita (1984), Miyazaki *et al.* (1984)
Vibrio sp.	*O. s* (Kuwait)	Lethargy, dark body colour, dermal necrosis, high mortality		Saeed (1993)
Streptococcus sp.	*O. n* (USA)	Hyperaemic gills, diffuse epithelial tissue proliferation, lesions, dermal haemorrhage		Bowser *et al.* (1998)
	O. m (Taiwan)	Haemorrhage, exophthalmia, corneal opacity, dark coloration, abscess of trunk muscles	Erythromycin, doxycycline	Tung *et al.* (1987)
	O. n × *O. a* (Saudi Arabia)	Erratic swimming, melanosis, exophthalmia, haemorrhage around the jaws and base of pectoral and pelvic fins, ascitic fluid in the abdominal cavity		Al-Harbi (1994)
Streptococcus iniae	*O. n* (USA)	Darkened skin pigments, bottom swimming, rising and falling, side swimming, loss of appetite		Evans *et al.* (2000)

	O. n (USA)	Dark skin pigmentation, abdominal distension, haemorrhage, erythema, eye lesion, lethargy, reduction or cessation of feeding, circular swimming, side swimming	Oxytetracycline (75–100 mg/kg feed)	Darwish and Griffin (2002)
	O. n × *O. a* (USA)	Loss of orientation, exophthalmia, corneal opacity, petechiae around the mouth and anus, fluid accumulation in the peritoneal cavity	• Tetracycline, oxytetracycline • Sulphadimethoxine–ormitoprim (5 : 1)	Perera *et al.* (1994)
Staphylococcus epidermidis	*O. a* (Taiwan)	Lesions of spleen and kidney, apoptosis in lymphocytes and macrophages, brain, liver, gonads, mesentery, stomach, intestines and skeletal muscles		Huang *et al.* (2000)
Mycobacterium marinum	*O. n* × *O. m* × *O. a* (USA)	Small visceral granulomas, high epithelial macrophages and peripheral lymphocytes		Wolf and Smith (1999)
Flexobacter columnaris	*O. n* (USA), Tilapia sp. (Korea)	Respiratory disorder, fin erosion, body discoloration, lesion in muscles and skin, heavy mucus secretion	Oxytetracycline, tetracycline, chloramphenicol, Amikacine erythromycin, kanamycin	Roberts and Sommerville (1982), Chun and Sohn (1985)
Fungi				
Branchiomyces	*O. n* × *O. m*, *O. n* × *O. a* (Israel)	Damage of gill tissue, high mortality		Paperna and Smirnova (1997)

O. a, *O. aureus*; *O. m*, *O. mossambicus*; *O. h*, *O. hornorum*; *O. n*, *O. niloticus*; *O. s*, *O. spilurus*; *T. z*, *T. zillii*; *S. g. Sarotherodon galilaeus*; GSI, gonadosomatic index.

in the epidermis of the skin and gills (Ewing and Kocan, 1992). The optimum temperature for the growth of the parasite is 20–25°C at a high stocking density.

Infection of tilapia with *I. multifiliis* has been widely reported in the Americas (Lightner *et al.*, 1988; Crosby, 2001; Straus and Griffin, 2001) and Asia (Lua *et al.*, 1999; Te *et al.*, 1999). Larval fish stages appear more susceptible than fingerling and adult stages. Subasinghe and Sommerville (1992) found that the effect of *I. multifiliis* on larval *O. mossambicus* was more severe than on fingerlings. Similar results have been reported with Nile tilapia, where the incidence of infection decreased with increasing fish size (Lua *et al.*, 1999). The parasites were found in large numbers in the nares, pharynx, gills and skin of tilapia fry, causing severe mortality (Roberts and Sommerville, 1982).

The confrontation of Ich and other parasites has received considerable attention in recent years. Potassium permanganate ($KMnO_4$) is widely used to control this disease throughout the world. Straus and Griffin (2001) determined the acute toxicity of $KMnO_4$ to *I. multifiliis* and the concentration required for preventing its infestation of juvenile blue tilapia (*Oreochromis aureus*) and channel catfish. They found that the lowest effective doses were 0.5 mg/l and 1.0 mg/l for tilapia and catfish. The protective immunity of tilapia against *I. multifiliis* has also been investigated. When *O. mossambicus* (120–200 g) were subjected to a trickle, sublethal infection of *I. multifiliis* tomites, they were free of the parasite 18 days after the initial infection, and remained free for 9 months in an infectious environment (Subasinghe and Sommerville, 1986). The survival of *O. mossambicus* fry obtained from adult females immunized against *I. multifiliis* and challenged with controlled infection was also better than the survival of fry of challenged non-immunized females (Subasinghe and Sommerville, 1989). Similarly, Sin *et al.* (1994) found that *O. aureus* fry obtained from broodstock previously vaccinated with ineffective live tomites of *I. multifiliis* exhibited 95% survival with mouthbrooding and 78.4% survival without mouthbrooding, compared to only 0 and 37.3% survival in non-vaccinated fish without and with mouthbrooding. The authors suggested that protective immunity against *I. multifiliis* in tilapia fry is both derived from the mother via eggs and acquired indirectly from the mouth cavity during mouthbrooding.

8.4.2. Other ciliates

Several other unicellular parasitic ciliates, including *Trichodina*, *Tricodinella*, *Epistylis*, *Ichthyobodo*, *Apiosoma* and *Ambiphrya*, are common among wild and cultured tilapia, especially when the fish are farmed at high stocking densities (El-Tantawy and Kazaubski, 1986; Lightner *et al.*, 1988; Bondad-Reantaso and Arthur, 1989; Ramadan, 1991). *Trichodina* is the most documented ciliate in Nile tilapia in the Philippines (Natividad *et al.*, 1986; Bondad-Reantaso and Arthur, 1989) and Vietnam (Lua *et al.*, 1999), in Mozambique tilapia, blue tilapia, tilapia hybrids and *Tilapia zillii* in the USA (Lightner *et al.*, 1988) and Egypt (El-Tantawy and Kazaubski, 1986) and in both Nile tilapia and Mozambique tilapia in Nigeria (Okaeme and Okojie, 1989; Bello-Olusoji *et al.*, 2000).

Trichodina is a saucer-shaped parasite (Fig. 8.1) that attacks fish skin and gills. The typical signs of the disease include skin and gill damage, respiratory distress, loss of appetite and loss of scales. Skin and gill damage caused by this parasite may lead to the entry of other pathogens, such as bacteria and fungi (Lightner *et al.*, 1988; Plumb, 1997). This parasite can be successfully treated by a formalin bath at about 250 ppm for 35–50 min or 5 ppm $KMnO_4$ for 10–15 min (Nguenga, 1988).

Infection by the above-mentioned parasites is generally affected by tilapia species, size and sex and season. Ramadan (1991) studied the infection of Lake Manzala (Egypt) fishes by

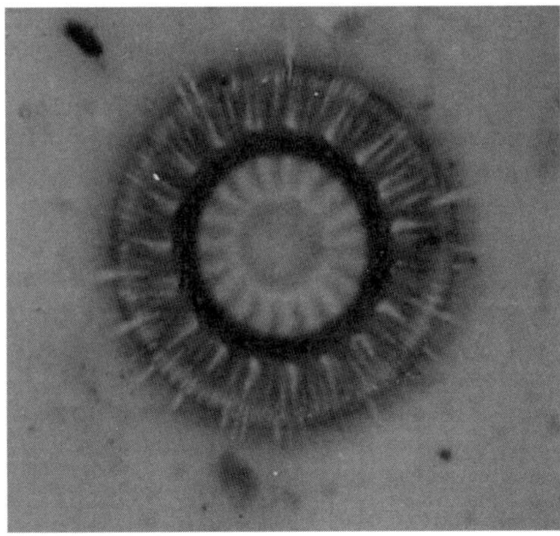

Fig. 8.1. *Trichodina,* an internal parasite isolated from Nile tilapia (photo provided by M.K. Soliman).

different parasites. The prevalence of infection was higher in larger fish than in smaller fish and in *T. zillii* than in Nile tilapia. Parasitism also increased with decreasing water temperature (i.e. the infection was higher in winter than in summer).

Most ciliate parasites can be treated with table salt (NaCl), $KMnO_4$ (Nguenga, 1988) and formalin (30–50 mg/l) (Lahav and Sarig, 1972; Natividad *et al.*, 1986; Plumb, 1997; Shoemaker *et al.*, 2000). Treatment dose is affected by water quality (temperature, salinity, dissolved oxygen, ammonia concentration, etc.).

8.4.3. Flagellated protozoa

A number of parasitic flagellated protozoans have been reported to infect tilapia in fish farms as well as in the wild. *Ichthyobodo necator* (previously known as *Costia necatrix*) is probably the most important flagellated parasite infecting tilapia (Plumb, 1997; Shoemaker, *et al.*, 2000). Crosby (2001) reported that *Ichthyobodo* represented a major problem for tilapia farmers in Virginia, USA. Heavy infection of *O. mossambicus* with *Ichthyobodo* in marine fish hatcheries in Hawaii has also been reported (Brock *et al.*, 1993). The general signs of the disease include flashing and scraping, listlessness and resting on the bottom, loss of appetite and the presence of a blue-grey sheen with thick mucous patches.

Cryptobia branchialis is another unicellular flagellate that has been reported to cause severe damage in tilapia in fresh water and brackish water (Natividad *et al.*, 1986; Kuperman *et al.*, 2002). This parasite is pear-shaped or elongated, anteriorly rounded and tapered posteriorly. Kuperman *et al.* (2002) reported an infestation of young *O. mossambicus* in Salton Sea (a highly saline lake in California, USA) by *C. branchialis*. Disease symptoms included deposition of thick mucus on the gill surface, swelling of gill filaments, reduction of respiratory lamellae and hypertrophy of respiratory epithelium. These changes led to depression in oxygen consumption. Tilapia fry were more susceptible to this parasite than adult fish.

The infection of tilapia by other flagellate protozoans has been recorded recently. Kuperman and Matey (1999) reported massive infestation of young *O. mossambicus* in Salton Sea by *Amyloodinium ocellatum* (Dinoflagellida). Outbreaks of the infestation and the subsequent mortality were recorded when water temperature and salinity were 40°C and 46‰. Paull and Matthews (2001) suggested that *Spironucleus vortens* is the flagellate that possibly caused a hole-in-the-head disease in cichlid fishes in the UK. The parasite was isolated from the kidney, liver, spleen and head lesions, as well as from the intestines. *Trypanosoma mukasai* has also been isolated and described from the peripheral blood of *O. mossambicus* in West Bengal, India (Sinha, 1986).

8.4.4. Sporozoan diseases

Myxosporidian sporozoans are a class of protozoans that may cause serious diseases in fish culture. They commonly occur as cysts, replete with spores, in the tissues of most wild tilapia (Okaeme and Okojie, 1989; Gbankoto *et al.*, 2001). Their pathological effects become higher in intensive earthen ponds, where the conditions facilitate the parasite's life cycle, since released spores require a period of initiation in pond mud prior to the infectious stage (Roberts and Sommerville, 1982).

Myxobolus sp. is the most important myxosporidian affecting tilapia in fish farms. Okaeme *et al.* (1989) reported that *M. ovariae* was highly prevalent among the tilapias *O. niloticus* and *Sarotherodon galilaeus* in Nigeria, and ovaries were the most affected organs. Histological changes of infected fish revealed an inflammatory reaction, mechanical pressure, reabsorption of ovary tissues, gradual replacement of gonadal tissues by the spores, gonadal atrophy and a reduced gonadosomatic index. The same authors (Okaeme *et al.*, 1989) recommended that infected fish should be culled and culture water should be sedimented and filtered to eliminate the spores. It was also found that *Myxobolus exiguus* infect the fry of different Nile tilapia strains (Thailand, genetically improved farmed tilapia (GIFT) and Vietnam strains) in North Vietnam (Lua *et al.*, 1999). *Myxobolus* sp. also infested tilapia (*Oreochromis* sp.) in Mexico (de Ocampo and Camberos, 1998). Disease symptoms included loss of appetite, increase in abdominal size, congestion and thickening of the medial and posterior intestines. Gbankoto *et al.* (2001) reported an occurrence of six myxosporidian parasites in the gills of *Sarotherodon melanotheron* and *T. zillii* from Lake Nokoué (Benin, West Africa). The two most common myxosporidians were *Myxobolus* sp. and *Myxobolus zillii*, which were located in the branchial filaments of *S. melanotheron* and *T. zillii*, respectively. Two other myxosporidians; *Myxobolus sarotherodoni* and *Myxobolus beninensis* were also recorded, but at lower levels, in the branchial arch cartilage of *S. melanotheron*. On the other hand, *Myxobolus dossoui* and *Myxobolus microcapsularis* were found in the branchial arch cartilage of *T. zillii* (Gbankoto *et al.*, 2001). Control of the disease includes killing the spores through culling infected fish, water filtration and partial drainage of the water, followed by addition of chlorine (at about 10 ppm) and air-drying of contaminated mud.

It appears, at least up until now, that there is no known effective treatment for myxosporidian diseases in tilapia. Avoiding the infective spores through water sedimentation and filtration, partial discharge, eventual pond drying and proper stocking densities have been the main controlling and protective measures. These methods have been tried with other farmed fish species, with varying degrees of success.

8.4.5. Metazoan parasites

MONOGENETIC TREMATODES. Monogenetic trematodes, including *Gyrodactylus*, *Dactylogyrus* (known as *Cichlidogyrus* in cichlid fishes; M.K. Soliman, Alexandria, 2005, personal communication) and *Cleidodiscus*, have been reported to parasitize tilapia, especially on gills and skin, in both farmed and wild fish (Ramadan, 1991; Omoregie *et al.*, 1995; Plumb, 1997; Shoemaker *et al.*, 2000). Roberts and Sommerville (1982) reviewed the occurrence of monogenetic and digenetic trematodes in cultured tilapia. Three species of monogenetic trematodes (*Cichlidogyrus sclerosus*, *Cichlidogyrus tilapiae* and *Gyrodactylus niloticus*) have been recorded in cultured tilapia in Vietnam (Te *et al.*, 1999). The infection of tilapia by monogenetic trematodes is affected by fish species, sex and size and environmental conditions. Ramadan (1991) found that *T. zillii* in Lake Manzala, Egypt, were more infected with *Dactylogyrus* than *O. niloticus*, and larger fish were also more infected than smaller fish. Furthermore, females were more susceptible to the disease than males, and the infection rate was higher in winter than in summer.

Gyrodactylus infection of pond-reared tilapia was recorded in Uganda, causing corneal damage (Fryer and Iles, 1972b) and Kenya (C. Sommerville and R.D. Haller, unpublished results). Infection is usually associated with poor handling and high stocking density. Nguenga (1988) reported an infection of Nile tilapia with *Dactylogyrus* (*Cichlidogyrus*) sp. in Cameroon. Infested fish showed rapid opercular movements and the opercula held open, thickened edges of the gills and destruction of branchial epithelium. *Cichlidogyrus* and *Gyrodactylus* have also been reported to infect tilapia in a number of fish farms in the eastern region of Saudi Arabia (FRRC, 2001). Infected fish suffered from difficulties in respiration, due to the damage of gills, haemorrhage, fin rot and increased mucus secretion.

Dactylogyrus was successfully treated with a single dose of formalin (250 ppm for 35–40 min) or two repeated applications of potassium permanganate at a dose of 5 ppm in static water with a 2-day interval (Nguenga, 1988). Table salt (sodium chloride) is also commonly used in tanks for preventive measures at a rate of 25 g/l of water (Nguenga, 1988). A number of other chemicals have been reported to control these parasites effectively in other fish species. The list includes a $KMnO_4$ bath, acriflavine, sodium hydroxide, methylene blue, formalin, magnesium sulphate, sodium perborate and many other chemicals (see Hoffman and Meyer, 1974, for details). The doses and methods of application depend on fish species and size and duration of treatment.

DIGENETIC TREMATODES. Digenetic parasitic trematodes can be very problematic to tilapia culture, and can cause heavy loss among fingerling and juvenile fish (Roberts and Sommerville, 1982). The life cycle of these parasites involves three hosts: a snail, a fish and a fish-eating vertebrate. Therefore, they may not cause any problems in closed systems, since one or more of the hosts may not be available. In open systems, they can infest the fish when the cercariae migrate to the target organ to develop into metacercariae (Shoemaker *et al.*, 2000).

A number of digenetic trematodes have been associated with disease infection in cultured and wild tilapia. *Clinostomum* sp., which is generally known as 'yellow grub' or 'white grub', is one of the most important digenetic trematodes that infect tilapia. The metacercariae cause bulging and distortion of the fish body profile, spoiling their appearance and make them more susceptible to handling (Roberts and Sommerville, 1982). *Clinostomum* sp. has been recovered from wild tilapia (*Oreochromis leucostictus*) in Lake Naivasha, Kenya (Aloo *et al.*, 1995) and *O. niloticus*, *S. galilaeus* and *T. zillii* in Lake Kompienga (Burkina Faso) (Coulibaly and Salembere, 1998). The prevalence of the parasite was highest in *O. niloticus* (56.2%), followed by *S. galilaeus* (45.79%) and *T. zillii* (18.21%). Omoregie *et al.* (1995) recorded *C. tilapiae* in Nile tilapia from a fish farm and petroleum-polluted water in Nigeria.

Another digenetic trematode, *Diplostimum* (eye fluke), was recorded in the same fish farm in Nigeria (Omoregie *et al.*, 1995). Okaeme and Okojie (1989) also found that feral tilapia, *O. niloticus* and *S. galilaeus*, in Nigeria were infected with *Diplostimim tregenna*. This parasite can cause complete blindness and loss of reflex and pigmentation control when the number of metacercariae infecting the eye is high (Roberts and Sommerville, 1982).

Black-spot disease, which is caused by *Neascus* metacercariae, has also been reported in Nile tilapia and *S. galilaeus* in Lake Kainji, Nigeria (Okaeme and Okojie, 1989). *Neascus* metacercariae stimulate the accumulation of the host capsule in the skin, resulting in obvious black spots, which can be easily observed (Roberts and Sommerville, 1982). The real effect of this parasite is that it makes the fish unmarketable if the infection is heavy. *Haplorchis* sp. is another digenetic trematode that is found in a wide range of freshwater fishes, including tilapia, in many countries, such as Egypt, Kenya, Israel, the Philippines, China and Japan (Roberts and Sommerville, 1982). Mass penetration of fish skin by large numbers of cercariae and the migration of metacercariae to accumulate at fin bases may lead to the loss of skin function and cause fish mortality (Roberts and Sommerville, 1982).

As far as I know, no specific drugs have been suggested for controlling digenetic trematodes in tilapia. However, di-*N*-butyl tin oxide has been effective in salmonids. Sand-gravel filtration of culture water has also been effective (Hoffman and Meyer, 1974). Above all, snail control remains the best method of controlling digenetic trematodes. In this regard, fish ponds should be dried and limed prior to stocking, to eliminate any trematodes that might have been present in pond water and/or mud. This practice is widely applied in the Philippines (Fig. 8.2).

PARASITIC NEMATODES, CESTODES AND ACANTHOCEPHALANS. Several nematode species have been reported in wild and cultured tilapia, but little information is available on their parasitic significance (Fryer and Iles, 1972b; Scott, 1977). The nematode *Contacaecum* sp. has been reported to cause pathological effects and growth

Fig. 8.2. Drying and liming earthen ponds in the Philippines to eliminate trematodes and other parasites that might have been present in the pond water and mud.

retardation in *Sarotherodon grahami* in Kenya (Scott, 1977). They have also been recorded in *O. leucostictus* in Lake Naivasha in Kenya (Aloo *et al.*, 1995), where males were more parasitized than females. This parasite can also be a problem to the consumer, since it occurs as large encysted worms throughout fish muscles.

Limited information is also available on the parasitic effects of cestodes and acanthocephalans on tilapia, despite the fact that they have been reported in wild and farmed tilapia (Fryer and Iles, 1972b; Ramadan, 1991; Aloo *et al.*, 1995; Omoregie *et al.*, 1995). Omoregie *et al.* (1995) isolated two cestodes, *Eubothrium tragenna* and *Polyonchobothrium* sp. from farmed Nile tilapia and oil-polluted water. The infection was concentrated in the intestines, stomach, liver and brain, while the muscles and gonads were significantly infected in adult and subadult fish. The acanthocephalan *Polycanthorhynchus kenyensis* was also recorded in *O. leucostictus* in Lake Naivasha, Kenya (Aloo *et al.*, 1995) and *T. zillii* and *S. galilaeus* in Lake Manzala in Egypt (Ramadan, 1991).

8.4.6. Parasitic crustaceans

Parasitic crustaceans, including the copepods *Ergasilus*, *Lernaea*, *Caligus* and *Lamperoglena*, the branchiurans *Argulus* and *Dolops* and the isopod *Alitropus typus*, are very common among wild and farmed tilapia. Most of these parasites can cause serious damage in cultured tilapias, and in turn lead to severe loss to tilapia farmers. They can also infect wild tilapia, as has been reported by Doueellou and Erlwanger (1994), who recorded an infection of the tilapias *Sarotherodon macrocephalus* and *Tilapia rendalli* in Lake Kariba (Zimbabwe) by *Lamperoglena monodi* and *Dolops ranarum*.

Parasitic copepods are among the most serious pathogens in fish culture. *Lernaea* spp. are sedentary parasites that have been found to cause varying losses of farmed tilapia in Nigeria, Israel, Malawi, Brazil and Indonesia (Roberts and Sommerville, 1982; Tavares-Dias *et al.*, 2001). Fryer and Iles (1972b) reported that *Lernaea tilapiae* are restricted to tilapia in Lake Malawi. The heads of *Lernaea* are deeply embedded in fish skin and muscles, causing severe inflammatory lesions, interference with feeding and mouthbrooding and reduction in fish growth and rendering the fish unmarketable. An infestation by *Ergasilus* sp. has been reported in polyculture of carp, mullets and tilapia in Israel (Sarig, 1971). However, the infection of tilapia was lower than that of the other two groups. Lin *et al.* (1996) described the developmental stages of *Caligus epidemicus* that infect cultured tilapia in Taiwan.

The fish louse (*Argulus* sp.) is a mobile branchiuran that has been found on tilapias, especially in stagnant water (Sarig, 1971; Tavares-Dias *et al.*, 2001). Experimental infection of Mozambique tilapia with hatched *Argulus americanus* larvae has been successful (Shimura and Asai, 1984). In addition to the damage that *Argulus* sp. can cause to the fish, the open wounds they cause are responsible for facilitating secondary infection by bacteria or fungi (Roberts and Sommerville, 1982).

Alitropus typus is a parasitic isopod that can cause extensive damage to tilapia culture. It feeds by sucking the blood of its host. A severe infestation of Nile tilapia cage culture by *A. typus* in central Thailand has recently been recorded (Chinabut, 2002). This parasite caused 50–100% mortality within 2–7 days after initial infestation, leading to economic loss estimated between US$234 and US$468/cage.

Hoffman and Meyer (1974) listed a number of chemicals that are widely used for controlling crustacean parasites (copepods) in many fish species. Sodium chloride, potassium permanganate, lindane, ammonium chloride, dichlorodiphenyltrichloroethane (DDT), Dipterex and benzene hexachloride were among the most effective drugs in controlling *Argulus* and *Lernaea*. Of course, the dose, method of application and treatment duration depend on fish species and size and drug concentration. However, the farmer should consult the authority in charge before using any of these drugs, because some of them may have been banned due to their carcinogenic effects and severe environmental impacts.

8.5. Bacterial Diseases

Bacteria are found naturally in the intestine of tilapia, and may become infective when environmental and culture conditions become favourable. Al-Harbi and Naim Uddin (2004) isolated and identified 17 bacterial genera from the intestine of hybrid tilapia (*O. niloticus* × *O. aureus*) reared in earthen ponds in Saudi Arabia. Despite the fact that many of these bacteria are pathogenic to tilapia, none of the sampled fish exhibited any bacterial disease symptoms or was dead. Nonetheless, increased attention has been paid to bacterial diseases of tilapia in recent years. Several studies have investigated the effects of bacterial infection of wild and farmed tilapia in different parts of the world. Both Gram-negative and Gram-positive bacteria have been associated with diseases of tilapia. The significance of bacterial infection depends mainly on water quality, organic load and culture intensification. The following groups of bacteria have been reported to infect wild and farmed cichlids.

8.5.1. Motile *Aeromonas* septicaemia (MAS)

The Gram-negative motile bacteria *Aeromonas* (Fig. 8.3) are probably the most common bacterial disease that infects wild and cultured tilapia.

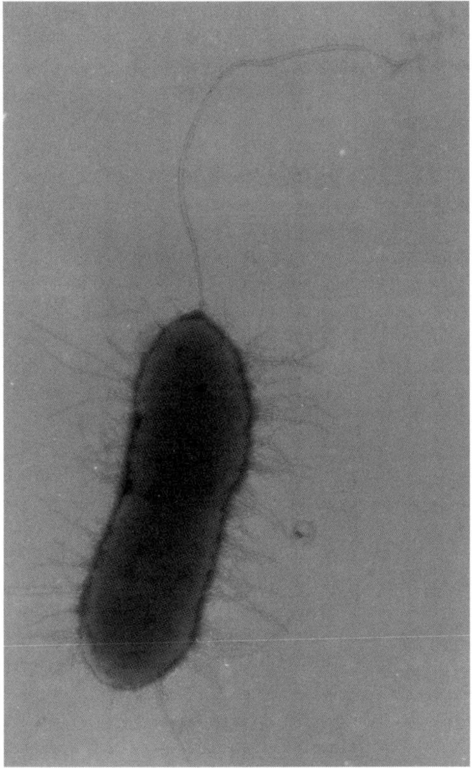

Fig. 8.3. *Aeromonas hydrophila* isolated from Nile tilapia (photo provided by M.K. Soliman).

They cause 'fin rot' or 'skin rot' diseases and may lead to heavy mortality in cultured tilapia (Roberts and Sommerville, 1982). Infected fish usually have a dark colour, lose their appetite and develop ulcers or hyperaemia at the bases of pectoral and pelvic fins (Scott, 1977), ascites and exophthalmia (M.K. Soliman, Alexandria, 2005, personal communication). Internal symptoms include a pale liver and the presence of many focal haemorrhagic necroses in the liver, heart and skeletal muscles and over the visceral and peritoneal surfaces. Massive losses of tilapia due to *Aeromonas* infection have been reported worldwide. Considerable attention has therefore been paid to this disease, for the sake of reducing its potential in tilapia farms.

Outbreaks of caudal fin rot caused by *Aeromonas hydrophila* have caused 80–100% mortality in farmed Nile tilapia in South China (Liu *et al.*, 1993). Infected fish showed slow movements, had rotted caudal fins, swam near the surface and had a poor appetite. Similar infection of Nile tilapia by *A. hydrophila* (biotype 1) was reported in wintering ponds in some districts of China (Wang and Xu, 1985). The major symptoms included the presence of circular or elliptic erosion of the skin ('rotten skin') and the dorsal fin. In early infection stages, the body surface was haemorrhagic. *Aeromonas hydrophila* were also isolated from Nile tilapia during fish disease outbreaks in various aquaculture farms and projects in the Philippines between 1994 and 1996 (Yambot, 1998). The occurrence of the disease was observed in low-volume and high-density aquaculture of Nile tilapia during the rainy season and cold months when water temperature was low. Infected fish suffered from skin lesion, ulceration, fin rot, body discoloration, mouth sore, eye opacity, exophthalmia, dislodged eyeball and sluggishness. High mortality was also observed in cage culture of Nile tilapia during the rainy season and cold months. The bacteria were isolated from the liver, kidney, spleen, gall bladder and opaque eyes of infected fish.

Garcia *et al.* (1999) isolated *A. hydrophila*, *A. schubertii* and *A. sobria* from cultured *O. mossambicus* and hybrid tilapia (*O. mossambicus* × *O. niloticus*) in Venezuela. The bacteria were isolated from the kidney, liver and dermal lesions of infected fish. *Aeromonas salmonicida* also caused mass mortality of wild common carp and *O. mossambicus* in Kalyani reservoir in India (Reddy *et al.*, 1994). The infection of blue tilapia with *A. hydrophila* was also recorded in Mexico (Constantino Casas *et al.*, 1997). Stress and lack of proper culture conditions were the major contributors to the development of the disease.

Experimental infection of Mozambique tilapia held in seawater with *A. hydrophila* via intramuscular injection has been tested (Azad *et al.*, 2001). Experimentally infected fish developed ulceration leading to open wounds, dermal necrosis and liquefying muscular degeneration, with infiltration of cellular and serum factors to the site of injection. The major histopathological manifestations were focal necrosis, loss of submucosa and sloughing of the intestinal microvilli. Similar experimental infection of tilapia by *Aeromonas* was reported by Liu *et al.* (1990) in Taiwan.

A number of studies have been conducted on the resistance of carp to *Aeromonas*, based on non-specific immunity functions, including intra-species and inter-species variations (Cai and Sun, 1994, 1995). Studies on potential heterosis for disease resistance of tilapia are almost lacking, despite the fact that resistance of tilapia to bacterial infection may vary from one tilapia species to another. Only very recently, Cai *et al.* (2004) studied the resistance of Nile tilapia, blue tilapia and their hybrids (male blue tilapia × female Nile tilapia) to *Aeromonas sobria*. Based on the median lethal dose (LD_{50}), a function of non-specific immunity, the authors found that the hybrids had the highest resistance to the disease, while blue tilapia were the least resistant.

Several methods of treatment of and protection from *Aeromonas* disease have been tested, with varying effectiveness. Jirawong (2000) studied the toxicity of copper sulphate ($CuSO_4$) to *A. hydrophila*. He found that a concentration of 0.5–1.0 ppm $CuSO_4$ killed 90–99% of the bacteria within 2–8 h, while 100% were killed within 2 h at 2 ppm. The immune response of tilapia (*Oreochromis* sp.) immunized with formalin-killed *A. hydrophila* by intramuscular injection (IM) and direct immersion (DI) was significantly increased when the fish were fed diets containing ascogen (5 g/kg feed) (Ramadan *et al.*, 1994). The survival of challenged fish after IM and DI vaccination was 89.80% and 40.82% in ascogen-fed groups compared to 75.51% and 10.20% in the vaccinated, non-ascogen-treated groups, respectively.

Protective vaccination against *Aeromonas* in other tilapias has also been successful. Ruangpan *et al.* (1986) found that Nile tilapia vaccinated with formalin-killed bacterial vaccines showed a 53–61% protection within 1 week after vaccination, while 100% of the fish were protected within 2 weeks. Non-specific immune systems of tilapia have been shown to be stimulated by different polysaccharides. Wang and Wang (1997) examined the efficacy of 11 polysaccharides in the protection of blue tilapia and grass carp against bacterial infection by *A. hydrophila* and *Edwardsiella tarda*. They found that four glycans, namely Bar (glycan extracted from barley), Krestin, scleroglucan and Zymosan, significantly increased the survival rates of tilapia (80, 60, 70 and 60%, respectively) and grass carp (60, 70, 90 and 60%, respectively). The non-specific immune response of Nile tilapia clones was also investigated (Sarder *et al.*, 2001). Serum lysozyme activity, phagocytosis and natural resistance to *A. hydrophila* infection by bacterial challenge were compared between fully inbred clones (IC) of Nile tilapia (produced by gynogenesis and sex reversal) and crosses between these lines (outbred clones). The results showed a positive correlation between the level of infection and the non-specific immune parameters measured. Cumulative mortality also showed that the cross between resistant IC and IC susceptible to *A. hydrophila* resulted in progeny with intermediate levels of resistance to those of their parents.

8.5.2. *Pseudomonas*

Pseudomonas are Gram-negative bacteria that have been reported to infect cichlid fishes in different geographical regions. *Pseudomonas fluorescens* were found to cause chronic mortality in farmed Nile tilapia in Japan (Miyashita, 1984). Infected fish were characterized by fine white nodules in the spleen and abscesses in the swim bladder. The infection occurred mainly in winter and spring, with peak mortality at low water temperatures (15–20°C). Miyazaki *et al.* (1984) also found that pond-farmed Nile tilapia infected with *Pseudomonas* in Japan suffered from exophthalmia, dark body coloration, nodular lesions, focal necrosis in the liver, spleen, kidney and gills, inflamed swim bladder, abscesses in the eyes, spleen and swim bladder and granuloma formation. *Pseudomonas* spp. have also been recorded in red tilapia, blue tilapia and their hybrids in brackish-water aquaculture in the Philippines, causing dermal lesions and increased mortality in infected fish (Aban *et al.*, 1999).

8.5.3. Vibriosis

Vibriosis, a bacterial disease caused by the Gram-negative bacteria *Vibrio*, can affect freshwater and saltwater fish. This bacterium could constitute a serious health hazard for cultured tilapia. Several incidences of *Vibrio* infections have been reported in tilapia, with varying levels of damage. The susceptibility of tilapia to vibriosis depends on tilapia species, bacteria species and strains, environmental conditions and culture systems. Sakata and Hattori (1988) isolated and characterized three strains of *Vibrio vulnificus* from infected, pond-cultured Nile tilapia in Kagoshima, Japan. Different *Vibrio* species, including *V. vulnificus*, *V. harveyi* and *V. mimicus*, have also been isolated from diseased red tilapia, blue tilapia and their hybrids in brackish-water floating cages in the Philippines (Aban *et al.*, 1999). Infected fishes showed body lesions and high mortality. This secondary infection was attributed to the great fluctuation of water salinity (18–35‰) at cage sites. In Saudi Arabia, it was reported that the outbreak of vibriosis among tilapia in fish farms in the eastern region was mainly due to stress and external parasite infection (FRRC, 2001). Disease symptoms included exophthalmia, scale loss, haemorrhage and increased mucus secretion. In a similar case, Saeed (1993) reported an outbreak of *Vibrio* sp. in *Oreochromis spilurus* reared in seawater tanks in Kuwait. Infected fish suffered from lethargy, dark coloration, dermal necrosis and mortality. Terramycin, added to the water at a concentration of 45 mg/l for 7 days, successfully controlled the disease and improved fish survival.

The susceptibility of Nile tilapia to a *V. vulnificus* strain with a high degree of virulence for eels has been investigated (Fouz *et al.*, 2002). Infected fish developed a septicaemia similar to eel vibriosis. This result suggested that when tilapia are co-cultured with eels they should be vaccinated against *V. vulnificus* (biotype 2) to prevent the occurrence of the disease and improve fish survival under adverse environmental conditions. In another study, the non-specific immunity of red and black strains of Nile tilapia was compared following challenge with the bacterium *Vibrio parahaemolyticus* (Balfry *et al.*, 1997). The results revealed a significant effect of tilapia strain on immune response (e.g. serum lysozyme and phagocytic activities). Phagocytic activity increased while lymphocyte numbers decreased following the bacterial challenge.

8.5.4. Streptococcosis

Streptococcosis is a disease caused by the Gram-positive, non-motile bacteria *Streptococcus* spp. This disease is a major cause of damage to the global aquaculture industry, with an economic loss exceeding US$150 million annually (Perera *et al.*, 1994, 1997; Shoemaker and Klesius, 1997). These bacteria are opportunistic pathogens that are widely spread in aquaculture environments because of their dependence on stress to assert pathogenicity (Bunch and Bejerano, 1997). They have been isolated from pond water, bottom mud, organic manure used for pond fertilization and contaminated fish (Bunch and Bejerano, 1997). *Streptococcus* gains access to the fish body by ingestion, injured skin and experimental injection (Evans *et al.*, 2000; McNulty *et al.*, 2003).

One other major concern about *Streptococcus* is that it can be transmitted from fish to humans. *Streptococcus iniae* has been reported to infect people who have handled fresh, whole fish (mostly tilapia) from fish farms in Toronto, Canada (Getchell, 1998). The patients, who were of Asian origin, cleaned and ate the fish, and some of them had injured their hands during fish cleaning, when the bacteria may have entered their bodies through the wounds. This particular accident indicates that tilapia infected with *Streptococcus* should be carefully handled and treated.

The infection of tilapia by *Streptococcus* sp. has been widely reported, but *S. iniae* is among the most serious pathogens affecting the tilapia culture industry. Stressful culture conditions, including low or high water temperature, high salinity and alkalinity ($pH > 8$), low dissolved oxygen, high nitrite concentration and high stocking density, increase the susceptibility of tilapia to streptococcal infection (Chang and Plumb, 1996a, b; Bunch and Bejerano, 1997; Perera *et al.*, 1997; Shoemaker *et al.*, 2000).

Streptococcal outbreaks have been recorded in farmed tilapia worldwide. Tung *et al.* (1987) reported epizootics of streptococcal infections in cage-cultured Mozambique tilapia in Taiwan, which resulted in heavy mortalities, reaching 50–60% within 1 month. The major pathological signs of the disease included haemorrhage, exophthalmia, with corneal opacity, and dark

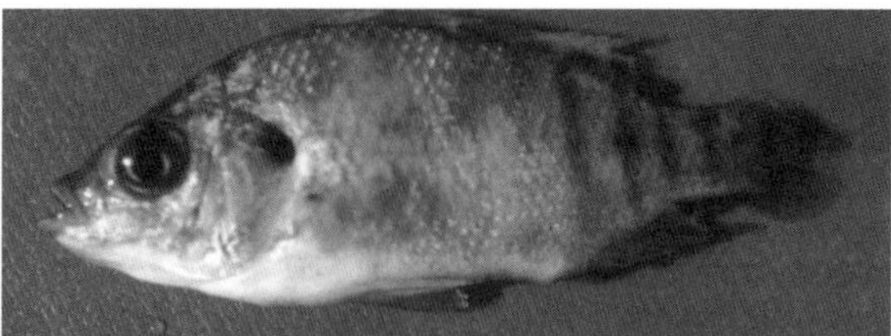

Fig. 8.4. Severe haemorrhage and fin erosion in Nile tilapia caused by streptococcal infection (photo provided by M.K. Soliman).

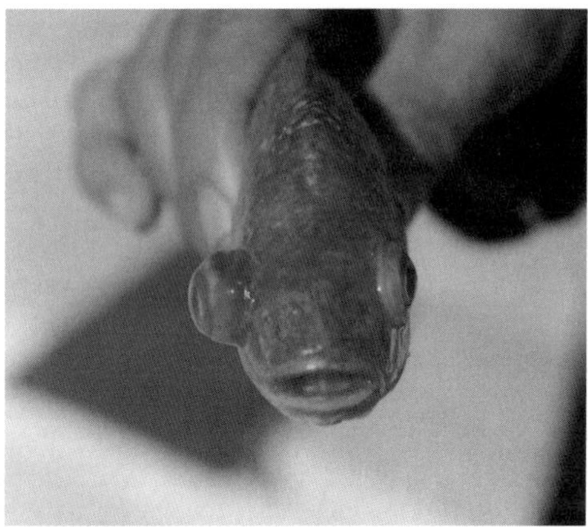

Fig. 8.5. Exophthalmia (pop-eye) in Nile tilapia caused by streptococcal infection (photo provided by M.K. Soliman).

body coloration, with nodular or abscess formation on the trunk and/or peduncle muscles (Figs 8.4 and 8.5). Similar disease symptoms and mass mortality were found in the same species infected with *S. iniae* in brackish-water ponds in India (Mukhi *et al.*, 2001). When healthy fish were injected with the pathogen (10^{5-8} cells/ml), 10% mortality was observed within 7 days of infection. Incidences of infection caused by *Streptococcus* sp. and *Enterococcus* sp. have also been associated with endemic mortality of tilapia hybrids in commercial freshwater farms in Colombia (Pulido *et al.*, 1999). In addition, Berridge *et al.* (1998) reported periodical outbreaks of streptococcosis caused by *S. iniae* and *Streptococcus difficile* in hybrid tilapia (*O. aureus* × *O. niloticus*) in Texas, USA. The authors observed that the rapid and marked mortality of market fish were associated with transferring fish from production units to market tanks.

McNulty *et al.* (2003) studied the haematological changes in Nile tilapia infected with *S. iniae* by naris inoculation. Infected fish had increased pigmentation, eye opacity, erratic swimming and lethargy. The authors suggested that two stressors may have resulted in the infection: the loss of the oxygen carrying capacity and an increase in iron levels.

Several drugs have been tested for the treatment of *Streptococcus*. Antibiotic treatment is generally ineffective, and the need for a proper vaccine has become a must (Klesius *et al.*, 2000). However, Darwish and Griffin (2002) found that oxytetracycline was effective in controlling *S. iniae* in blue tilapia (*O. aureus*). Oxytetracycline was incorporated into the feed at 0, 25, 50, 75 and 100 mg/kg body weight. The 75 and 100 mg dose significantly increased the survival of the infected fish from 7% in the infected non-medicated to 85 and 98%, respectively.

In addition, recent studies have shown the success of passive immunization in Nile tilapia by vaccination with anti-*S. iniae* whole sera (Klesius *et al.*, 2000, 2001; Shelby *et al.*, 2002a). Immunized fish developed significant antibacterial responses and a sharp reduction in abnormal behaviour and morphology. Furthermore, Shelby *et al.* (2002b) found that Nile tilapia intraperitoneally injected with anti-*Streptococcus* whole sera developed a secondary antibody response and immunity to *S. iniae*, with 100% survival after challenge with *S. iniae*.

8.5.5. Staphylococcosis

Staphylococcus epidermidis is another bacterium that has been reported to infect tilapia. Huang *et al.* (1999) described the epizootiology and pathogenicity of *S. epidermidis* in cultured tilapia (*Oreochromis* spp.) in Taiwan. Diseased fish showed splenomegaly with diffusion of several white nodules and lesions in the spleen and anterior kidney. When blue tilapia (*O. aureus*) were challenged with viable

S. epidermidis and its supernatant, apoptosis was predominantly detected in the lymphocytes and macrophages in the spleen and kidney, and occasionally in the brain, liver, gonads, mesentery, stomach, intestine and skeletal muscles (Huang *et al.*, 2000). This particular study indicated that the pathogenicity of *S. epidermidis* for tilapia is due to the toxicity of the bacterial product, which induces the apoptosis.

8.5.6. Mycobacteriosis

Mycobacteriosis or 'fish tuberculosis' is a chronic bacterial disease caused by *Mycobacterium* spp. The disease can infect a wide range of freshwater and marine fish. The infection of wild and cultured tilapia by *Mycobacterium* is well documented. Three pathogenic agents, *M. marinum*, *M. fortuitum* and *M. chelonae*, have been reported to cause the disease (Chen *et al.*, 1998). *Mycobacterium fortuitum* has been reported in the tilapias *Sarotherodon andersonii* and *Tilapia sparrmanii* from the Okavangu swamp in Botswana (Roberts and Matthiessen, 1979). The infection of farmed Nile tilapia in intensive culture in Kenya with *M. fortuitum* (Roberts and Sommerville, 1982) and the infection of tilapia hybrids (*O. niloticus* × *O. mossambicus* × *O. aureus*) with *M. marinum* (Wolf and Smith, 1999) was also reported. Affected fish showed focal granulomata in the liver, spleen, kidney and viscera, a high proportion of epithelial macrophages and more peripheral lymphocytes.

The effects of the extracellular products (ECP) of *Mycobacterium* spp. on the non-specific immune response of Nile tilapia have been examined by Chen *et al.* (1998). The fish were immunized by injecting their swim bladders with ECP, Freund's complete adjuvant (FCA) and Freund's incomplete adjuvant (FIA). The results revealed that ECP from *Mycobacterium* spp. and the adjuvants used in the study provided a good stimulation of the non-specific immune response in Nile tilapia.

8.5.7. Edwardsiellosis

Edwardsiellosis is a bacterial disease caused by *Edwardsiella tarda*, which are short-rod, Gram-negative bacteria belonging to the family Enterobacteriaceae. *Edwardsiella tarda* have been isolated from reptiles, fishes, amphibians and freshwater and integrated aquaculture (Muratori *et al.*, 2000). The pathogenicity of *E. tarda* to fish increases under stressful conditions, especially at a high water temperature (> 30°C) and high organic contents (even at a low water temperature).

The infection of tilapia by *E. tarda* has been reported by a number of authors. Miyashita (1984) recorded chronic mortalities in Nile tilapia in some farms in Japan due to infection by *E. tarda* and *Pseudomonas*. Muratori *et al.* (2000) also isolated *E. tarda* from the skin, gills, fins, intestines and muscles of Nile tilapia reared in an integrated fish farm fertilized with pig manure. External surfaces (skin, gills and fins) were more affected than other organs. Pathological signs of the disease include haemorrhage, corneal opacity (Fig. 8.6) and chronic mortality.

Fig. 8.6. Corneal opacity of Nile tilapia caused by *Edwardsiella* infection (photo provided by M.K. Soliman).

The resistance of tilapia to *E. tarda* infection has also been investigated, with contradictory results. When Nile tilapia were intraperitoneally injected with a protein-bound polysaccharide preparation (PS-K), at a dose of 0.1 mg/g body weight, maximal resistance was developed in the fish 1 week after the injection (Park and Jeong, 1996). PS-K-injected fish showed increased phagocytic activity of the pronephros cells. These results suggested that PS-K activated the non-specific immune system of injected tilapia. Similarly, Wang and Wang (1997) evaluated the effectiveness of 11 polysaccharides in the protection of blue tilapia against *E. tarda*. They found that four glycans (Bar, Krestin, Scleroglucan and Zymosan) improved fish resistance and survival rates. In conflict with the above results, Lio-Po and Wakabayashi (1986) found that Nile tilapia immunized with formalin-killed *E. tarda* through intraperitoneal injection were not effectively protected from infection by *E. tarda*.

8.5.8. Columnaris disease

Columnaris disease or 'gill rot' is caused by the aquatic myxobacterium *Flexobacter columnaris*, which can affect cold-water and warm-water fish worldwide. *Flexobacter columnaris* is the most common myxobacterial disease in tilapia culture, and is usually associated with stressful culture conditions, especially high or low water temperature and high ammonia concentration (Roberts and Sommerville, 1982; Amin *et al.*, 1988). *Flexobacter columnaris* has been isolated from Nile tilapia (Amin *et al.*, 1988) and *Tilapia* spp. (Chun and Sohn, 1985). The disease symptoms included respiratory disorder, body discoloration, heavy mucus secretion, fin and gill erosion and lesions in the skin and muscles.

Improving the quality of culture water is the best way of protecting the fish from columnaris disease. Meanwhile, *F. columnaris* is susceptible to antibiotics such as oxytetracycline, chloramphenicol, erythromycin (Amin *et al.*, 1988), kanamycin, tetracycline and amikacine, but resistant to sulpha drugs (Chun and Sohn, 1985).

8.5.9. Other bacterial diseases

Incidences of bacterial infection by other species of bacteria have been reported in tilapia. Paperna *et al.* (1996) found that overwintered tilapia hybrids (*O. aureus* × *O. niloticus*) had focal white lesions in the liver, filled with amoeboid organisms inside and outside the host cells. The lesions were granulomata comprised of degenerating epithelial cells surrounding an expanding necrotic core. The susceptibility of Mozambique tilapia to the toxins (types A–E) produced by *Clostridium botulinum* (an anaerobic, spore-forming, Gram-positive bacillus) was also investigated by Lalitha and Gopakumar (2001). All five toxins were found to be toxic to the fish. The authors emphasized that proper hygiene should be maintained in fish farms to minimize bacterial contamination.

8.6. Fungal Diseases

Fungal diseases are considered 'secondary diseases' because they are generally opportunistic, taking advantage of necrotic tissues associated with injuries, bacterial or parasitic lesions, dead and decaying eggs and poor culture conditions (Pillay, 1990). Several fungal species have been isolated from wild and farmed tilapia and organic fertilizers used for pond fertilization. El-Sharouny and Badran (1995) recovered 17 fungal species from Nile tilapia and *S. galilaeus* in Egypt. Similarly, Okaeme and Olufemi (1997) isolated 18 fungal species in tilapia culture ponds in Nigeria, eight of them having originated from the organic manure used for pond fertilization. Although many of these fungi are infectious to tilapia, the following fungal diseases are the most important and also the most documented.

8.6.1. Saprolegniasis

Saprolegniasis is a disease caused by *Saprolegnia* and *Achlya* spp. and appears as white cotton-like patches on the necrotic tissues of infected fish, mainly skin, gills, fins, eyes and mouth. The clinical signs of saprolegniasis include haemorrhagic ulceration, erosion of skin, fins, gills and muscles, systemic mycosis of the liver, spleen, eyes and kidney and massive mortality (Okaeme and Olufemi, 1997). Low water temperature is a main reason for the spread of this disease. Lightner *et al.* (1988) reported 100% mortality of *O. mossambicus*, *O. aureus* and tilapia hybrids (*O. mossambicus* × *Oreochromis urolepis hornorum*) caused by *Saprolegnia* due to low water temperature (< 15°C). The tilapia hybrids were the most susceptible to the fungal infection, followed by *O. mossambicus*, while *O. aureus* was the most resistant because of its higher cold tolerance. Severe infection caused by *Saprolegnia* was also reported in Nile tilapia eggs and crowded fish in Kenya (Arrumm, 1987) and *T. zillii* in freshwater ponds in Nigeria (Ogbonna and Alabi, 1991). Similarly, El-Sharouny and Badran (1995) found that Nile tilapia and *S. galilaeus* infected with *Saprolegnia ferax* and *Saprolegnia parasitica* suffered from high mortality, but *S. galilaeus* was more susceptible to fungal infection than Nile tilapia.

8.6.2. *Branchiomyces*

Branchiomyces spp. are generally known as 'bad-management disease' since they flourish in poor-quality water with high levels of organic matter. As expected, water reuse systems (closed systems) are more susceptible to this infection than other culture systems. The fungus invades the gill via branchial vessels or the epithelium and can cause massive destruction of respiratory surfaces. *Branchiomyces*-like fungus spp. have been isolated from red tilapia (*O. niloticus* × *O. mossambicus*) and green tilapia (*O. niloticus* × *O. aureus*) in intensive farming in Israel (Paperna and Smirnova, 1997). Severe infection resulted in the loss of 85% of the fish.

8.6.3. *Aspergillus*

Aspergillus flavus produces aflatoxins that are toxic to fish. Poor storage of fish feeds leads to the growth of this mould. Farmed tilapia fed with feed contaminated with aflatoxins B_1 and B_2 showed severe haemorrhage in the branchial muscles below the dorsal commissure of the operculum, intestinal haemorrhage, haemopoiesis and massive accumulation of haemosiderin in splenic and renal melanomacrophage centres (Roberts and Sommerville, 1982). The genotoxic potential of the spores isolated from *Aspergillus niger*, *Aspergillus ochraceus* and *Helminthosporium oryzae* in intraperitoneally injected tilapia (*O. mossambicus*) was found to be positive for the frequency of chromosome aberration in gill epithelia and primary spermatocyte metaphase (Manna and Sadhukhan, 1991).

Different chemicals, such as malachite green, potassium permanganate, quicklime, copper sulphate, formalin, mercurochrome and sodium chloride, have been effective in treating fungal disease. However, improving water quality and culture conditions and developing good management practices are the key factors in preventing fungal diseases.

8.7. Viral Diseases

Despite the damage that viral diseases may cause to infected fishes, information available in the literature on their effects on cichlid fishes in general, and tilapia in particular, is very limited. Only a few studies have considered the viral diseases of cichlid fishes. The limitation of viral infection in tilapia may be attributed to the fact that tilapia are reared in warm-water environments, which are not favourable for viral multiplication (M.K. Soliman, Alexandria, 2005, personal communication).

Iridovirus (Bohle iridovirus, BIV), which causes lymphocystic diseases, was first reported and described in wild tilapia from East African lakes in the early 1970s (Paperna, 1974). This virus has been suggested to have caused 100% mortality of *O. mossambicus* fry over a period of 60 days (Ariel and Owens, 1997). Infected fish showed rapid corkscrew-like swimming patterns ('spinning'), a symptom that renders them more susceptible to cannibalism. An outbreak of viral diseases, including an iridovirus-like agent, has also been reported in cultured Nile tilapia in Canada (McGrogan *et al.*, 1998). Infected fish were characterized by a dark colour, lethargy, slow swimming or resting on the bottom, exophthalmia, marked pallor of the gills and visceral organs, particularly the liver, erythema of the submandibular skin, severe abdominal ascites and haemorrhages in the liver. In addition to these symptoms, Smith *et al.* (1998) found that Nile tilapia fingerlings infected with this virus exhibited necrosis and inflammation of several organs, including the choroid of the eye, heart, intestines, kidney, liver and spleen. The most severe change was necrosis of the haematopoietic tissue of the spleen, kidney and heart. Similarly, iridoviral disease caused gill pallor, degeneration of tissue cells, the presence of intranuclear inclusion bodies and progressive weakness of the chromide cichlid *Etroplus maculatus*, leading to the death of the fish (Armstrong and Ferguson, 1989).

Rhabdovirus is a viral agent of cichlid fishes that has been isolated and described from the cichlids *Cichlasoma cyanogutatum*, *Cichlasoma nigrofaciatum* and *T. zillii* (Lautenslager, 1986). This agent causes fatal diseases of these hosts. Skliris and Richards (1999) studied the susceptibility of Mozambique tilapia to another viral agent, nodavirus, isolated from naturally infected European sea bass, *Dicentrarchus labrax*. The authors suggested that nodavirus might be transmitted to seawater-reared tilapia. Another virus, birnavirus, was described and isolated from Mozambique tilapia in Taiwan in the early 1980s (Hedrick *et al.*, 1983). The virus was widespread and may have led to secondary effects on the fish, such as increasing their susceptibility to other infectious agents. In addition, Avtalion and Shlapobersky (1994) reported that the larvae of blue tilapia, Nile tilapia and Galilee tilapia were highly susceptible to 'unidentified' whirling viral disease. Disease symptoms included whirling, followed by darkening of the skin and anorexia.

There are no reported effective treatments for viral diseases of tilapia except the removal and destruction of infected fish. The best protective treatment is the quarantine of larval fish and the avoidance of introducing fish with viral disease histories into culture facilities (Shoemaker *et al.*, 2000).

8.8. Non-infectious Diseases and Disorders

Tilapia, like other farmed aquatic animals, may be subjected to unfavourable conditions, leading to health problems and non-infectious diseases. Unfortunately, the information that is available on non-infectious disorders of tilapia is very limited. Only nutritional disorders, gas bubble disease and the disorders caused by pollutants are reasonably well documented. The following sections briefly discuss these disorders.

8.8.1. Gas bubble disease

Excessive aeration in intensive aquaculture systems may result in supersaturating culture water with dissolved gases. When these gases are absorbed by the fish during respiration, they cause a change in partial pressure, leading to gas accumulation (gas bubbles) at different sites of the fish body. In tilapia fry, gas bubbles are concentrated in the yolk sac, while in older fish they are found in the gills and under the skin (Roberts and Sommerville, 1982).

Outbreaks of gas bubble disease have been reported in farmed saltwater juvenile tilapia, *O. spilurus* (52.5 g), in Saudi Arabia (Saeed and Al-Thobaiti, 1997). About 50% of fish populations were affected, resulting in 30% mortality. Another outbreak occurred in adult Nile tilapia (270 g) in brackish water, affecting about 40% of the fish population, with 40% mortality. Affected Nile tilapia were heavily infected with monogenetic trematodes. In all cases, the total gas was higher than 111% saturation, nitrogen was supersaturated and oxygen undersaturated. The disease symptoms were overcome by reducing the gas pressure through splashing the water and using water with lower gas pressure.

8.8.2. Nutritional diseases

The deficiency of certain nutrients in the diets of farmed fish may lead to disease symptoms and health problems. The effects of dietary vitamin deficiency on farmed tilapia are well documented (see Chapter 6 for more details). Blue tilapia fed pantothenic acid-deficient diets exhibited poor growth, haemorrhage, sluggishness, anaemia, hyperplasia in epithelial cells of the gill lamellae and high mortality (Soliman and Wilson, 1992a). Anorexia, low growth, fin erosion, loss of body colour, cataract and mortality were also observed in fish fed with riboflavin-deficient diets (Soliman and Wilson, 1992b). In addition, low vitamin E contents in the diets resulted in poor growth, low feed conversion ratio (FCR), skin haemorrhage, muscle dystrophy, impaired erythropoiesis and abnormal skin coloration (Roem *et al.*, 1990). Similar deficiency symptoms have been reported in tilapia hybrids fed test diets deficient in niacin, pyridoxine and thiamine (Lim and Leamaster, 1991; Shiau and Suen, 1992; Shiau and Hsieh, 1997). In the meantime, Lim and Klesius (2001) reported low growth and reduced red blood cells and total cell counts in Nile tilapia fed diets deficient in folic acid.

On the other hand, the addition of certain feed elements to tilapia feeds may play a significant protective role in the fish, especially when they are cultured at a high stocking density and under deteriorated water quality. Schlechtriem *et al.* (2004) studied the effects of dietary L-carnitine on tilapia hybrids (*O. niloticus* × *O. aureus*) reared under intensive pond-culture conditions. The fish were fed diets containing either 150 or 450 ppm of L-carnitine. Histological examinations revealed that fish fed 150 ppm carnitine showed the lowest permeability to fluorescein in gills, gut and skin epithelia, the highest activity in the system of active transport for organic anions and the highest levels of the multixenobiotic resistance transporter (MXRtr) for lipophilic/amphiphilic xenobiotics, compared to the control and the 450-ppm treated fish. The MXRtr activity in the liver bile canaliculi and renal proximal tubules of this group of fish was much higher than that of the other groups except those from the indoor tank. The intralysosomal accumulation of neutral red in their head–kidney macrophages was also significantly higher. The authors suggested that even a low level of L-carnitine enrichment can provide considerable protective effects in fish reared under intensive pond-culture conditions.

8.8.3. Disorders caused by pollutants

Aquatic environments receive a wide variety of agricultural runoffs, including pesticides, herbicides, fungicides, fertilizer residues, heavy metals, etc. Considerable amounts of various industrial wastes, human wastes, molluscicides and oil compounds are also discharged into aquatic environments. These organic and inorganic compounds are very likely to cause environmental impacts such as bioaccumulation, eutrophication, environmental degradation and imbalance of aquatic biodiversity. Farmed and wild fish, including tilapia, are therefore expected to be adversely affected by these pollutants. Such impacts depend on fish species and size, type and concentration of pollutant, duration of exposure and environmental conditions. However, little attention has been paid to the impacts of environmental pollution on farmed tilapia. Despite the fact that hundreds of chemical compounds are likely to be hazardous to these fish, even at low concentrations, only a few compounds have been considered.

Phenolic compounds are among the most serious organic pollutants in the aquatic environments. These chemicals are highly toxic to aquatic animals and can cause severe ecological and economic losses. For example, Hart *et al.* (1998) found that Nile tilapia subjected to the carcinogenic polycyclic aromatic hydrocarbon (PAH) 7,12-dimethylbenzanthracene (DMBA) suffered from reduced spleen, pronephros and total white blood cell counts. The fish also exhibited reduced swimming activity and feeding activity and increased skin pigmentation and mortality. In another study, Mehrim (2001) evaluated the maximum tolerable level of phenol for fingerling Nile tilapia, and the effects of chronic exposure on growth performance and feed utilization efficiency. He found that 30 ppm was the maximum tolerable level, and beyond that level fish exhibited a respiratory manifestation and hyper-irritability, followed by lethargy, lesions, increased mucus secretion, skin darkness, fin erosion, gill and liver congestion and distension of the gall bladder. Fish exposed to the maximum tolerable level (30 ppm) had significantly reduced growth, survival and feed utilization efficiency. The addition of Biogen as a feed supplement to the diets has significantly overcome the symptoms mentioned.

Bayluscide® is another phenolic compound that is widely used as a molluscicide. It is one of the chloronitrophenol derivatives (niclosamide ethanolamine salt) (5,2-dichloro-4-nitro-salicylic-anilide). This compound is frequently used for eradication of the intermediate host snails of schistosomiasis (bilharzia) and fascioliasis in Egypt. It is very toxic to snails, but it can be toxic to fish as well. Acute toxicity of Nile tilapia intoxicated with Bayluscide included erratic and nervous swimming, continuous opening of mouth and gill cover, haemorrhage under the scales and at the base of the fins and degenerative and necrotic changes in the liver, kidney, spleen, heart and gills (Nafady *et al.*, 1986; Marzouk and Bakeer, 1991; Khalil, 1998). Khalil (1998) found that 0.3 mg/l was lethal to Nile tilapia, while 0.1 mg/l was sublethal. Fish exposed to 0.15 mg/l (50% of the median lethal concentration (LC_{50})) suffered from nervous and respiratory manifestations, corneal opacity, a decrease in red and white blood cell counts, haemoglobin concentration and phagocytic activity, and a high accumulation of Bayluscide in the gills, liver and muscles.

Several other organophosphorus compounds, such as Metrifonate, Fenthion, Diazinon,

Foschlor, Endosulfan, Malathion, Nuvalron and Dichlobenil, are widely used as insecticides, molluscicides, herbicides and therapeutic agents for many diseases. As expected, these compounds can have a wide range of impacts on cultured and wild tilapias. Rajavarthini and Michael (1996) found that exposure to Nuvalron caused dose-dependent suppression of antibody response in Mozambique tilapia. *Tilapia rendalli* exposed to Endosulfan suffered from hepatic lesion, encephalitis, meningitis, oedema and inflammatory infiltrate of eosinophilic granular cells in the brain (Makhiessen and Roberts, 1982). When Nile tilapia were exposed to Diazinon, they exhibited lethargic behaviour, loss of coordination, dark red gills, a pale liver, a distended gall bladder with dilated bile ducts, dilated arterioles and lamellar capillaries (El-Kateib and Afifi, 1993). Similarly, Nile tilapia exposed to a high level of Metrifonate showed a respiratory manifestation, loss of pigmentation, dark body coloration, protrusion of the anal opening and haemorrhage (El-Gohary, 2004). Chronic symptoms included darkness of body colour, haemorrhage, decrease in phagocytic activity and necrotic and degenerative changes in the kidney, liver, spleen, heart and brain.

The above results clearly indicate that water pollution can have harmful impacts on tilapia. The use of the different chemicals and drugs for controlling pests, insects, snails, herbs, etc. should be carefully regulated and managed. In addition, culture water should be analysed prior to, and continuously monitored during, culture practices to investigate whether it contains any traces of pollutants and whether the levels of these pollutants are hazardous to farmed tilapia. In the late 1990s, a sudden, tragic, mass mortality occurred in Nile tilapia, cage-cultured in a waterway close to El-Salam Canal in Lake Manzala, Egypt, within the 'Cages- for-Graduates Project', which was funded by the Social Development Fund (A.-F.M. El-Sayed, Mansoura, 1999, personal observation). This massive mortality was attributed to increased levels of pesticides in the canal from land runoffs. In fact, the owners of these cages faced a serious problem because they had to pay back the loans they had taken from the Social Development Fund. Only governmental interference saved them from prison. This particular example stresses the necessity of regular monitoring of culture water, especially in areas where the water may receive pollutants from various sources.

8.9. Closing Remarks

1. Stressful conditions adversely affect tilapia and make them more susceptible to different diseases. An essential cause of chronic stress in tilapia is social interactions and hierarchies. Reducing stress is an important tool for disease control and improvement of immunity in cultured tilapia.

2. Tilapia have been reported to be subjected to parasitic, viral, fungal and bacterial diseases, in addition to a number of non-infectious diseases.

3. Most ciliate parasites can be treated with table salt and formalin (30–50 mg/l).

4. *Gyrodactylus* infecting tilapia was successfully treated with a single dose of formalin (250 ppm for 35–40 min) or two repeated applications of potassium permanganate at a dose of 5 ppm in static water with a 2-day interval.

5. Black-spot disease, which is caused by *Neascus* metacercariae, makes tilapia unmarketable if the infection is heavy.

6. The digenetic trematodes *Haplorchis* spp, which are found in freshwater fishes, including tilapia, represent a risk to human health if the affected fish are not cooked adequately.

7. Little information is available on the parasitic effects of cestodes and acanthocephalans on tilapia, despite the fact that they have been reported in wild and farmed tilapia. More research is needed along this line.

8. Parasitic crustaceans, including the copepods *Eragasillus*, *Lernaea*, *Caligus* and *Lamperoglena*, the branchiurans *Argulus* and *Dolops* and the isopod *Alitropus typus*, are among the most serious pathogens in tilapia culture.

9. The significance of bacterial infection of tilapia depends on water quality, organic load and culture intensity. Most infectious bacteria can cause massive losses of farmed tilapia. Several methods of treatment, including copper sulphate ($CuSO_4$), vaccination against the bacteria with formalin-killed bacterial vaccines, antibiotics and passive immunization by vaccination with antibacterial whole sera, have been used, with varying levels of effectiveness.

10. Despite the damage that viral diseases may cause to infected tilapia, very limited information is available on their effects. There are no effective treatments for these diseases except the removal and destruction of the infected fish. The quarantine of larval fish and the avoidance of introducing fish with viral disease histories into culture facilities are the most effective protective treatment.

11. Organic and inorganic compounds discharged into aquatic environments are very likely to have adverse environmental impacts on farmed and wild tilapia. Controlling the use of these compounds, together with regular monitoring of culture water, is necessary for controlling such impacts.

9

Harvesting, Processing and Economics

9.1. Introduction

As indicated in Chapter 1 of this book, the production of farmed tilapia has been rapidly expanding in the past few years. Most of this production is used directly for local markets, mainly as a cheap, good-quality, animal protein source in rural areas. Because tilapia were considered to be a low-value product, little attention has been paid to their harvesting and handling for many years. Tilapia have also not been identified as a trade item according to the International Standard Statistical Classification (ISSC). As a result, information on tilapia exports and imports has been very limited. However, recently, the international trade in tilapia and tilapia products started to gain increased global attention. Thus, it is essential to shed some light on the factors that may affect tilapia production, costs, trade and overall economics, with emphasis on harvesting, handling, processing and marketing, in order to adopt the best cost-effective production strategies.

9.2. Harvesting

Tilapia culture is dominated by small-scale rural farms, especially in developing countries, and most of the production is consumed in local areas. There is no specific harvesting method for farmed tilapia. The information available on tilapia harvesting methods is also very limited. Moreover, polyculture with other aquatic species makes it more difficult to adopt specific harvesting techniques for tilapia. For these reasons, small-scale farmers in different regions may adopt different, but simple, harvesting techniques.

Harvesting methods depend on pond size, culture systems and levels of technology applied. For example, small-scale tilapia farmers in many parts of Africa and Asia cannot easily get harvesting nets and other equipment required for complete harvesting of their ponds. To overcome this problem, many farmers adopt partial harvesting techniques, using locally available gear (Fig. 9.1) (Brummett, 2002a). Large-scale tilapia producers adopt more advanced harvesting tools, such as winches, because they generally prefer batch harvesting (Fig. 9.2). Partial harvesting is usually designed to remove large fish and provide smaller fish with more space for growth (Hepher and Pruginin, 1981). Yet many questions regarding this system remain unanswered: (i) Is this system preferable to batch harvesting methods? (ii) At what size and harvesting intervals should the fish be harvested? (iii) How much fish (as a per cent of total density) should be harvested each time? (iv) Should harvested fish be replaced with new fish? (v) What is the best fish size that will replace the harvested portion? (vi) What are the best harvesting methods for this system? To the extent that these questions are not answered, the system may not lead to the outputs that the farmers hope to get and may also fail to increase overall yield (Knud-Hansen and Lin, 1996). Therefore, finding answers to these questions is necessary in order to evaluate and improve partial harvesting methods.

The limited information available may suggest the use of partial harvest as a more cost-effective

Fig. 9.1. Manual harvesting of tilapia from earthen ponds in the Philippines.

Fig. 9.2. Mechanical harvesting of tilapia from floating cages at Lake Harvest Farm, Lake Kariba, Zimbabwe (photo provided by R. Brummett).

harvesting method than using single harvest techniques. Nitithamyong *et al.* (1990a) compared partial harvest and batch harvest of Nile tilapia reared in monoculture systems or in integration with pigs. Marketable-size fish were harvested monthly (multiple harvest) from one monoculture system and from the integrated system, while another monoculture system was harvested only once at the end of the study. Total fish production from the monoculture and integrated systems using the multiple harvesting techniques was higher than that from the monoculture system with a single harvest. However, despite higher yields, the integrated system with multiple harvest provided a significantly lower return than the monoculture system, probably because of the higher labour cost. This result indicated that the supposed benefits to be gained from employing partial harvesting may be misleading, and therefore it should be evaluated biologically and economically.

McGinty and Alston (1993) reported that the annual yield of all-male tilapia reared in polyculture with juvenile freshwater prawns (*Macrobrachium rosenbergii*) and partially harvested monthly was similar to that from a batch harvest system. The authors suggested that partial harvesting offers good potential for small farmers who need to supply markets on a regular basis. This system also facilitates meeting the demands for fish fry since all the fish for stocking are not required at the same time. Again, evaluating tilapia harvesting methods must consider technical, biological, economic and socio-economic factors.

In many rural areas, fish harvesting methods are one of the major bottlenecks in the development of rural aquaculture. In Malawi, for example, most fish farmers have been completely dependent on the government because of borrowing harvesting nets from the Fisheries Department (Kaunda and Costa-Pierce, 1993). In this context, cheap and locally available harvesting tools are an important prerequisite for the success of rural aquaculture. In a recent study, Brummett (2002a) tackled this issue and evaluated the effectiveness of a variety of local fishing gear used in partial harvesting methods in small tilapia (*Oreochromis shiranus*) ponds in Malawi. The fish (13.7 g) were stocked in 200 m^2 ponds (20 m × 10 m × 0.6 m = 120 m^3) at a density of 2.5 fish/m^3 and fed local feed sources (maize, bran and vegetable leaves). After 122 days, the ponds were assigned to: (i) weekly seining with a reed fence; (ii) weekly fishing for 4 h with hook and line by a single fisherman; (iii) weekly fishing for 9 h with two basket traps; and (iv) no fishing (control). The partial harvesting methods continued for 20 weeks. The author found that hook-and-line fishing and fishing with a reed fence partially harvested significantly greater weights of fish than trapping. Partial removal of tilapia juveniles had the highest impact on increased yield. The partial removal of tilapia juveniles (not fry or adults) is probably the main limiting factor in increasing productivity of fish ponds.

9.3. Handling

Tilapia is generally marketed in rural areas and local markets in developing countries, either fresh or iced, with little handling and processing. However, the global tilapia market is growing sharply, not only in producing regions (mainly South-east Asia, the Far East and Africa), but also in many non-producing regions, such as the USA, Europe and Australia. More attention should therefore be paid to tilapia handling and processing in order to maintain quality and meet the quality standards required by external markets. Poor handling and holding conditions, inadequate processing and the use of inappropriate processing methods can seriously affect the quality of tilapia and increase postharvest losses. For example, Enujiugha and Nwanna (1998) reported that more than 20% of Nile tilapia and African catfish yields were lost in Nigeria as a result of inadequate handling and processing.

Live tilapia have become an important product in the market in many parts of the world where consumers prefer to buy live rather than iced or frozen fish (Singh and Daud, 2001). It is becoming common to find live tilapia in display tanks and aquaria in seafood restaurants and supermarkets in countries like Singapore, Thailand and Malaysia (Singh and Daud, 2001). However, live food systems require effective support systems, including live holding containers, specially equipped trucks, live holding centres and other infrastructure components. Consumers prefer to pay higher prices in order to get live fish. In Malaysia, for example, the price of live red tilapia is 37–40% higher than that of chilled tilapia (Singh and Daud, 2001). The price of live tilapia in Egypt, particularly in large cities such as Cairo and Alexandria, is 50–75% higher than that of iced fish of similar sizes (A.-F.M. El-Sayed, Alexandria, 2004, personal observation).

Singh and Daud (2001) described the different steps and processes involved in the handling and marketing of tilapia in Malaysia, including transport, distribution, display and holding facilities. A modern technology has been developed by the Malaysian Agricultural Research and Development Institute (MARDI) for the transport of live fish such as tilapia in a minimum quantity of water (Wan Johari *et al.*, 1999). The system involves transferring the fish into tanks at a temperature of 21°C for about 10 h to reduce stress. The water in the tanks is gradually cooled to 11°C, with the addition of ice, to reduce fish activity. The fish are then packed in plastic bags with water and oxygen at a fish : water ratio of 1 : 1. The bags are tied and placed in polystyrene boxes, which are sealed and transferred to refrigerated trucks maintained at 15°C and transported to their final destination. On arrival, the fish are stocked in holding facilities and left to regain their normal physiological functions as the temperature gradually rises. By using this technique, live tilapia can be transported for a journey of 12 h without any mortality on arrival (Singh and Daud, 2001).

9.4. Processing

Fish processing is becoming increasingly important, due to the increasing demand for processed fish in global markets. Traditionally, processed freshwater fishes, supplied in various simple forms, have increased consumer choice. Normally, small processors sell their products directly to consumers. In many provinces of Thailand, for example, processed fish, mostly in dried and salted form, are sold along highways to serve domestic travellers. On the other hand, large fish processors usually sell their products to the wholesale fish markets in Bangkok and to retailers as well. Sometimes, processed fish, such as fish balls and minced fish, are resold in supermarkets.

Marketing of processed tilapia is also gaining high popularity in some South-east Asian countries. Indonesia and Thailand, for instance, export filleted tilapia to the US market (Guerrero and Guerrero, 2001). Other products of processed tilapia, such as dried, smoked and canned tilapia, fish balls, sausages, surimi, fish flour, fish burgers, fish fingers, fish noodles, sashimi and various breaded, battered and marinated products are now well accepted by Asian consumers (Irianto and Irianto, 1997). This simply means that tilapia can be an ideal raw material for producing value-added products. This approach is likely to create more jobs, generate more income and improve the standard of living, especially in rural areas.

Typically, the processing of tilapia includes one or more of the following techniques: (i) cooling (icing, refrigeration, freezing, etc.); (ii) filleting; (iii) radiation; and (iv) industrial processing (drying, smoking, canning, fish balls and sausages, etc.). A number of precautions must be taken into consideration during processing in order to ensure a good-quality product. These are:

1. Assurance of the total removal of off flavours in the flesh either during the culture of the fish or during processing.
2. Proper purging and bleeding.
3. Careful heading to avoid damage to the fish and subsequent contamination by internal organs such as the gall bladder.
4. Fillets must not contain excess fat.

9.4.1. Cooling

Icing, chilling and refrigeration are the most common methods used for the storage of tilapia. However, cool storage is usually done improperly and without considering scientific, quality and hygienic standards. This generally leads to reducing the quality and shelf life of the fish and increasing health risks. It is therefore essential that the proper temperature and processing methods are adopted for the storage of tilapia.

The effect of storage temperature on the quality and shelf life of tilapia (*Tilapia* spp.) in Sri Lanka has been evaluated by Rajaratnam and Balasundaram (2001). Gut-retaining and gutted fish and shrimp (*Penaeus monodon*) were stored at 0, 4 and 28°C (room temperature). The authors found that freezing significantly reduced the microbial populations, while the initial bacterial population increased by 300% after 72 h of storage at room temperature. Spoilage rate was also more rapid in gut-retaining than in gutted fish and headed shrimp. It was concluded that shelf life and quality acceptance could be extended up to 25 days when the products are stored at 0 and 4°C. It was also suggested that tilapia and shrimp should be gutted and headed before refrigeration. Similarly, Irianto and Irianto (1997) reported that icing is an effective method for extending shelf life of tilapia and improving their quality.

9.4.2. Filleting

Among commercial aquaculture species, tilapia is the most difficult to process and has a relatively low fillet yield (33%) compared to salmon (> 50%), catfish (> 38%), striped bass (> 40%), sea bream, sea bass and others (Snir, 2001). The main product of tilapia processing is the fillet (Fig. 9.3). Filleting yield depends on machinery efficiency, operator experience, fish shape and filleting method. Souza and Macedo-Viegas (2000) evaluated the effects of four different filleting methods on the processing yield of Nile tilapia in Brazil. The filleting methods were: (i) skinning and filleting the whole fish; (ii) filleting the whole fish and then removing the skin; (iii) skinning and filleting headless fish; and (iv) filleting headless fish and then removing the skin. The best fillets and total eatable yield (36.59 and 42.15%, respectively) were obtained with skinning the whole fish before filleting, followed by

Fig. 9.3. Tilapia processing plant in Brazil (photo provided by K. Fitzsimmons).

filleting and skinning headless fish. Skinning and filleting the whole fish also produced thicker and longer fillets compared to other methods. The type of head cut used during processing may also affect the dressout and yield of tilapia, as reported by Souza *et al.* (2000). These authors found that contour and oblique head cuts produced better dressout and yield than straight head cuts.

9.4.3. Rigor mortis

It is important to throw some light on the rigor mortis phenomenon in order to ensure that the quality of the product is retained during production, processing and storage. Rigor mortis is the rigidity or stiffness of the fish body that occurs after death. The onset of rigor mortis may vary from about 10 min to several hours after death. This natural phenomenon may therefore influence the appearance and structure of fish muscles and render it unsuitable for filleting. Jarding *et al.* (2000) examined the occurrence of rigor mortis in farmed Nile tilapia at a commercial processor in Zimbabwe. Live fish were killed by cranial fracture or bleeding for 10 min after severance of the gills and/or the ventral aorta, and rigor mortis was then measured. The fish were left at an ambient temperature (20 ± 3°C) or iced at 2–5°C. The occurrence of rigor mortis took place within 1–1.5 h and was faster for fish stored on ice than for those stored at ambient temperature, while the effects of killing methods were not different. It has also been reported that stressed fish go into rigor mortis more rapidly than unstressed fish and also generate more tension in their muscles post mortem (Nakayama *et al.*, 1992). The rapid rigor mortis may render filleting tilapia difficult or even impossible, due to the stiffness of the muscles and their inability to appear in a relaxed stage after entering the stiffening phase (Jarding *et al.*, 2000).

It is clear that filleting tilapia during rigor mortis is not a good idea and it is not recommended since it has an adverse effect on fillet quality. Tilapia stay in rigor mortis for at least 1 week, leading to decreased shelf life. Therefore, they must be processed before rigor mortis because waiting for filleting after the end of rigor mortis may lead to fish spoilage. It is also necessary to freeze the fish prior to rigor mortis because they usually develop gaping when they are frozen. It appears from this discussion that the low fillet yield of tilapia (35–43%) and the problem of rigor mortis make the fillet market difficult to enter and develop.

9.4.4. Radiation

Radiation can be an effective tool for maintaining high fish quality and enhancing the shelf life of processed fish. Radiation is generally combined with cooling. Cozzo-Siqueira *et al.* (2003) evaluated the effects of combining ionizing radiation and refrigeration with minimal processing on the shelf life of processed Nile tilapia in Brazil. The physical, chemical, nutritional and microbiological characteristics of the fish were studied in eviscerated samples and in commercial cuts. The fish were irradiated with 0, 1.0, 2.2 and 5 kGy and stored at temperatures ranging from 0.5 to 2°C for 20 and 30 days. During storage, the level of moisture in the non-irradiated samples decreased and the levels of protein and lipid increased, while the irradiated samples remained stable. Muscle amino acids and fatty acids remained stable in the irradiated samples, but decreased in the non-irradiated samples. The levels of total volatile base (TVB-N) and non-protein nitrogen (NPN) increased in the non-irradiated samples but tended to remain stable in the irradiated fish samples. The microbiological content in the irradiated samples was below the levels established by Brazilian seafood legislation, whereas the non-irradiated samples had a higher microbiological content. In addition, lipid oxidation tended to increase with increasing irradiation dose.

9.5. Marketing and Economics

Typically, tilapia marketing is controlled by the following factors: (i) production systems; (ii) operating costs; (iii) competition among producers; (iv) degree of processing; (v) production scale; and (vi) consumer demand. For example, production costs in the USA and Canada are higher than in Central America, Latin America, the Caribbean and other tropical regions, due to the high cost of labour, energy and operating costs. Importing tilapia from these regions would be more profitable than farming them in North America (Watanabe *et al.*, 2002). This gives tropical producers a competitive advantage over US and Canadian producers. However, this disadvantage can be overcome by optimizing resource utilization, such as using fertilization for the production of tilapia fingerlings, and improving production management (Hargreaves and Behrends, 1997; Watanabe *et al.*, 2002).

Since the production system is probably the most important factor affecting tilapia marketing and economics, it is discussed further in the following section. The cost-effectiveness of some farming systems is also described under various culture conditions, in order to help tilapia farmers and farm operators and managers adopt the most economically sound farming system.

9.5.1. System economics

Tilapia are cultured mostly in semi-intensive systems, where operating costs are low and culture inputs are minimal. Intensive tilapia culture with high investment and maximum inputs is also practised in many countries, but on a smaller scale. The expected outputs of culture practices are therefore a function of the culture systems adopted and the inputs to those systems. A number of studies have been carried out to evaluate and analyse the economic returns of different farming systems in different geographical locations. Such analyses would certainly help fish farmers and farm managers adopt the most sustainable and economically feasible systems.

Samonte *et al.* (1991) compared the economic feasibility of the polyculture of tiger shrimps (*P. monodon*) with Nile tilapia in brackish-water ponds in the Philippines. A stocking combination of 6000 shrimps and 4000 tilapia/ha produced the highest total yields and net income compared to monoculture of tilapia or shrimp. Two crops per year provided a 70% return on investment and 1.2 year payback. Sensitivity analysis revealed that shrimp/tilapia polyculture was profitable up to a 20% decrease in the selling price of both species.

The feasibility of small-scale hapa culture of Nile tilapia as an additional income source for rice farmers in north-east Thailand has also been evaluated (Middendorp and Verreth, 1992), using different production scenarios. Rice farmers stocked tilapia at an initial weight of 50 g to reach marketable size after 85 days. Investment costs were about 5% of total operating costs. Fingerling costs (about 25% of total costs) could be reduced significantly when fingerlings were collected from backyard ponds or from the hapas. Feed costs, which represented about 70% of total costs, could also be reduced by optimizing the feeding regime, using home-mixing of fish feed and extending the rearing period to 100 days. A density of 14 fish (50 g)/m^3 was suggested to reach a market size of 216 g in 100 days. If smaller sizes (< 50 g) are stocked, two rearing cycles (about 200 days) will be needed to reach market size. In this way, rice farmers may increase their household income by about 20%.

The economics and profitability of artisanal and commercial tilapia culture in Malawi are another example of the importance of system analysis (Stewart, 1993). Considering the local support services, including fry and fingerling supply and local market development, it was found that polyculture of Nile tilapia and African catfish (*Clarias gariepinus*) in earthen ponds was more cost-effective than other culture systems.

The economic and engineering evaluation of tilapia production in a small recirculating culture system using a computer simulation model has been described in the USA (Losordo and Westerman, 1994). Sensitivity analysis indicated that, among all operational variables tested, reducing feed cost and improving the feed conversion ratio (FCR) were the most effective variables in reducing the production costs. Increasing productive capacity or decreasing the investment cost also increased system profitability. Head *et al.* (1996) found that feed, processing and distribution and sex-reversed fry represented the highest variable costs in commercial-scale, seawater pond production of Florida red tilapia in Puerto Rico, while salaries and depreciation accounted for the highest fixed costs. At a stocking density of 2 fish/m^2, the proposed production was not economic, while increasing the density to 3.5 and 4 fish/m^2 increased the profitability of the operation. Using locally prepared feed, together with the integration of hatchery and grow-out operations, is also suggested for reducing operational costs and increasing returns.

Similarly, Head and Watanabe (1995) evaluated the economics of a commercial-scale, recirculating, brackish-water hatchery for Florida red tilapia (*Oreochromis urolepis hornorum* × *Oreochromis mossambicus*) in the Bahamas, using previously collected production data. Salaries and benefits were the largest fixed costs. Air-freight shipping to Miami, Florida, represented the largest variable cost when air charters were used. The study indicated that the system is economically feasible if fry can be sold year-round, while seasonal market demand may limit the economic return. The authors also reported that more economic success can be achieved if more economical building construction, modifications in rearing tank design and vertical integration of hatchery and grow-out operations, with external sales of excess fry and broodstock, are adopted. When the economic and technical feasibility of tilapia cage culture in Ghana was investigated, using least-cost, easily available local inputs (Oduro-Boateng, 1998), the sensitivity analysis indicated that the internal rate of return was 58%, meaning that the investment in tilapia cage culture in Ghana is highly profitable, especially in rural areas, where the system can be easily adopted by rural farmers.

These results clearly indicate that system analysis is necessary for successful rural aquaculture practices. This is simply because each geographical location has its own unique characteristics. In other words, the culture inputs that may be limiting in one place may not be so in another. Accordingly, system analysis would provide the farmer with different culture scenarios, so that he/she can choose the most cost-effective scenario.

9.5.2. Risk analysis

It is evident from the above examples that economic analyses applied to tilapia culture enterprises are specific to the culture system. This means that each system has its own risks, and therefore enterprises that generate higher returns must accept higher risks. Thus, economic analyses that indicate a higher economic return must also report the higher risk, so that the producer can be aware of this risk. Nevertheless, most economic studies concerning tilapia culture are directed to technical and economic practices and incentives (as mentioned in the previous sections), with little attention to economic risk. It is essential that economic analyses of aquaculture projects consider not only economic benefits but also the downside risk and risk potential.

In an earlier study, El-Sayed (1990) evaluated feeding Nile tilapia with cottonseed meal (CSM) containing different protein levels compared with a fishmeal (FM)-based diet. Fish fed hulled CSM (25% crude protein (cp)) resulted in the lowest fish performance. However, cost/benefit analyses indicated that this diet was economically the best, followed by the dehulled CSM (40% cp), while the FM-based diet had the lowest economic return. Similarly, Hatch and Falck (2001) evaluated alternative production systems of tilapia in Honduras, in terms of profitability and risk. They developed 'user-friendly' tools to assess the economic performance of these systems and the level of risk to tilapia producers. Economic indicators of profitability and risk were determined using baseline production data, such as nutrient inputs, stocking rates, etc. in addition to the prices of nutrient inputs and sale prices of fish. The authors concluded that this evaluation procedure will allow researchers to use their limited resources more efficiently through directing their research to production strategies that are likely to be profitable at an acceptable risk level.

Along the same line, Veverica *et al.* (2001b) compared the performance and economic benefits of polyculture of Nile tilapia and African catfish in earthen ponds in Kenya, under different feeding and fertilization regimes, using low-cost, locally available resources. In addition to fertilization, the fish received rice bran (RB), pig finisher pellets (PFP) or a test diet (20% cp) for 180 days. The RB produced the lowest growth, while PFP resulted in the highest performance. Relative profitability analysis indicated that PFP was economically the best, followed by RB. However, RB had the lowest break-even price and the least investment cost. This result revealed that using low-cost ingredients such as RB in semi-intensive, polyculture of tilapia could be highly profitable with minimal risks, especially in developing countries, where resources are limited.

In another study, the polyculture of Nile tilapia and African catfish in earthen ponds in Kenya was reassessed under other feeding and fertilization regimes (Omondi *et al.*, 2001). The fish received the following combinations of RB and fertilizer in order to realize the least-cost combination: (i) urea and diammonium phosphate (DAP) to provide 16 kg of N and 4 kg of P/ha/week; (ii) urea and DAP to provide 8 and 4 kg of N and P/ha/week, respectively, in addition to RB fed at 60 kg/ha/day; (iii) RB fed at 120 kg/ha/day; and (iv) RB at 120 kg/ha/day plus urea and DAP at 8 kg of N and 4 kg P/ha/week. The economic analyses indicated that the net profits of polyculture were lower than those of monoculture of Nile tilapia and catfish, but were higher for treatment (i) followed by treatment (ii) and significantly reduced in treatments (iii) and (iv), because of the high cost of feed and fingerlings. This study indicated that the inclusion of RB was less cost-effective than fertilizer alone. This result looks contradictory to that of Veverica *et al.* (2001b), who recommended the use of RB as a low-cost ingredient for tilapia polyculture systems. This controversy could be attributed to the differences in the cost of culture inputs (seeds, feed, fertilizer, transportation, labour, etc.) between the studies. Once again, this controversy means that economic analyses are system-specific, and each system has its own risk.

Lin and Yi (1998) assessed the economic feasibility of different culture inputs of small-scale pond culture of Nile tilapia. Production inputs included: dry chicken manure (CM), triple superphosphate (TSP), CM + urea or CM + urea + TSP, urea + TSP, urea + TSP + feed or feed only. The highest return was achieved with supplemental feeding beyond 100 g/fish; however, the best net income was obtained when CM (75 kg/ha/week) + urea (28 kg N/ha/week) + TSP (7 kg P/ha/week) were used.

These results clearly demonstrate that the practice that may produce the best yield may not necessarily be the most cost-effective. This emphasizes the need for economic and risk analyses for tilapia culture projects concomitant with biological and technical analyses, prior to the start of the project. Such analyses will help the farmer use the available resources in a sustainable manner and adopt the most efficient production strategy.

9.6. Domestic Markets

Tilapia culture is practised mainly by small-scale farmers in rural areas in developing countries. Consequently, the fish are marketed in rural and local markets, either fresh or iced. As expected, transportation, distribution and marketing channels in these areas are not well managed, due to the scattering of the farms in remote areas and the lack of technical and extension assistance. This situation is generally perpetuated by poor quality control, lack of corporate resources and limited market information (Young and Muir, 2000). The development of domestic tilapia marketing is therefore becoming of prime importance. Yet very few studies have considered this subject, mainly in certain Asian regions.

The early studies on tilapia economics and marketing potentials in the Philippines indicated that the substitution of *O. mossambicus* by the introduced *O. niloticus* led to rapid development of tilapia culture in various parts of the country (Torres and Navera, 1985; Bimbao and Smith, 1988). This has narrowed the margins between wholesale and retail prices of farmed tilapia, and resulted in more competition due to the increase in production. These studies suggested that the future market potential of farmed tilapia will depend on the future income and purchasing power of the consumers and the ability of producers to reduce production costs.

Today, the increased production of farmed tilapia, together with better income and increased purchasing power in many South-east Asian countries, has shifted the consumer preference from the traditional, small-sized, unprocessed, fresh tilapia to large-sized tilapia and other tilapia products. Guerrero and Guerrero (2001) reviewed tilapia domestic marketing practices in the Philippines, Thailand and Vietnam. They reported that tilapia is an important food commodity commercially produced in many South-east Asian countries, with Nile tilapia being the most important species marketed in the region. Consumer preference has shifted to large-sized tilapia, which are generally sold fresh, chilled or frozen. Typically, tilapia producers sell their products to traders, who in turn sell them to wholesalers and/or retailers. As stated earlier, processed tilapia are also gaining increased marketing momentum in the region.

In a comprehensive review, Ferdouse (2001) discussed the recent past, current and future potential of tilapia in Asian markets. She reported that South-east Asia and the Far East, namely mainland China, the Philippines, Thailand and Indonesia, are the most important markets for tilapia, where a majority of the supply is consumed domestically. Market acceptance is also gradually increasing in other Asian countries, such as Malaysia, Singapore, Hong Kong, Taiwan, Bangladesh and India, especially in growing urban populations. The fish are sold fresh and/or chilled for household consumption in the cities. She indicated that better local and regional tilapia marketing opportunities have been created in recent years due to the following:

1. Increased fish consumption in Asian countries.
2. Availability and sustainability of supplying high-quality tilapia.
3. Better market opportunities within and outside South-east Asia.
4. Consumer perspective towards healthy food.
5. Economic changes in the region since the mid-1990s.

The author stressed that tilapia is no longer considered a low-value species in Asian markets.

In Africa, tilapia marketing and distribution channels are very simple, with minimal interference from governmental authorities. In countries such as Egypt, Zambia and Nigeria, most of the tilapia produced from aquaculture or capture fisheries are directed to urban markets in large cities, in addition to the amounts that are consumed locally. The increased demand for tilapia in many African countries has increased the prices of these fish. In Egypt, for example, the retail price of large-sized Nile tilapia (> 200 g) grown in fish ponds is about US$2/kg, while the prices of small fish (100–< 200 g) range from US$1 to US$1.5/kg (A.-F.M. El-Sayed, Alexandria, 2004, personal observation). The prices also fluctuate from one season to another (higher in winter than in summer) and from one area to another. In some other countries, like Malawi, freshness rather than fish size affects the retail price of tilapia (Brummett, 2000). In that country, the demand for tilapia is very high and fish produced in rural areas are generally absorbed by local markets and rarely reach urban markets (Jamu, 2001). It is no surprise that these fish are sold out the day before harvest.

In South America, tilapia marketing is positively responsive to population growth and advertising, which are the two main factors affecting consumer demand. Tilapia marketing in Costa Rica is an example of how advertising affects the demand for tilapia. There was market resistance in the country to Mozambique tilapia because of its dark skin and the use of pig manure for pond fertilization. Nile tilapia was then introduced and given an attractive name 'St Peter's fish' in order to convince consumers that this species was not tilapia. In this way, tilapia was successfully marketed and has gained increasing popularity. Similarly, red tilapia (*Oreochromis* spp.) was introduced to the country in 1984, and was more acceptable because it resembles the colour of red snapper, which is a popular marine species.

The domestic tilapia market in Honduras is also growing very well. Many supermarkets and restaurants sell tilapia, and the percentage that offer the fish is expected to increase (Fúnez *et al.*, 2001). In a survey of buyer attitudes towards tilapia in Honduran markets, Neira and Engle (2001) found that over 41% of supermarkets and 25% of the restaurants sold tilapia. An additional 24% of supermarkets used to sell tilapia, but stopped sales because of the lack of fish supply, low demand, lack of freshness, storage problems and lack of seafood sections in the supermarket. However, the survey indicated that 46% of restaurants and 50% of supermarkets would probably begin adding tilapia to their menus and shelves in the next year.

9.7. Global Markets

Despite the fact that most produced tilapia is consumed in domestic markets in production areas, especially in rural Asia, Africa and South America, the demand for tilapia is growing in non-traditional, non-producing countries. Thus, global trade in tilapia products has witnessed an impressive flourishing in the last two decades, and this is expected to continue (Vannuccini, 2001). Among the major advantages of tilapia in global markets are their competitive prices and their white flesh, which make them an important substitute for white fish species that face shortages in supply (Vannuccini, 2001).

Despite these attributes of tilapia marketing, the picture of their world market is not clear, since very little information is available on global trade figures. Tilapia are generally not identified as a separate commodity item in the statistics of most countries, but are included with 'other' freshwater fish species. For this reason, tilapia trade is generally negligible in international fish markets. One other important factor that makes global tilapia trade more difficult is that the bulk of produced tilapia is consumed locally, generally without any handling and processing, at low prices and with little recording of their trade statistics.

However, the available information, despite being limited, shows that the global market for tilapia is expanding, and the number of countries exporting and importing tilapia is increasing. Vannuccini (2001) reviewed global tilapia marketing, including imports, exports and major markets (Figs 9.4 and 9.5). This review shows that the last few years have witnessed a growing acceptance for

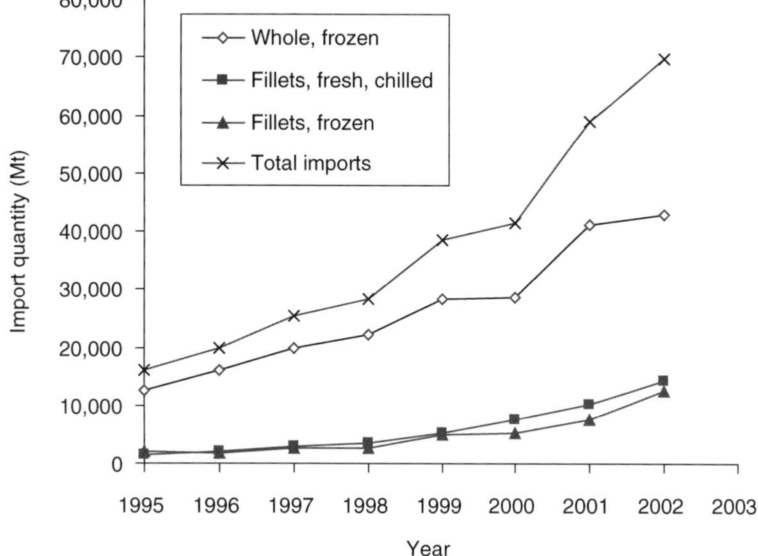

Fig. 9.4. Global tilapia imports during 1995–2002.

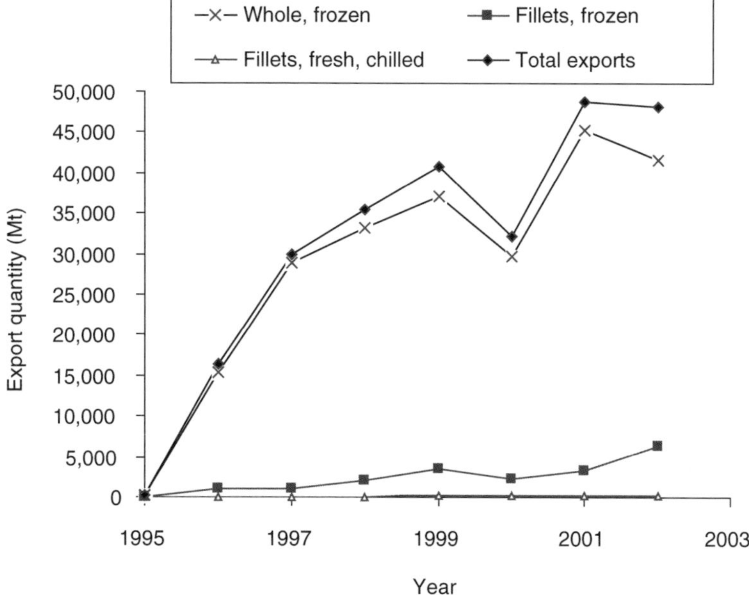

Fig. 9.5. Global tilapia exports during 1995–2002.

and consumption of tilapia in non-producing countries, especially the USA, Europe, Central America and South America. As a result, large-scale production and international trade in tilapia products are flourishing at an exceptionally high rate.

9.7.1. The US market

Domestic production of tilapia in the USA has increased significantly over the past few years, so that it has become the fastest-growing species cultured in the country. In spite of this, the US market remains the largest market for tilapia import in the world. A great variety of tilapia products, including live fish, fresh fish and frozen fillets, are currently marketed. According to the US Department of Agriculture (USDA) Economic Research Service (2004), tilapia import quantities have increased sharply in recent years. These quantities have jumped from 56,334 Mt in 2001 to 90,206 Mt in 2003, with an increase of over 60% in just 3 years (Fig. 9.6). The value of imported tilapia has also increased from about US$128 million to over US$241 million during the same period, with an increase of over 88% (Fig. 9.7). The price of tilapia also increased at higher rates than that of shrimp and salmon imports. The USA imported tilapia and tilapia products from 33 countries during 2001–2003 (Table 9.1). Surprisingly, some exporters are not traditionally tilapia producers, such as South Korea, New Zealand, Japan, the UK and Canada. These countries may import tilapia and then re-export it to international markets, including the US market, mainly as fresh and/or frozen fillets.

The volume of imported tilapia increased in all product categories; however, much of the increase has been attributed to the imports of frozen fillets, which increased by 358%, from 5081 Mt in 2000 to 23,250 Mt in 2003. Frozen whole fish rose by 76.5% and accounted for 54.4% of all tilapia imports in 2003. Imported fresh fillets totalled 17,951 Mt in 2003, 140% higher than in 2000 and 32.6% higher than in the previous year. The value of this category was also increasing at a very high rate, from about US$61 million in 2001 to reach US$102 million in 2003, with an increase of about 68%. It also represented about 42% of total value of imported tilapia in 2003. The volume and value of this category of the tilapia market continue to expand rapidly. The estimated amount of foreign tilapia production required to supply the US market is 173,778 Mt of live fish.

Going back to 1993 statistical data, the first full year that tilapia imports were reported separately, one will find that the value of tilapia imports increased from US$18 million in 1993 to US$241.2 million in 2003, an increase of 1240%. This huge increase has been attributed mainly to a higher volume of imports.

Asia is the main tilapia exporter to the US market. About 80% of imported tilapia in 2003

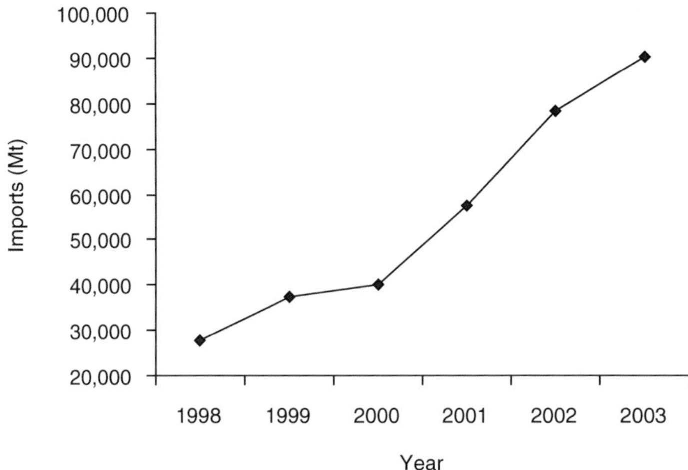

Fig. 9.6. Tilapia imports to the US market during 1998–2003.

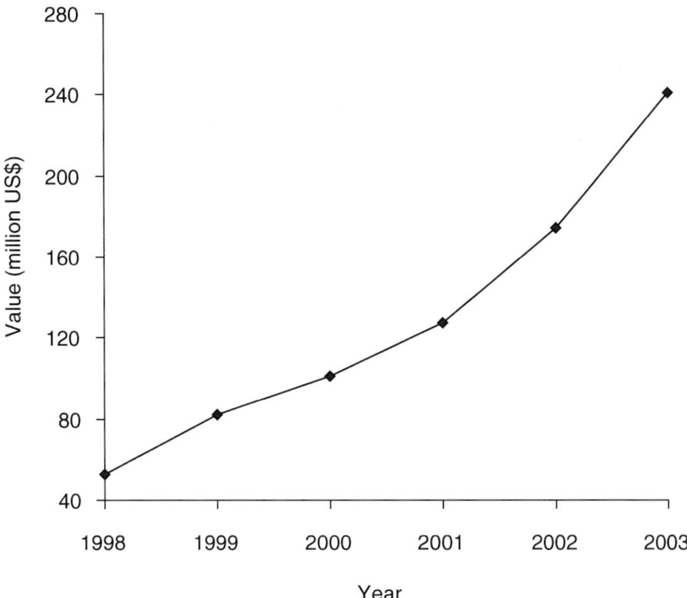

Fig. 9.7. Value of tilapia imported to the US market during 1998–2003.

came from Asia. The USA has been the largest market of Taiwanese red tilapia. Out of the 37,000 Mt of Taiwanese tilapia exported in 1999, 64% went to the US market, in the form of low-priced frozen whole fish (Chiang, 2001). Currently, mainland China is becoming the biggest competitor in the US tilapia market, and is now the dominant tilapia supplier. As a result, tilapia imports from Taiwan are declining, while those from mainland China are increasing (Table 9.1). The export of Taiwanese tilapia to the USA declined from 29,809 Mt in 2001 to 22,415 Mt in 2003. At the same time, tilapia imports from mainland China jumped from only 13,590 Mt in 2001 to 45,477 Mt in 2003. More than 63% of tilapia imports from China in 2003 were in the form of frozen whole fish, while about 35% were frozen fillets. To get back into the US market and to sustain the growth of Taiwan's tilapia exporting industry, changing the product from whole frozen fish to fillets and adopting promotional campaigns have been suggested. Other Asian countries, such as Indonesia and Thailand, are also becoming important tilapia exporters to the US market.

Tilapia imports from South America and the Caribbean region are also increasing. Fourteen countries from these regions are currently exporting tilapia and tilapia products to the US market. Ecuador is the largest tilapia exporter from the region, followed by Costa Rica and Honduras. Tilapia imports from Ecuador increased from 3378 Mt in 2000 to 9726 Mt in 2003, with an increase of 188%. Over 96% (9397 Mt) of tilapia imports from Ecuador in 2003 were in the form of fresh fillets. Costa Rica and Honduras are also major tilapia fresh fillet exporters to the US market. The three countries accounted for over 90% of all imported fresh fillets in 2003. Some aquaculture producers in Ecuador have switched from shrimp production, due to falling prices in the shrimp market, to tilapia production. This means that imports of tilapia from Ecuador are expected to increase at the expense of shrimp. Other countries, such as Panama and Nicaragua, are becoming important tilapia exporters to the US market.

It is expected that the increase in tilapia imports will continue in 2004 (Harvey, 2004). Total tilapia imports are expected to reach about 100,000–104,000 Mt, on a product weight basis (198,000–204,000 Mt on a live-weight basis). The value of these imports is also expected to reach US$265–275 million. On the other hand, the prices for the individual types of tilapia imports

Table 9.1. Tilapia imports to the US market (Mt) during 2001–2003 (from Economic Research Service, USDA) (www.ers.usda.gov).

	Whole, frozen			Fillet, fresh			Fillet, frozen			Total		
Continent/country	2001	2002	2003	2001	2002	2003	2001	2002	2003	2001	2002	2003
Asia												
China	10,869.8	19,616.0	28,763.1	191.0	191.0	856.6	2,529.0	6,026.0	15,857.0	13,589.9	26,485.8	45,476.7
Indonesia	39.0	2.5	5.4				2,179.0	2,572.4	3,582.7	2,218.0	2,575.0	3,588.1
Malaysia	17.3									17.3		
Hong Kong		39.9	135.2								39.9	135.2
Philippines	50.8		18.0				2.1			52.9		18.0
Taiwan	27,599.3	20,659.7	19,663.6	76.1	246.5	281.5	2,133.3	2,760.9	2,469.7	29,808.7	23,667.0	22,414.7
Thailand	48.8	249.6	121.3	1.6	26.9	6.6	209.2	338.5	939.8	259.6	615.0	1,067.7
Vietnam	6.9	0.9	41.4		1.5	17.0	52.9	106.3	73.2	59.8	108.7	131.6
Burma								2.4	19.1		2.4	19.1
Japan						0.5			18.0			18.5
Pakistan			22.0								22.0	
Subtotal	38,631.9	40,568.6	48,770.0	268.7	465.9	1,162.2	7,105.5	11,806.5	22,959.5	46,006.2	53,515.8	72,869.6
S. America												
Colombia		7.9	3.0	32.2				4.0		32.2	11.9	3.0
Ecuador	95.4	16.3	143.2	4,924.2	6,615.5	9,396.9	139.5	271.6	186.4	5,159.1	6,903.4	9,726.5
Panama	2.4	1,496.9	103.8	350.2	147.4	95.6		48.2	41.5	352.6	1,692.5	240.9
Venezuela		5.9						3.8			9.7	
Brazil					111.8	208.3	8.0	49.0	26.9	8.0	160.8	235.2
Cayman Isl.					1.8	8.2					1.8	8.2

Chile				14.9	12.4	34.6			0.5	14.9	12.4	35.1
Costa Rica			1.6	3,108.9	3,206.0	3,996.1		2.3	2.5	3,108.9	3,208.3	4,000.2
El Salvador					77.7	188.6					77.7	188.6
Honduras				1,437.7	2,873.6	2,856.6				1,437.7	2,873.6	2,856.6
Jamaica				91.6	16.4		27.5	19.2	18.1	119.1	35.6	18.1
Nicaragua				3.8	0.9	2.9	16.9	24.6	8.4	20.7	25.5	11.3
Grenada						1.0						1.0
Subtotal	97.8	1,527	251.6	9,963.5	13,063.5	16,788.8	191.9	422.7	284.3	10,253.2	15,013.2	17,324.7
Others												
Canada					0.8	0.6	8.2	5.1		8.2	5.9	0.6
Uganda					3.9						3.9	
New Zealand							19.8	170.9		19.8	170.9	
South Korea							20.0		1.2	20.0		1.2
Tanzania							26.4			26.4		
UK								18.0			18.0	
Mauritius			5.3						4.5			9.8
Subtotal			5.3		4.7	0.6	74.4	194.0	5.7	74.4	198.7	11.6
Grand total	38,729.7	42,095.6	49,026.9	10,232.2	13,534.1	17,951.6	7,371.8	12,423.2	23,249.5	56,333.8	68,727.7	90,205.9

and the average import price are expected to decrease slightly in 2004, due to strong competition among producers and from other seafood products. Ecuador and China are expected to put more downward pressure on prices of exported tilapia in order to expand their shares of the US market.

In order to promote tilapia marketing in the USA, a generic marketing campaign is being conducted (Schramm, 2001). This campaign is organized and implemented by the Tilapia Marketing Institute (TMI), which is a consortium of tilapia producers and suppliers of goods and services supporting the tilapia culture industry. The TMI is a non-profit, self-funded organization, aimed at increasing US consumer awareness of tilapia. TMI works closely with the media across the USA in order to obtain full advertising and a coverage of tilapia stories in these media. TMI also sponsors events attended by chefs and food writers to draw their attention to tilapia as an important food fish in the US market.

9.7.2. Other international markets

The demand for tilapia is growing in many European countries, especially among particular ethnic groups, such as in African, Chinese and Asian communities in big cities. Taiwan is the main tilapia exporter to Europe. European imports from Taiwan increased from 889 Mt in 1996 to 3711 Mt in 1999, with the UK being the major importer (Vannuccini, 2001). Tilapia are also marketed in France, Germany, Belgium, the Netherlands, Austria, Italy, Hungary, Switzerland, Malta, Denmark and Sweden. Large-sized, whole Nile tilapia are generally preferred in Germany, while the UK market favours fresh red tilapia imported from Jamaica (Vannuccini, 2001).

Arab countries also represent a huge market for tilapia imports. Despite the high tilapia production in some countries of the region, like Egypt, local markets usually absorb this production, leaving little or no room for exportation. Arab countries import considerable quantities of tilapia, mostly whole frozen, from Taiwan. The Arabian Gulf States are the major tilapia importer in the region. The tilapia market has also grown substantially in Canada during the last few years. The live tilapia market is considered to be the most important, especially in certain cities like Toronto and Vancouver. Canada imports fresh and frozen tilapia from Costa Rica and Jamaica (Vannuccini, 2001).

9.8. Closing Remarks

1. There is no specific harvesting system for farmed tilapia; partial harvest may be the most cost-effective, but this technique should be evaluated biologically and economically. Cheap and locally available harvesting tools should also be adopted.

2. Poor handling, inadequate processing and holding conditions and inappropriate processing methods may cause the quality of tilapia to deteriorate and increase postharvest losses. Proper handling, transportation and displaying should therefore be used.

3. Tilapia is traditionally marketed as whole fresh and/or iced fish. Marketing of processed tilapia and tilapia products is now gaining great popularity in many parts of the world.

4. Cooling (icing, refrigeration, freezing, etc.), filleting, radiation and industrial processing (drying, smoking, canning, fish balls and sausages, etc.) are commonly used for processing tilapia.

5. Tilapia guts should be removed before refrigeration, because the spoilage rate is more rapid in gut-retaining than in gutless fish.

6. Tilapia are difficult to process and have a relatively low fillet yield. Skinning the whole fish before filleting may produce the best fillet and total eatable yield. Contour and oblique head cuts may also produce better dressout than straight head cuts.

7. Rigor mortis takes place within 1–1.5 h after death and is faster for fish stored on ice than for those stored at ambient temperature. It may influence the appearance and structure of fish muscles and render it unsuitable for filleting. Stressed fish go into rigor mortis more rapidly than unstressed fish.

8. Rigor mortis reduces the fillet quality of tilapia, therefore, filleting during rigor mortis is not recommended. The fish must be processed before rigor mortis. It is also necessary to freeze the fish prior to rigor mortis.

9. Economic analyses of tilapia culture enterprises are specific to the culture system, and cannot be generalized to all systems. Economic analyses that indicate a higher economic return must report a higher risk, so that the producer can be aware of such a risk.

10. Economic analyses of tilapia culture projects must consider both economic benefits and downside risk and risk potential.

11. The production and global trade in tilapia and tilapia products have witnessed an impressive flourishing during the past two decades. Global markets for importing and exporting tilapia are also expanding.

12. The US market is the largest tilapia market in the world. A great variety of tilapia products, including live fish, fresh fish and frozen fillets, are currently marketed in the USA. Taiwan and mainland China are the major tilapia exporters to the US market.

10

The Role of Tilapia Culture in Rural Development

10.1. Introduction

Rural development is a continuous process facilitated by governments, non-governmental organizations (NGOs) and international donors, to sustain the growth of rural economies, improve the livelihoods of rural communities and promote food security through the improvement of food supply, employment and income (Edwards *et al.*, 2002; Halwart *et al.*, 2003). Despite the fact that several sectors can play different roles in rural development, the major contribution to poverty alleviation and food supply is very likely to come from the agricultural sector, simply because agriculture is the main activity in rural areas.

Aquaculture can play an important role in national economies in countries practising aquaculture activities. This role includes increasing fish supply for domestic consumption, generating more foreign currency, raising producers' incomes and creating employment opportunities. As an agricultural subsector, aquaculture is also expected to play a core role in rural development, especially in developing countries in the tropical and subtropical regions of the world. It can contribute significantly to the alleviation of poverty and malnutrition and improving the welfare of rural households, due to the following main reasons:

1. Aquaculture is one of the fastest-growing food-producing systems in the world (Ahmed and Lorica, 2002). It is even the fastest-growing subsector in many countries (Halwart *et al.*, 2003).
2. More than 85% of world aquaculture production comes from developing countries, and mainly from low-income food-deficit countries (LIFDCs).
3. Aquaculture is becoming an important component of rural livelihood and poverty alleviation in situations where increasing population pressure, environmental degradation and/or loss of access will limit dependence on natural fisheries.

In rural areas, aquaculture is generally integrated with other farming activities, and should not be examined in isolation from other sources of livelihood. At the same time, it is important to understand the role of households not only as producers, but also as consumers. Thus, to evaluate the role of aquaculture in rural development, one must consider the contribution of rural aquaculture to household economies, in the case of both smallholder fish farmers and subsistence farmers, and the interaction with the non-aquaculture economic activities of farmers, as well as the household's food/protein requirements and actual consumption levels.

Rural aquaculture is also a complex and diverse practice, including various aspects and interrelationships, such as technical aspects, environmental aspects, economic and socio-economic aspects, extension services and policy issues. All these issues should be fully considered and integrated for sustainable rural aquaculture development. The benefits of aquaculture in the development of rural livelihoods are also related to health, nutrition, employment, income, reduction of vulnerability and farm sustainability.

In small-farmer systems, aquaculture provides good-quality animal protein and other nutrients, especially for nutritionally vulnerable groups, at low prices. It also provides jobs and 'own enterprise' employment, including work for women and

Fig. 10.1. A rural woman and her son sell tilapia in a local fish market at Hamool, Kafr El-Shaikh, Egypt.

children, generating additional income through the harvesting and sale of their products (Fig. 10.1). On larger farms, income can be increased through employment in seed supply networks and in marketing chains and through manufacture/repair functions. Indirect benefits include an increased availability of fish in local rural and urban markets and a concomitant reduction in price, allowing savings to be spent on other income-generating farm products. Aquaculture can also offer a benefit from the utilization of common resources, particularly for the landless, through cage culture, culture of molluscs and seaweeds and enhanced fisheries in communal water bodies (FAO, 2002).

One other important benefit of rural aquaculture is its relevance to integrated agriculture–aquaculture systems, which increase farm efficiency and sustainability. Agricultural by-products, such as crop residues, livestock manures and green manures, can be used as feed/fertilizer inputs for small-scale commercial aquaculture systems. Similarly, farming fish in rice fields has several economic, biological and environmental benefits (discussed in Chapter 4).

In rural areas, aquaculture is likely to be more advantageous than other agricultural activities, such as cash crops and livestock production, due to the following:

1. It can be easily integrated into other agriculture and animal production activities.
2. The use of low-cost inputs and technologies by using locally available on-farm sources.
3. The limited investment needs.
4. The low levels of risk.
5. The low labour requirements.

As a result of the recent recognition and increasing role of aquaculture in rural development, both governments and governmental and non-governmental organizations have started paying more attention to the subject. Scientific conferences, workshops, training courses and extension services on rural aquaculture development are spreading worldwide. The scientific conference that was held in Thailand in 1998 on rural aquaculture, with emphasis on Asian experience, is just one example. The papers presented at the conference were published in a book entitled *Rural Aquaculture* by CAB International in 2002. Many Food and Agriculture Organization (FAO) 'Technical Papers' and 'Reports' have also been published on the subject.

Several other research studies have been conducted on the role of aquaculture in rural development. Ahmed and Lorica (2002) evaluated the role that aquaculture plays in attaining food security in Asian developing countries. They reported that pond culture along with cash crops and other on-farm activities contribute between 5 and 10% of total income in rural areas in Bangladesh. The effects of aquaculture adoption on household labour use and employment have also been significant (Ahmed *et al.*, 1995). Veerina *et al.* (1993) found that 71% of traditional crop farmers in Andhra Pradesh, India, considered aquaculture as their primary activity, and about 49% of the farmers depended completely on aquaculture. These figures clearly indicate that aquaculture can play an important role in rural development.

Rural aquaculture can also increase the consumption of aquatic food by both producers and domestic consumers by increasing fish availability in rural and urban areas. Per capita consumption of fish in various Asian countries has been reported to be significantly higher for fish-producing households in the rural areas, compared with both non-producers and the national average consumption (Dey *et al.*, 2000b; Ahmed and Lorica, 2002). Not only that, but there is a wide variation in per capita consumption and types of fish consumed among various economic and social strata. Consumer demand for major fish species, such as carp and tilapia, shows a price-elastic and income-elastic demand in India, Bangladesh, Thailand and the Philippines, but not in China (Dey, 2000). Price elasticity and income elasticity are the changes in consumer demand according to changes in fish price and consumers' income. This means that the degree by which consumers' demands will change because of the changes in fish price and consumer's income is lower in China than in the other four countries surveyed. Generally, low-value fishes, such as carp and tilapia, have lower price elasticity, whereas high-value products, such as crustaceans, sea bass and sea bream, have higher price elasticity. It has also been found that price elasticity of demand for big fish (e.g. carp) in Bangladesh is higher in the urban areas than in rural areas for all income groups, and the level of household income is inversely related to the income elasticity of demand for fish (Dey, 2000).

A considerable part of aquaculture production in rural areas is consumed on farm, especially when aquaculture becomes part of subsistence activities on the farm (Ahmed and Lorica, 2002). On-farm household consumption is directly related to pond production of low-price fish, such as tilapia and carp (Gupta *et al.*, 1999). Increased on-farm fish availability reduces the dependence of household consumption on purchased fish (i.e. fish supply produced from family farms covers most of the needs of the household for fish). Consequently, as pond production increases in rural areas a large proportion of this production is likely to be sold in local markets, in addition to the major amounts consumed by the households themselves. In the meantime, high-price fish crops are sent to the market and sold at higher prices (Ahmed and Lorica, 2002).

10.2. Tilapia and Rural Development

Tilapia have many attributes and advantages that make them an excellent candidate for aquaculture (see Chapter 1). As a result, tilapia culture has witnessed a rapid global expansion during the past two decades, especially in developing countries, where the bulk of farmed tilapia is produced. Moreover, tilapia culture is becoming more popular and well established in many smallholders' ponds in rural areas, even in countries that do not have any history of tilapia culture and have traditionally relied on their native species, such as Vietnam, India, Bangladesh and Sri Lanka.

In Bangladesh, for example, where carp are considered the major cultured fishes while exotic fish culture is rare in Gazipur District (Ahmed *et al.*, 1993), Barman *et al.* (2002) found that 56% of pond farmers in north-west Bangladesh raised tilapia either in monoculture systems or, more commonly, in polyculture systems with carp. Tilapia are also an important contributor to farmers' risk avoidance in riverine, flood-prone areas in the region. Farmers favour tilapia culture in these areas because tilapia do not need to be stocked or managed carefully and also require very low culture inputs, compared to carp, for which seeds are generally purchased. These advantages have led to the relative popularity of tilapia culture in other areas of Bangladesh (Ahmed *et al.*, 1996; Hossain and Little, 1996). The tendency to favour tilapia culture to reduce cash costs was also noticed in north-east Thailand (Little *et al.*, 1996).

Tilapia production systems are currently categorized into three distinctive categories: (i) smallholder; (ii) commercial; and (iii) industrial (Little, 1998). Little (1998) proposed a framework for categorizing factors affecting those production systems in relation to the market and the various beneficiaries (Table 10.1). It would be useful here to throw some light on these production systems.

I. SMALLHOLDERS' SYSTEM. The smallholders' system is widely established in much of rural Asia, Africa and Latin America, where smallholders' ponds, ditches and rice fields, typically individually owned, are often used for tilapia culture.

Table 10.1. Framework for the main types of tilapia production systems (from Little, 1998, with permission).

Category	Smallholder	Commercial	Industrial
Main descriptors	Rural, few/no inputs, organic spread, natural recruitment	Peri-urban, fertilized pond, hatchery-supplied seed	Isolated, vertically integrated, seed production on site
Inputs	None, fertile runoff, incidental fertilization	Wastes/fertilizers, supplementary feeds	Complete feeds
Systems	Small ponds, ditches, rice fields, community ponds	Large ponds	Raceways, cages, intensive ponds
Importance of self-sustained seed	XXX	XX	XXX
Importance of linkages with hatcheries by trading networks	X	XXX	X
Primary reasons for farmers' self-sufficiency	Few options, minimal costs	Reduced costs	Assured supply, maintain and upgrade quality
Relevance of monosex	X	XX	XXX
Market	Poor, rural producers and consumers	Poor and middle-income, urban	Primarily export, domestic, urban better-off
Main beneficiaries	Producers, may be poorest within households, immediate community, local traders	Traders, landowners	Shareholders, processors, ice/equipment suppliers, transporters, governments, grain farmers, importers, contract farmers
Other stakeholders	Fish seed traders	Transporters, harvest teams	Co-users of water resource/waste water
Major constraints	Erratic seed supplies	Poor-quality seed, land-related costs, nutrient availability	Off flavours, water availability, cost of feed and labour

X, least important; XX, medium importance; XXX, most important.

The farmers generally use very few inputs, if any (kitchen waste, incidental fertilization), with poor management. Consequently, the yield of this system is generally low, but can still significantly improve the nutrition and livelihoods of rural households.

II. COMMERCIAL PRODUCTION SYSTEM. The commercial production system is generally practised semi-intensively, typically in densely populated peri-urban areas, where culture inputs such as fertilizers and supplemental feed sources are available. These areas are also characterized by growing appreciation of the overall value and production profitability of tilapia culture. Culture inputs, such as seed, feeds, fertilization, harvesting, transportation and marketing, and overall management planning are generally higher than in the smallholders' system.

III. INDUSTRIAL PRODUCTION SYSTEM. The industrial production system of tilapia is usually practised by large companies, which generally establish operation mechanisms to best suit the market needs. Extensive investments are usually pumped into this production system. This system also requires sophisticated production, marketing and overall management policies. In many Asian countries, tilapia produced by industrial production systems are exported to foreign markets, particularly the US market.

The following sections throw light on the role of tilapia culture in rural development in the major producing continents, namely Asia, Africa and Latin America.

10.3. Tilapia and Rural Development in Asia

Developing countries in Asia are the leaders of world aquaculture production, with China alone producing over 71% of the total global production (36,576,341 Mt in 2002) (FAO, 2004). In addition, the top ten aquaculture producers are LIFDCs (FAO, 2004). These figures reflect the importance of aquaculture in developing countries in general and in rural areas in particular.

Asia is also the major producer of farmed tilapia, with a production of 1,191,611 Mt in 2002, contributing about 80% of global tilapia production. Most of this production comes from small-scale, rural farms (Fig. 10.2). Therefore, small-scale, family-owned systems in much of Asia are becoming more competitive with large-scale, commercially managed tilapia farms (Little, 1998). However, little information is available on the importance of tilapia to households and their role in the overall economies in these rural areas. This gap is mainly because small-scale tilapia farms are scattered in remote and isolated areas that are difficult to approach. In addition, tilapia is cultured mainly in polyculture systems or in integrated systems with other plant/animal species and separate information and statistical data on their production, value, etc. are not generally available in many countries. Only in recent years has increasing attention been given to the role of tilapia culture in rural development in some tilapia-producing countries in Asia.

Most of the cultured tilapia in Asia is produced in China, the Philippines and Thailand. These three countries produced over 78% of total farmed tilapia in Asia in 2002 (see Chapter 1). It would be appropriate to discuss the role that tilapia culture plays in rural development in these countries. A considerable amount of information has also been gathered on tilapia culture in rural Bangladesh in recent years. It would be useful, therefore, to throw light on Bangladesh experiences. It is a pity, however, that no information is available on the role of tilapia culture in rural development in China, despite China being the world's top producer. Although I have contacted many personnel and organizations in and outside China in order to provide me with any available information on the subject or to direct me to the sources of this information, I could not get any response. The following section will therefore be limited to the Philippines, Thailand and Bangladesh as examples of tilapia culture in rural Asia.

10.3.1. The Philippines

The Philippines are the third largest tilapia producer in the world, after China and Egypt, with a production of 122,390 Mt in 2002 (FAO, 2004). The tilapia culture industry in the Philippines is constrained by a number of social, economic,

Fig. 10.2. Tilapia culture in earthen pond in rural area in Vietnam (photo provided by P. Edwards).

technical and environmental factors (e.g. high input costs, fish mortality, water supply and typhoon damage) (Mair *et al.*, 2002). Limited supply of quality seed at reasonable prices remains a major constraint. The performance of farmed tilapia in the Philippines (mainly Nile tilapia) has been declining compared to newly introduced pure strains, presumably due to the loss of genetic variation through founder effect and introgression with Mozambique tilapia (*Oreochromis mossambicus*) (Mair *et al.*, 2002). The early sexual maturation and unwanted reproduction of tilapia have also been another problem facing optimization of tilapia marketing. It appears, therefore, that genetics-based technology can be one of the important alternatives to resolve these problems.

Accordingly, Mair *et al.* (2002) examined the effects of genetically male tilapia (GMT) technology on the livelihood of poor households in the Philippines. The effects on the producers, consumers and employment opportunities were evaluated. Analyses of current dissemination strategies indicated that ongoing 'passive' mechanisms do not reach many poor, small-scale farmers. These farmers can improve their livelihood from appropriate genetics-based technologies. The authors found that strategic, supported partnership with NGOs and people's organizations (POs) are the most effective means of targeting the technology at poor farmers. Providing improved broodstock for small-scale hatcheries, together with the suitable technical, extension and financial assistance, is also another way for improving the livelihood of poor, small-scale tilapia farmers. Moreover, the use of genetically male tilapia (GMT) can increase the yield by 30–40% and profitability by over 100% compared to mixed-sex tilapia in the Philippines. The following recommendations for further research and dissemination activities related to GMT in the Philippines have been proposed by Mair *et al.* (2002):

- Since the number of farmers practising subsistence-level forms of tilapia culture in more rural and isolated areas in the country is likely to increase, there is a need for research to improve our understanding of the role of subsistence aquaculture in Philippine farming systems and the potential for these farmers to benefit from genetics-based technologies.
- There is a need for further quantification of the role of tilapia in the diet of the poor and for a better understanding of the pathways by which the poor obtain their tilapia. Economic models can be used for prediction of the impacts of improved fish and more efficient production upon the price and availability of tilapia in this sector of urban and rural societies.
- With regard to monitoring the impacts of dissemination of GMT, the feedback mechanism should be improved for a better understanding of the magnitude and nature of impacts and for profiling the socio-economic status of beneficiaries in order to characterize the short- and medium-term influence of the technology upon their livelihoods.
- Phil-Fishgen, a fish dissemination organization, should adopt a more proactive dissemination strategy specifically targeted at small-scale, poor farmers in order to better target the technology to the benefit of poor producers. Improved dissemination strategies should be supported by NGOs and POs to target, and possibly subsidize, small- and medium-scale hatcheries. This policy should be supported by efficient extension services and improved access to affordable credit.

Although the development of genetically improved tilapia can be a breakthrough in the tilapia aquaculture industry in the Philippines, the use of GMT broodstock and the production of quality fry and fingerlings depend mainly on the existing socio-economic and biophysical profiles of existing tilapia hatchery operations (Bimbao *et al.*, 2000). This means that this new technology cannot be applied without knowing how much it will cost to produce fry/fingerlings compared to the existing hatchery technologies. Knowledge of the existing levels of technical efficiency in tilapia hatcheries is vital for examining policies to increase productivity and decrease production costs, and also for maximizing the dissemination strategies of improved tilapia strains among tilapia hatcheries.

A comprehensive, micro-level analysis of socio-economic and production efficiency of tilapia hatchery operations in the Philippines has been made by Bimbao *et al.* (2000). The main objective was to provide enough information on socio-economic characteristics, farming environment, hatchery practices and management, marketing practices, problems associated with land-based tilapia hatchery operations and the existing costs and returns and farm-specific technical efficiency of hatchery operators. Such information was vital prior to the adoption of the genetically improved tilapia strain. The authors found that tilapia hatchery operators have the necessary education, tenure security, experience and willingness to adopt new technologies and use better tilapia strains when these become available to them at reasonable prices. However, the estimated technical efficiency of the hatchery operators tested (76 hatcheries) was only 48%. By operating at full technical efficiency, tilapia hatcheries can increase their production from 748,000 to 1,558,000 fingerlings production/cycle/ha. This would result in a significant reduction in production costs and, in turn, make hatchery operations more profitable. These results show the significant lack of technical efficiency among tilapia hatchery operators in the Philippines. Substantial efforts should therefore be made to improve the productivity of hatchery operations and reduce the production cost of tilapia fry and fingerlings. There is also a need to identify improved management practices that reduce the average cost of fingerling production and ensure better broodstock quality.

It is clear that the adoption of any new aquaculture technology should be preceded by careful analysis of economic, socio-economic and technical factors that might affect the success or failure of this technology. In the meantime, if the exciting aquaculture technologies are not efficiently used, improving their efficiency becomes more cost-effective than introducing new technologies. Therefore, evaluation of the efficiency of existing production technologies used for tilapia production is necessary to maximize alternative options for increasing farm productivity (Dey *et al.*, 2000b). In this regard, the technical efficiency of tilapia growout operations in ponds in the Philippines has been examined (Dey *et al.*, 2000c), using data from a project entitled 'Dissemination and Evaluation of Genetically Improved Tilapia Species in Asia' (DEGITA), conducted by the International Center for Living Aquatic Resources Management (ICLARM), in collaboration with the Bureau of Fisheries and Aquatic Resources (BFAR). This study showed extremely important results that would clearly help in developing the efficiency of tilapia farms. The results can be summarized as follows:

1. The total farm area, education and age of the farmers were among the factors affecting technical efficiency. Farmers with a larger farm area, higher age and higher education level attained higher technical efficiency.
2. The introduction of genetically improved farmed tilapia (GIFT) into the Philippines provides a new technology that is likely to improve the productivity and output of tilapia farming.

Once again, the introduction of any new technology into rural areas should consider the age and educational background of the selected candidates who would apply this technology. In addition, capacity building and continuous training and extension services should be provided for farmers and/or farm operators. To support this approach, a research project on educational development activities in support of tilapia aquaculture in the Philippines has been conducted at Central Luzon State University, in collaboration with Florida International University, USA (Brown *et al.*, 2002b). This activity aims at the exchange of students, staff, principal investigators and executives. It also provided an improved aquaculture facility, including experimental units, teaching laboratories and information technology, so that it can benefit a large number of students.

10.3.2. Thailand

Thailand is currently the third top tilapia producer from aquaculture in Asia, only after China and the Philippines, with a production of 100,576 Mt in 2002 (FAO, 2004). Most of this production comes from freshwater aquaculture. About 80% of freshwater fish culture in ponds and 20% of rice fields are devoted to Nile tilapia and red tilapia (Guerrero, 2001). It is no surprise, therefore, that tilapia production is at the top of freshwater fish culture in Thailand. Typically, tilapia is cultured in private/family-owned, small ponds (< 1 ha), in polyculture systems with carp and silver barb (Guerrero, 2001; Pant *et al.*, 2004). Integrated agriculture–aquaculture systems (IAAS) (fish–chicken/duck–pig), in addition to fish-cum-rice, are very common. Integrated culture of tilapia can play a significant role in improving households' livelihood, by providing them with a high-quality, cheap protein source, generating additional income and rehabilitating the soil through better on-farm nutrient recycling (Alsagoof *et al.*, 1990).

Fish is the major traditional animal protein source, even for the poorest people, in the north-east region of Thailand. The demand for fish in the region has exceeded the local supply in recent years, mainly because of the decline of natural fish stocks due to population pressure and environmental degradation (Edwards *et al.*, 1991). This situation has led to the promotion of aquaculture in the region. The prospects of IAAS farming for sustainability of agriculture development in north-east region of Thailand have been evaluated by Pant *et al.* (2004). A survey of IAAS households with varying lengths of experience in four different agroecologies – one irrigated and three rain-fed ecologies (drought-prone, rain-fed lowland and rolling land) – was carried out. This survey collected an extensive amount of information on culture systems, fish species cultured, stocking densities, pond inputs, yield, gross return and the role of ponds in farming system diversification. The main findings of the study were as follows:

1. The fish culture system of the IAAS varied with agroecology but not with the length of farmers' experience in IAAS farming.
2. The size of the pond holding in the rain-fed lowland was smaller than in the other three agroecologies.
3. The highest gross fish yield of 2.3 Mt/ha was recorded in the irrigated agroecologies, and was significantly higher than in the rain-fed ecologies.
4. Over 75% of households raised tilapia and carp in a polyculture system, with nearly 90% in irrigated agroecology, compared with about 50% in the rain-fed lowland. In the drought-prone and rolling land agroecologies, about 75% of households practised tilapia/carp polyculture.
5. The average income of IAAS households from fish varied across agroecologies.
6. Farm ponds played a significant role in the diversification of farming systems in rain-fed areas.
7. Improved management practices and input supply are key factors for increasing fish production in IAAS.
8. Intensification of fish culture through pond fertilization with inorganic fertilizers can improve fish production in IAAS in the region.

It appears from these results that tilapia/carp polyculture makes a significant contribution to the improvement of households' livelihood in north-east Thailand. On the other hand, only about 10% of the households in the region raise tilapia in monoculture in IAAS. The tilapia/carp polyculture system appears appropriate and profitable in many parts of the region because pond inputs, such as rice bran, crop by-products, broken rice and cattle and buffalo manure, are available, mostly on farm, at low cost.

Communal ponds may also have a great potential for small-scale tilapia farmers in north-east Thailand. These ponds are large and deep ponds constructed in north-eastern Thailand for rainwater storage. They are characterized by low fish production and little management, if any, by local people. Hapa culture is an appropriate method to increase the fish production of village ponds, by privatizing parts of the common resource. Middendorp (1988) evaluated the use of communal ponds for small-scale culture of tilapia in hapas. At a recommended stocking density of 150 fish (50 g)/hapa, about 25 kg of tilapia could be produced in 100 days, using commercial catfish pellets (30% crude protein (cp)). The benefit/cost ratio was greater than 67%, with a monthly internal rate of return of about 16%. This clearly indicated that tilapia culture in hapas in communal ponds can provide an additional income for rural villagers.

Small-scale culture of tilapia in hapas in village ponds can be an additional source of income for rice farmers in north-east Thailand. The availability and low cost of culture inputs can significantly reduce investment costs. When rice farmers stocked Nile tilapia in hapas (8 m^3) at an initial weight of 50 g to reach marketable size after 85 days, with a mean specific growth rate of 1.46% and feed conversion ratio of 1.4, investment costs were about 5% of total operating costs (Middendorp and Verreth, 1992). Fingerling costs (about 25% of total costs) could be considerably reduced when fingerlings were collected from backyard ponds or from the hapas ('wild spawning'). Feed costs, which represented about 70% of total costs, could be reduced by optimizing the feeding regime and home mixing of fish feed. Adopting shorter rearing periods was preferred by local rice farmers. However, a 100-day rearing period was suggested, with a stocking density of 14 fish/m^3, to reach a market size of 216 g. If smaller fingerlings are stocked, two rearing cycles will be needed to reach market size. In this way, rice farmers may increase their household income by about 20% (Middendorp and Verreth, 1992). The importance of this type of research is that it provides different culture scenarios for rural villagers so that they can choose the scenario that fits with their potential needs and the available inputs.

10.3.3. Bangladesh

Mozambique tilapia (*O. mossambicus*) and Nile tilapia (*O. niloticus)* were introduced into Bangladesh from Thailand in 1954 and 1974. Both species reproduced rapidly and were distributed in governmental and private fish farms throughout the country. Information has been gathered on the production of these fish in ponds, sewage-fed ponds and lakes, in cages in Kaptai Reservoir and in open inland waters. Tilapia is considered a low-cost operation with good financial returns. This incentive has encouraged poor rural farmers to culture these fishes (mainly Nile tilapia) in small seasonal ponds, ditches and micro-impoundments, which are of little benefit. The progressive expansion of tilapia culture in these rural areas has created an urgent need for technical and extension assistance on tilapia culture. This assistance is essential because farmers in these remote areas generally lack the basic culture principles and adopt primitive methods, which lead to extremely low production.

Unused or underused seasonal ponds and ditches in areas like Mymensingh can be an economic source of food fish and additional income in rural areas in many parts of Bangladesh, by using them for farming tilapia. Farming Nile tilapia in these ponds is simple, requiring very low inputs and labour, and can be undertaken by women and even children. Selling a proportion of produced fish and fingerlings can meet the operational costs and sustain the operation. Gupta *et al.* (1992) reported that a pond of 169 m^2 yielded an average 23.6 kg of fish, with a return on investment of 343%, which is almost equivalent to the national annual consumption of low-income rural households with six family members. As expected, 70% of fish produced is consumed on farm, improving the nutrition and health of farming families. It is clear that tilapia culture in this region is economically viable and 90% of the farmers are happy with this business and willing to expand their operations. Similarly, Gupta (2001) found that the culture of short-cycle species, such as silver barb (*Puntius gonionotus*) or Nile tilapia (*O. niloticus*), in these seasonal ponds was profitable. In addition, landless farmers can benefit from this technology by culturing these fish in common-property roadside ditches.

As mentioned earlier, tilapia culture in Bangladesh is used mainly for subsistence. Increased use of commercial inputs, as suggested by the economic efficiency criteria, may not be realistic, bearing in mind the subsistence nature of tilapia farms and the limited resources available for small-scale rural farmers. Therefore, assured availability of commercial inputs and cash have been suggested as the major prerequisites for increasing the production and income from tilapia aquaculture by small-scale farmers.

The extension service in rural aquaculture has not been given the necessary attention despite the crucial role it plays in aquaculture development. Extension programmes can provide small-scale farmers with services, such as farm visits, technical advice, training and demonstration. Ahmed *et al.* (1995) evaluated the effects of adopting a new strategy of extension for the improvement of polyculture of Indian, Chinese and common carp, culture of Nile tilapia and culture of silver barb within the existing farming systems of Bangladesh. The main objective was to assist farmers to adopt fish culture in small ponds and ditches. Farmers responded positively in respect of stocking density, species ratio and input use. By adopting new aquaculture practices, the farmers could produce much more fish (9.89 kg/40 m^2) than the baseline production (2.2 kg/40 m^2) and earned more income per unit area of their ponds. The average net profit–cost ratio was estimated at 2.44. The authors concluded that extension services can increase the production and benefits of farmers who already have ponds and/or ditches.

Alongside this line, the culture of GIFT has been tried in Bangladesh as a means of increasing tilapia production and economic return. The results showed that the GIFT yield was much higher than that of red tilapia and other existing strains. In order to propagate the culture of GIFT in Bangladesh, Dey *et al.* (1997) conducted a survey of the socio-economic status of small-scale farmers, marketing systems and consumption patterns. They found that both actual and potential fish farmers are generally small-scale operators. Farmers with farms of less than 0.4 ha have lower productivity and profitability than those with larger farms. Nevertheless, all farmers can make a profit regardless of farm size. When the demand for fish was analysed in Bangladesh using an econometric estimation procedure, the estimates for fish demand were essential in assessing the impact of technology advancement, such as genetic improvements (Dey *et al.*, 1997). As mentioned in Section 10.1, the demand for fish was found to be both price- and income-elastic. An inverse relationship occurred between total expenditure elasticity and income level, whereas a positive relation was found between price elasticity and average income. It is expected therefore that a technological breakthrough in small-scale tilapia culture will improve the livelihood of poor consumers at a higher rate compared to the rest of the population (Dey *et al.*, 1997).

10.3.4. Other Asian countries

Tilapia culture started to gain popularity in some other Asian countries, and is expected to play a noticeable role in aquaculture development in these countries. Little *et al.* (2000) reported that semi-intensive aquaculture in Sri Lanka in on-farm ponds depends on hatchery-produced tilapia. Cage-based fattening of tilapia was also introduced as a means of poverty alleviation in some rural areas in Sri Lanka (Pollock and Little, 2002). In Vietnam, Pekar *et al.* (2002) reported that tilapia was the third most important fish species stocked in integrated systems in the Mekong Delta (13–21%), after silver barb (26–34%) and common carp (19–26%). Tilapia are also more attractive in Hanoi because they reach their marketable size faster than carp. Tilapia culture is also spreading in many parts of India (Nandeesha, 2002). In Kolkata, for example, tilapia are the fish most in demand by both poor and rich consumers. In a recent review on tilapia farming in India, Natarajan and Aravindan (2002) reported that tilapia has been successfully cultured in sewage-fed ponds, rice fields and brackish-water ponds in various parts of India. Tilapia culture has also become well established in Bheries in West Bengal.

10.4. Tilapia and Rural Development in Africa

Aquaculture practices in Africa are very limited, despite the fact that experimental trials were conducted on fish farming in countries like Egypt in the early1930s. The contribution of aquaculture to food security, poverty alleviation and overall economic development in the continent is small. This is despite the high potential for aquatic genetic resources and water adequacy in many parts of Africa (Kapetsky, 1994). About 56–70% of Africa's lands have the highest production potential for warm-water fish such as tilapia, carp and catfish (Aguilar-Manjarrez and Nath, 1998). The FAO has also reported that 37% of sub-Saharan Africa is suitable for small-scale fish farming, and by the year 2010, 35% of Africa's fish needs could be met by small-scale aquaculture (Aguilar-Manjarrez and Nath, 1998). Nevertheless, there are several constraints that hamper the development of aquaculture in Africa. ICLARM–GTZ (1991) and Brummett and Williams (2000) listed these constraints as follows:

- Severe political instability in many African countries limits the development of agricultural sectors, including aquaculture. Such political problems also threaten the economic viability of aquaculture enterprises and security of investment.
- Poverty of consumers, leading to small local markets and reliance on external markets.
- The poor infrastructure, including poor storage, poor marketing channels and poor maintenance of farming facilities.
- Poor quality of governmental services, such as extension.
- Sociocultural constraints.
- Shortage of good-quality seed.
- Shortage of experienced labour.
- Shortage or lack of governmental budgets allocated for aquaculture.
- Lack of local expertise and technical experience, which may increase the risk of enterprise and limit technical options.

One other major constraint facing the development of rural aquaculture in Africa is its dependence, to a large extent, on external donations and technologies. In most cases, the farmers do not have the necessary experience and training in these new technologies, and the governments do not provide the necessary support (Moehl, 2002). The situation is getting worse because external funds have become very limited. As a result, the numbers of fish farms and farmers have increased dramatically, but production has not grown as progressively as planned for. In fact, the yield has declined in many cases, making the enterprises a big loss. Also, despite the fact that several aquaculture research and extension stations have been built in many African countries, a large part of them have been abandoned due to the lack of experience and funds. Therefore, it is unwise to adopt any new aquaculture technology in rural areas in Africa before 'setting the stage'. This means that appropriate training and extension programmes should be prepared first.

Small-scale African fish farmers are better off, and better educated, than non-fish farmers (Sen *et al.*, 1996). To alleviate the poverty of these rural fish farmers, appropriate extension and technology supports should be provided for them. The technology adopted and the level of extension services, rather than the financial situation or the level of formal education, are the driving force for the success of small-scale tilapia culture in rural, poor Africa. For example, Van der Mheen-Sluijer (1991) found that small-scale aquaculture adopters in the Eastern Province of Zambia had a lower income than non-adopters. The author attributed this to the extension service, which had focused on distributing aquaculture information to all small-scale farmers rather than to better-off farmers only.

Despite the constraints and difficulties facing rural aquaculture development, smallholder fish farms are increasing in many African countries and are integrated in the farmers' existing farming systems. These small-scale farms rely mainly on locally available inputs, with little investments, and that only in the form of land, water and labour (Brummett and Williams, 2000). They play an important role in increasing production efficiency, generating extra income, poverty alleviation and food security and improving overall household livelihoods in rural areas. It is estimated that more than 95% of African aquaculture production comes from these small-scale farms, with tilapia and/or African catfish being the dominant farmed species. Relatively large-scale, commercial farms are also raising tilapia as a cash crop in some African countries, such as Egypt, Kenya, Nigeria, Zambia and Zimbabwe.

It should be emphasized here that the decision making regarding rural aquaculture development must consider the social factors and not only economic factors. Ignoring these non-financial factors, such as family context, social status and food habits, may lead to the failure of rural aquaculture development initiatives. For example, the assessment of the potential of mixed-species productivity profit in temporary wetland ponds in northern Nigeria revealed a conflict between economic gain and community acceptance of the products (Thomas, 1994). Another striking example is common carp culture in Egypt, where local acceptance of this fish is extremely low, despite the fact that it is widely cultured with tilapia in semi-intensive systems (A.-F.M. El-Sayed, Alexandria, 1999, personal observation). As a result, the price of carp has sharply declined, causing heavy economic losses to many farmers, who have completely abandoned carp culture.

10.4.1. Tilapia and food security

Freshwater aquaculture dominates aquaculture systems in Africa. Tilapia, especially Nile tilapia, are the most important farmed fish in Africa, despite their slow expansion (about 2.5%/year) (Jamu, 2001). They contribute over 30% of total aquaculture production. Tilapia culture in small ponds (200–500 m^2) remains the most common culture practice in Africa (Fig. 10.3), contributing about 38–93% of total tilapia production in many countries (FAO, 1995a). Other farming systems, including tanks, cages (Fig. 10.4) and raceways, are also adopted in some areas, though at much smaller scales. Thus, tilapia farming has a great potential for improving livelihoods in rural Africa. The available information, despite being very scarce, has shown that tilapia culture has significantly improved the health of rural households and increased their income in various parts of the continent.

Rural aquaculture in Rwanda is a good example of how rural households depend on their own fish production for obtaining their fish needs. Over 90% of Rwandan farmers consumed fish (mostly tilapia) from their own ponds, and 28% of their total fish production is generally used for home consumption, whereas 56% is sold. This means that farmed fish is used for family consumption and also as a cash crop.

Raising fish in agricultural areas can also be more profitable than traditional land crops. Research has been conducted in Rwanda to assess the feasibility of small-scale tilapia culture compared with the production of land crops, namely sweet potato, Irish potato, cassava, taro, sorghum,

Fig. 10.3. Tilapia culture in small earthen pond in rural Kenya (photo provided by R. Brummett).

Fig. 10.4. Tilapia culture in small cages in agricultural drainage canals in a rural area in Behaira Governorate, Egypt.

maize, red beans, soybeans and cabbage (Hishamunda and Jolly, 1998). Tilapia culture was the second most expensive way of protein production after cassava, while soybean was the cheapest source of protein for rural farm households. Tradewise, tilapia farming was competitive with the other enterprises for the use of the scarce resources in Rwanda, and was the most promising enterprise for improving the welfare of rural households in a subsistence economy, like Rwanda. The authors came up with the following important conclusions:

- Tilapia farming was the most profitable enterprise in terms of income above variable costs and net return to land, labour and management.
- Tilapia culture was the most expensive way of supplying farmers with protein after cassava.
- Fish culture is an important activity in increasing farm cash income and improving households' welfare.

Under Rwandan conditions, and similar conditions in other African countries, tilapia are better farmed as a cash crop, where the cash generated from their sale can be used for buying the necessary commodities, including cheaper protein sources. Fish farming is also a strong competitor with horticulture for land fertilization in Rwanda.

Enterprise analyses indicated that fish farming yielded the highest net return to land, labour and management compared to land crops (sweet potatoes and soybean). Once again, these analyses support the argument that tilapia is better used as a cash crop.

The role of tilapia farming in food security has also been documented in other African regions. In North Western Province in Zambia, for instance, more than 50% of farmed fish is sold, making the fish a major source of cash (Wijkstrom and Larsson, 1992). In Eastern Province, cash income from tilapia farming has attracted the attention of many crop farmers towards fish farming (Van deer Mheen-Sluijer, 1991). In Ruvuma Region, Tanzania, fish farming has rapidly grown, to the extent that it was ranked as the third most important source of income by farmers with fish ponds (Seki and Maly, 1994). Similarly, Vincke and Schmidt (1991) suggested that economic return from tilapia farming was higher than that generated from coffee, dairy or poultry.

10.4.2. Aquaculture/agriculture integration

For small-scale tilapia culture to be sustainable and profitable, it should be integrated with broader rural development initiatives, especially with agricultural farming activities. This type of integration can also lead to a significant improvement in the cash income of farmers' households in Africa. If properly integrated, tilapia culture can lead to more fish production, better diversification and enhancement of the farming systems and retaining environmental and social sustainability (Brummett and Williams, 2000). Research in Malawi (Brummett and Noble, 1995; Brummett, 2002b) has shown that fish ponds integrated into farm activities, such as rice–fish integration, in such a way that they recycle wastes from other agricultural and household enterprises can increase production and profitability. Fish ponds integrated with vegetable gardens in Malawi significantly increased fish and vegetable productivity (Brummett and Noble, 1995). Vegetable by-products are generally used as pond inputs, whereas pond water is used for irrigating garden vegetables. With continuous practice and appreciation of the system, farmers become able to further improve productivity and profitability. The success of small-scale integrated system generally encourages other farmers to adopt it. This generally leads to a rapid spread of this system from farmer to farmer.

Switching from smallholder tilapia production (i.e. production for subsistence) to a more commercial practice requires additional inputs and new technologies for the storage and/or processing of feed ingredients. This will add more operating costs, but can still be highly profitable. In Malawi, for example, when commercial production systems in rural areas are properly conducted, fish production can be more profitable than most other cash crops (Chimatiro and Janke, 1994).

Although it can be highly profitable, small-scale aquaculture/agriculture integration is faced with many constraints that may render the system unable to achieve its objectives (Koffi *et al.*, 1996). This is mainly because farmers are generally unable to innovate in respect of their choice of economic strategies. Moreover, the socio-economic constraints of the farmers are totally unknown, and the internal economic system of the rural development projects often fails to propose production units that could be accessible to the farmers.

Despite these constraints, a number of research projects have considered the integration of fish culture with other agriculture activities in Ivory Coast (Copin and Oswald, 1993; Koffi *et al.*, 1996), Malawi (discussed above) and Ghana (Ruddle, 1996). These projects have shown that rural aquaculture integrated with agriculture and the use of agricultural wastes as pond inputs were highly profitable. Liti *et al.* (2002) evaluated the growth performance and economic efficiency of Nile tilapia/African catfish raised in polyculture in fertilized ponds in Kenya for 80 days. Feed inputs included commercial pig finisher pellets (12.5% cp), rice bran (6.5% cp) or a test diet (12.5% cp). Rice bran exhibited the poorest growth and lowest production. But the cost/benefit analyses indicated that rice bran had the lowest break-even price and the least investment cost and, in turn, was economically better than the other diets.

An interesting approach regarding reconsidering rural fish farming development in Africa and how tilapia culture can be well integrated in smallholder farming systems has been described by Oswald *et al.* (1997). These authors reported that the process that aimed at improving fish farming efficiency in the early 1990s had failed and the productivity of smallholders who had adopted fish farming was lower than that of smallholders who had not attempted such diversification. After communication with fish farmers, they realized that the stocking density of Nile tilapia must be decreased, since the quality requirements of the market could only be met at low fish densities. This finding has imposed a new pond planning model, 'pond-dam', and the initiation of a new innovation process. The pond-dam model was well adapted to the local economy and integrated with spontaneous swamp rice cultivation in fish ponds. It is clear from this particular example that fish farming methods should be compatible with local agricultural development processes.

10.5. Tilapia and Rural Development in Latin America and the Caribbean

Latin America and the Caribbean region have a wide variety of ecosystems with strong potential for developing aquaculture activities. The inland and coastal waters of Brazil and Mexico are by far the largest and most promising ecosystems for aquaculture development. Thus, the potential for aquaculture in the region is high.

Studies that have analysed the contribution of rural aquaculture to household economies in Latin America and the Caribbean are scarce (Charles *et al.*, 1997). Only general estimates, based on gross figures of total production or employment (by species or production system), were found. The impact of aquaculture on rural livelihoods is rarely evaluated. Meanwhile, the impacts of aquaculture on other non-aquaculture economic activities are generally made in terms of effects on specific factors, such as water and soil quality, sedimentation and diseases (Molnar *et al.*, 1996).

According to Martinez Espinosa (1995), rural aquaculture in Latin America can be divided into two categories: (i) the 'poorest of the poor' aquaculture, which is characterized by very low input and cost and very low output as well; and (ii) the 'less poor' aquaculture, characterized by low/medium inputs and low/medium outputs. The second type of aquaculture is believed to have an excellent potential for rural development. This category has already received assistance from governments and international organizations such as FAO, for the formulation of general plans for the development of rural aquaculture in the region. Martinez Espinosa (1994) evaluated the possibilities of the development of aquaculture in the rural environment, in the framework of the AQUILA II development project. Two pilot cases were selected for the proposed study, one in Barrancabermeja in Colombia and the other in San Cristobal in Venezuela. Various strategies for the formulation of a plan of development were suggested and discussed. The results indicated that small-scale farmers were present with adequate characteristics to carry out this type of aquaculture. It was also concluded that rural aquaculture has a high potential if technical, economic and financial support is provided.

As a result of the above-mentioned development projects, rural tilapia culture in small-farm ponds in south-west Colombia has been improving (Sere, 1988). Nile tilapia is the dominant species in both small-scale farms and in commercial farms. Extensive efforts have also been paid to improving seed production and controlling unwanted breeding.

Tilapia has accidentally become among the most highly valued culture species in Ecuador, as a result of unique and unexpected circumstances. The Ecuadorian National Chamber of Aquaculture believes that the tilapia culture industry is a consequence of the two major pathological crises that have occurred in the shrimp industry in recent years. The first occurred in 1993 in the province of Guayas, which is a major shrimp production centre, where taura syndrome attacked the farms and caused massive losses. The second crisis occurred in May 1999, when the white spot syndrome virus (WSSV) hit the coasts of Ecuador and infected all the shrimp farms, causing great loss. Since then, tilapia has been considered as a potential candidate for culture in Ecuador (Watanabe *et al.*, 2002).

In Venezuela, many tilapia farms are integrated with small farms or ranching operations, while some farmers grow tilapia with *Clossoma macropomum* in a polyculture system (Watanabe *et al.*, 2002). Tilapia is cultured mainly for domestic consumption, but small amounts of processed products are exported to Colombia, France and the USA.

Among all fish species suggested for rural aquaculture in Latin America and the Caribbean, the semi-intensive culture of tilapia seems to be the most profitable. Alceste *et al.* (2001) reported that small-scale as well as commercial, large-scale tilapia culture is currently practised in many Latin American countries. Tilapia appears very advantageous in countries like Costa Rica, Brazil and Venezuela. In other countries, with limited resources, such as Guatemala and Panama, tilapia culture is practised mostly for local consumption and for the diversification of rural activities related to agriculture and animal husbandry (Alceste *et al.*, 2001). Mixed-sex Nile tilapia have been cultured in family- and communally managed fish ponds in Panama and Guatemala since the 1980s for family consumption and as a cash crop (Lovshin, 2000b). Tilapia culture systems were originally designed to allow farmers without any experience in fish culture to produce their own fish from 'egg to market size', using cheap and locally available inputs, such as household by-products and animal manure.

Family-scale Nile tilapia culture in Guatemala is becoming more popular in rural areas. Small-scale farmers raise tilapia using low-quality nutrient inputs, including agricultural and food wastes; organic and chemical fertilizers are commonly used. This type of culture is very profitable for rural families. Many fish farms are integrated with chicken broilers or layers. A 100 m^2 pond with 48 broilers can produce 38 kg of fish in 6 months, while 17 layers can produce 29 kg during the same period (Popma *et al.*, 1995). Popma *et al.* (1995) also reported that annual fish production from a typical 120 m^2 pond was 48 kg of fish, of which 20 kg were sold and 23 kg were consumed by the producer's family. As a result, the average annual per capita consumption of fish among producers' families increased from 0.5 kg to 3.3 kg/year. On-farm income has also increased by 18% as a result of selling farmed fish, despite the fact that fish ponds occupied only 2% of total land.

In contrast, a dim picture of tilapia culture in Panama and Guatemala has been reported by Lovshin (2000b). This author evaluated the success and sustainability of tilapia culture in these two countries in 1998. He found that, in Guatemala, the mixed-sex culture system introduced to Guatemalan pond managers in the 1980s is still the only system employed by tilapia farmers. Forty-three per cent of pond projects were abandoned between 1989 and 1998. Seventy-two per cent of the farmers produce their own small tilapia for pond stocking or obtain them from neighbours. The number of predator fish (*Cichlasoma managuense*) used to control seed production in grow-out ponds increased from 14% in 1989 to 36% in 1998. There has always been a deficiency in seed supply by governmental hatcheries. Even when the fingerlings were available, it was difficult for the Guatemalan government to transport them to countryside; moreover, the majority of farmers lack the means to travel to get fingerlings from these hatcheries. Two out of three governmental hatcheries visited were closed due to inadequate funds; only one hatchery remains in operation.

In Panama, pond managers were more flexible and started to grow either mixed-sex tilapia with guapote tiger as a predator (29% of the farmers) or monosex (male) Nile tilapia (71%). However, 48% of the tilapia projects were abandoned, and 36% of the existing farms culture mixed-sex tilapia with a predator and 64% produce all-male fish. Most pond managers did not produce their own seeds, since 82% of fish pond projects receive their tilapia fingerlings from the government.

It appears from these results that rural tilapia culture in ponds in Guatemala and Panama was not successful or sustainable. Lovshin (2000b) attributed this failure to the following:

1. Lack of technical assistance.
2. Lack of adequate feed and fertilizers.
3. Lack of permanent water source and/or leakage from ponds leading to seasonal drying.
4. Theft (Guatemala).
5. Social problems among pond management members (Panama).

Attempts to promote small-scale, rural tilapia culture in the rest of Central America (Belize, Costa Rica, El-Salvador, Honduras and Nicaragua) have also ended in failure (Meyer, 2001). For over three decades, governmental and private organizations have been promoting tilapia culture in the region as a means of producing low-cost, high-quality animal protein for the poor. However, the promotion has faced several limitations and difficulties that made developmental processes very slow. Meyer (2001) diagnosed these limitations as follows:

- *Poor extension*: Tilapia culture promotion has always been implemented by extension agents and technicians with little or no knowledge of tilapia biology and culture techniques. This problem was tackled in the early 1980s by the Pond Dynamics/Aquaculture Collaborative Research Support Programme (PD/CRSP), Oregon State University. This programme played an important role in developing reliable information and technical bases for successful tilapia culture in the region.

- *Poor site selection*: Many ponds have been constructed without adequate conditions for the success of tilapia culture. For example, many ponds were built at very high elevation, where water temperature is generally lower than the optimum range for tilapia culture.
- *Inadequate water sources*: Many ponds were constructed in areas with limited water resources. It is known that many streams in the region are temporary, where water flow diminishes or stops during the dry season and increases during the rainy season. The lack of water resources and water management generally limit aquaculture and agricultural development in the region.
- *Inadequate technical assistance and training*: There is an acute shortage of technical assistance and training provided for tilapia farmers in rural areas. Most non-governmental organizations that are committed to aquaculture promotion lack the necessary resources and knowledge to adequately transfer the technology to the farmers effectively.
- *Fish theft and predation*: Because Central American republics are generally poor countries, theft of agricultural crops, including farmed fish, is very common. Theft can cause a great loss for tilapia farmers. Several local and migratory avian species are also known to prey on tilapia, especially red tilapia fingerlings because of their distinguishable colour. This can cause complete elimination of tilapia fingerlings if the ponds are not protected (e.g. by netting covers).
- *Limited seed availability*: As in many other Latin American regions, seed availability at reasonable cost remains one of the major factors that limit tilapia culture development. The principal suppliers of tilapia seeds are national hatcheries. As mentioned earlier, these hatcheries lack the necessary funds and technical support. Therefore, their seed production does not meet the increasing demands of tilapia farmers.

To avoid, or at least minimize, the above limitations, the development of tilapia culture programmes in rural areas in Latin America and the Caribbean (as in Asia and Africa) should be preceded by careful analyses of social status of rural households, available culture resources and inputs, technical and training assistance, funds and donations and extension services.

10.6. Closing Remarks

1. Rural aquaculture is a multidisciplinary practice, including technical, economic and socio-economic aspects, extension services and policy aspects. It is necessary that all these factors be considered for efficient and sustainable development of rural aquaculture.
2. If the existing aquaculture technologies are not efficiently used, improving their efficiency becomes more cost-effective than introducing new technologies.
3. Rural aquaculture development should fully consider social factors such as family context, social status and habits.
4. The introduction of any new technology into rural areas should consider the age and educational background of the selected candidates who would apply this technology. Capacity building and continuous training and extension services should be provided for farmers and/or farm operators.
5. Rural aquaculture development in Africa relies mainly on external donations and technologies, with little local experience and/or governmental assistance. As a result, several enterprises have failed to achieve their objectives.
6. The adoption of any new aquaculture technology should also be preceded by careful analysis of economic, socio-economic and technical factors that might affect the success or failure of this technology. Well-prepared extension programmes should also be used to help farmers adopt these new technologies.
7. Tilapia culture in rural areas is practised mainly for subsistence. Increased use of commercial inputs may not be realistic, because of the subsistence nature of tilapia farms and the limited resources available for these rural, small-scale farmers.
8. Although the development of small-scale tilapia culture in rural, poor African regions is affected by many economic and social factors, such as family context, social status and habits, it seems that the technology adopted and the level of extension services play a crucial role in the success or failure of rural aquaculture in these regions.

11

Recent Technological Innovations

11.1. Introduction

Tilapia culture is expanding globally at a very high rate, with a 12.2% average annual increase in production during the past decade. Currently, tilapia are cultured in more than 100 countries all over the world. As a result, aquaculture in general and tilapia culture in developing countries in particular are expected to face many challenges over the coming years, in order to meet the increasing demand of growing populations. Improvements in tilapia broodstock, reproductive performance and seed production, hatchery and grow-out technologies, stress and disease control, water quality management and nutrition and feeding mechanisms will be the major challenges facing tilapia farmers, researchers and farm managers.

Modern technologies, particularly biotechnology, will play the core role in confronting these challenges. Biotechnology can be used for growth enhancement, improvement of reproduction and early development of farmed organisms, enhancement of ecological sustainability, satisfying new markets for farmed products and conserving aquatic biodiversity (Subasinghe *et al.*, 2003). Recent technologies in molecular biology and genetics have already been involved, and will continue to be highly involved, in tilapia culture development, with an ultimate goal of food security, poverty alleviation and income generation in developing countries. This chapter throws some light on the recent technological innovations in tilapia culture, with emphasis on the following areas: genetics and reproduction, including transgenesis and selective breeding, disease management, feed and feeding and farming systems.

11.2. Transgenesis

The application of genetic principles in aquaculture, for increasing the production of farmed aquatic animals, lags far behind their application in the plant and livestock sectors. Very little attention has been paid to the genetic improvement of most farmed aquatic species despite the great potential of biotechnology in this field. However, tilapia are one of few fish groups that have been the subject of reproduction and genetic improvements during recent years. Extensive research has been carried out on the genetic development of farmed tilapia in many parts of the world during the past two decades. Transgenesis is a modern genetic biotechnology that has received great attention in recent years.

Transgenesis is the introduction of any foreign gene of interest into the genome of plants and animals by gene transfer technology. When the introduced gene is integrated, inherited and expressed, the transgenic organisms acquire new genotypes and phenotypes, depending on the nature of the gene introduced and the specificity and strength of the promoters driving the expression (Hew and Fletcher, 2001). Transgenic technology is one of the most promising technologies for generating rapid genetic improvements. This technology is expected to become more beneficial to aquaculture in developing new fish strains than traditional breeding programmes. Transgenesis is also expected to witness continuous improvement and development in order to produce less costly and more affordable fish, and will be applicable in the developing countries. According to Hew and

Fletcher (2001), transgenesis in aquaculture can be used for the following:

1. Enhancement of growth, feed conversion ratios and survival.
2. Increasing disease resistance against pathogens.
3. Controlling sexual maturation, sterility and sex differentiation.
4. Improving fish adaptability to extreme environmental conditions, such as cold resistance.
5. Altering the biochemical characteristics of the flesh to enhance its nutritional qualities.
6. Altering biochemical or metabolic pathways to improve food utilization.

In addition, transgenic tilapia can be used as biofactories for producing pharmaceutical products. Growth enhancement and cold resistance of farmed transgenic tilapia have received more attention than the other uses.

11.2.1. Growth enhancement

The improvement of growth and feed conversion efficiency of farmed fish has been the principal objective of transgenic studies. Transgenic fish, including tilapia (Rahman and Maclean, 1999; Rahman *et al.*, 2001), salmonids (Delvin *et al.*, 1994), common carp (Chatakondi, 1995) and channel catfish (Chitminat, 1996), containing an exogenous growth hormone gene construct, exhibited better growth and feed efficiency than their non-transgenic siblings. The increase in the growth rates of transgenic fish ranges from 20 to 1000%, compared to non-transgenic controls (Rahman *et al.*, 2001).

Tilapia have many attributes that make them ideal candidates for transgenic studies. These include: (i) frequent, aseasonal reproduction (i.e. eggs are constantly available for research studies); (ii) short generation time (5–8 months); (iii) easy reproduction in captivity; (iv) embryos and fry can be easily reared; (v) embryos are semi-transparent, so that some reporter genes can be detected without sectioning or sacrifice; and (vi) eggs can be fairly easily injected via the micropyle. Therefore, extensive research has been carried out on transgenic tilapia over the past few years. As a result, some of the lines of growth-enhanced tilapia have been commercially marketed (de la Fuente *et al.*, 1999).

Transgenic tilapia have been produced in many countries since the late 1980s (Brem *et al.*, 1988). However, the Centre of Genetic Engineering and Biotechnology in Havana, Cuba, and the Division of Cell Sciences, School of Biological Sciences, University of Southampton, UK, have been taking the lead for many years until they have become the chief centres researching transgenesis in tilapia. Some of the studies from these two laboratories, together with other studies concerning tilapia transgenesis, are discussed in this section.

Rahman and Maclean (1999) studied the growth performance of transgenic tilapia containing a Chinook salmon growth hormone (GH) gene construct spliced into the ocean pout antifreeze gene regulatory sequence. The initial transmission rate from generation G0 to G1 was less than 10%, with mosaic distribution of the transgene in the germ cells, but transmission rates from G1 to G2 followed the expected Mendelian ratio. Meanwhile, transgenic tilapia lines exhibited dramatic growth enhancement. The average weight of the G1 and G2 transgenic fish was three times heavier than that of their non-transgenic siblings. Long-term and short-term evaluation of growth and feed efficiency of transgenic Nile tilapia containing an exogenous fish growth hormone gene have also been investigated by Rahman *et al.* (2001). In a long-term trial, the authors found that transgenic tilapia showed a 2.5-fold increase in growth compared with non-transgenic siblings. After a rearing period of 7 months, the average weight of transgenic tilapia was 653 g compared with 260 g for non-transgenic fish. Feed conversion efficiency was also more than 20% greater in the transgenic fish. In a shorter-term trial (4 weeks), the growth rate of transgenic fish was about four times higher than that of non-transgenic siblings. Furthermore, apparent protein and energy digestibility were higher in the transgenic fish. This means that transgenic tilapia utilized the feed more efficiently than non-transgenic fish. Similar growth enhancement of transgenic tilapia has been reported by Maclean *et al.* (2002). These authors emphasized that farmed transgenic fish should be sterile. Sterility can be achieved through triploidy, gene knockout of crucial hormone-encoding genes via homologous recombination, and knock-down of the function of the same genes via ribozyme or antisense technologies.

Growth enhancement and feed efficiency in transgenic tilapia (*Oreochromis* sp.) carrying a single copy of a homologous complementary DNA (cDNA) growth hormone (GH cDNA) have also been evaluated by Martinez *et al.* (2000). Transgenic tilapia consumed about 3.6-fold less feed than non-transgenic siblings. Subsequently, the feed conversion efficiency was 290% higher in transgenic tilapia than in the control group. Growth efficiency, synthesis retention, anabolic stimulation and average protein synthesis were also higher in transgenic than in non-transgenic tilapia. Clear metabolic differences were also found in transgenic juvenile tilapia, including differences in hepatic glucose and in the level of enzymatic activities in target organs. The authors concluded that GH-transgenic juvenile tilapia showed altered physiological and metabolic conditions and were biologically more efficient than the control fish.

The growth hormones of different animals have been used for induction of transgenesis in tilapia. Hernandez *et al.* (1997) characterized transgenic tilapia lines with different ectopic expression of tilapia growth hormone (tiGH) cDNA. The tilapia GH was used to construct chimeric genes expressing different levels of tiGH *in vitro* and *in vivo*. These constructs were, in turn, used to generate different lines of transgenic tilapia. Various patterns and levels of ectopic expression of tiGH were, therefore, detected in transgenic tilapia. The lines with lower ectopic tiGH mRNA levels were the only ones that showed growth acceleration. This result suggested that the expression of ectopic tiGH promoted growth only at low expression levels. Another technique used for producing transgenic tilapia was the expression of recombinant common carp growth hormone (rcGH) in yeast (*Pichia pastoris*) (Li *et al.*, 2003). The growth of juvenile tilapia injected with purified rcGH was 24.5% and 53.1% higher than the control at a dose of 0.1 and 1.0 μg/g body weight/week. However, the chemical composition of fish muscles was not significantly affected by the rcGH treatment.

More recently, the poultry genome has been used to induce transgenesis in Nile tilapia (Helmy, 2004). Poultry DNA was incorporated into Nile tilapia broodstock (80–100 g average weight) at 10 and 20 μg/fish. The growth and feed conversion ratio (FCR) of transgenic siblings, fed a 30% crude protein (cp) diet, were exceptionally higher than those of non-transgenic controls. Transgenic fish were 88 and 42% heavier at 10 and 20 μg/fish than the control counterparts.

11.2.2. Cold resistance

Several cold-water marine and freshwater fish inhabiting ice-laden environments are well adapted to freezing water. To avoid freezing, these fish are able to produce a unique group of proteins, known as antifreeze proteins (AFPs) or antifreeze glycoproteins (AFGPs), which can interact with ice crystals and effectively lower the freezing temperature (Ewart *et al.*, 1999). The administration of AFPs in temperate and tropical fish species may improve their cold resistance during cold seasons (e.g. during overwintering). This assumption has been tested by Wu *et al.* (1998), who studied the effect of AFP on cold tolerance in juvenile Mozambique tilapia and milkfish (*Chanos chanos*) exposed to low temperature. The AFP was administered via anal injection or by feeding at 0 (control) or 20 μg AFP/g body weight every 2 days for six doses. The fish were later subjected to a cold-tolerance test. In the injection trial, the mortality of tilapia 24 h after the transfer from 26°C to 13°C was 53.3% in the control and only 14.3% in the AFP-treated group. In the feeding trial, the AFP was included in the feed at 0 (control), 100 or 1000 μg AFP/g body weight and fed to tilapia juveniles at a rate of 20% body weight/day for 12 days. The mortality at 24 h after the transfer to 13°C was 60% in the control, 41.9% in the 100 μg/g AFP group and 3.4% in the 1000 μg/g AFP group. Milkfish juveniles were given the AFP via anal injection of 100 μg AFP/g body weight, 100 μg bovine serum albumin (BSA)/g body weight or saline every 2 days for six doses. After injections, milkfish were treated by gradually decreasing temperature within 4–5 days from 26°C to 16 or 13°C. The mortality of milkfish was 66.7–100% in the controls and 13.3–33.3% in the AFP group. These results clearly demonstrated that AFP is able to enhance the tolerance of tilapia and milkfish juveniles for low temperatures.

11.2.3. Producing pharmaceutical products

There is a great possibility that transgenic tilapia can be used as biofactories for producing valuable pharmaceutical products (Maclean *et al.*, 2002). For example, Wright and Pohajdak (2001) developed transgenic tilapia containing a humanized insulin gene. Humanized insulin encodes human insulin α and/or β chains while using fish-preferred codons and regulatory sequences. Because tilapia and human insulins are different in certain amino acids, the authors cloned, sequenced and modified the tilapia insulin gene by site-directed mutagenesis, resulting in a tilapia insulin gene that codes for humanized insulin while still maintaining the tilapia regulatory (non-coding) sequences. These humanized genes are expressible in fish islet cells, providing transgenic fish with islet cells containing and capable of expressing humanized insulin genes. These islet cells can be harvested, encapsulated and transplanted into diabetic patients. In this way, normoglycaemia can be achieved in these recipient organisms.

11.2.4. Risk assessment

It appears from the previous discussion that transgenesis would offer several advantages for tilapia culture (high growth rates, cold resistance, etc.). However, the rate of genetic change in transgenic tilapia is such that their phenotypic and behavioural properties cannot easily be predicted (Mair, 2002). Marketing of these fish also faces many constraints. Consumer response to such genetically modified fish is negative in many countries, to the extent that their culture could be an economic risk. At the same time, they may have adverse ecological effects when they are introduced into new environments. Therefore, food safety assessment is necessary before consuming these fish on a commercial scale (de la Fuente *et al.*, 1996). The environmental impacts and risks associated with the adoption of transgenic tilapia for aquaculture activities should also be adequately and carefully assessed. Necessary protective measures must also be taken before introducing these fish into national aquaculture programmes (see Chapter 12 for details).

In this regard, Guillen *et al.* (1999) evaluated the behaviour of transgenic tilapia in comparison with wild tilapia as a way of assessing their environmental impact in Cuba. They also evaluated the safety of consuming transgenic tilapia as food. Transgenic tilapia had a lower feeding motivation and dominance status than wild tilapia. Food safety assessment indicated that growth hormone contained in transgenic fish has no biological activity when administered to non-human primates. No effects were detected in human volunteers after the consumption of transgenic tilapia. It would appear from these results that transgenic tilapia have no environmental impacts, and they are also safe for human consumption. However, more long-term work is needed along this line before a final conclusion can be drawn regarding the impacts of transgenic tilapia on human health and on the aquatic environment.

11.3. Genetically Improved Farmed Tilapia (GIFT)

One of the most recent technological developments in tilapia culture is the production of genetically improved farmed tilapia (GIFT) through a selective breeding project. The GIFT project was a multidisciplinary research and development programme implemented by the International Center for Living Aquatic Resources Management (ICLARM) (the name has recently been changed to World Fish Center) and co-partners (such as the United Nations Development Programme (UNDP)) to develop a selective breeding programme for Nile tilapia (Eknath and Acosta, 1998). It was one of the longest international research projects, since it started in 1988 and ended in 1997. The project was based in the Philippines, using a combined selection methodology on a synthetic base population developed from tilapia populations newly introduced from Africa and domesticated in Asia. The GIFT project focused on growth studies of tilapia, in addition to studying the genetic parameters of other important traits, such as size at first maturity, survival, disease resistance, skin colour, body conformation and cold tolerance. The following sections discuss the studies conducted on some of these traits.

11.3.1. Growth and survival

The genetic gain of GIFT has been documented by a number of authors. An estimated heritability of 0.24 for growth in Nile tilapia has been reported by Eknath *et al.* (1993). Further investigations on genetic gains of Nile tilapia over successive generations have also been reported (Eknath and Acosta, 1998). These investigations revealed that GIFT achieved 13% genetic gains over five generations, providing an estimated cumulative increase of 85% in growth rate compared to the base population from which they were selected. The average genetic gain across five generations of selection using a combined selection was 12–17% per generation. When within-family selection for growth improvement in Nile tilapia at 16 weeks was carried out (Bolivar and Newkirk, 2002), about 12% genetic gain per generation was achieved.

The GIFT strains appear to have better growth and metabolic performance than their non-GIFT counterparts. Dey *et al.* (2000c) analysed the comparative performance and nature of the GIFT strain on an average farm as well as on an efficient farm in five Asian countries (Bangladesh, China, the Philippines, Thailand and Vietnam). The growth performance of the GIFT strain was superior to that of the non-GIFT strains on both average and efficient farms. On an average farm, the body weight of the GIFT strain was 18% heavier at harvest in China to 58% heavier in Bangladesh. Chiayvareesajja *et al.* (1999) studied the genetic variations in lytic activities of blood serum of Nile tilapia in relation to survival and body weight. The lysozyme activity was measured at 15 and 30°C incubation temperatures. The estimated heritability for lysozyme activity was relatively high (0.6–0.7) at 30°C and intermediate at 15°C (0.3). There was also a negative correlation between lysozyme activity and fish survival. This finding suggests that blood lysozyme activity can be used as a useful trait for indirect selection for the improvement of the survival rate of tilapia.

11.3.2. Early and late maturation

One of the disadvantages of tilapia culture is their early maturation and reproduction at small sizes. Generally, uncontrolled reproduction leads to overpopulation, stunting and poor growth rates. However, tilapia show great plasticity in growth and maturation. Some tilapia species can grow to a large size, and both males and females grow to and mature at similar sizes (Longalong *et al.*, 1999). Under certain conditions (e.g. trapping in lagoons), these fish become very small and mature at a small size, but the females are much smaller than males at first maturity. Females that do not mature at an early age follow the growth trends of the males (Bolivar *et al.*, 1993). Therefore, delaying the onset of sexual maturation in tilapia females by selection can be an effective tool for reproduction management of these fish.

The possibility of delaying the onset of sexual maturation in Nile tilapia has been investigated in the laboratory using a breeding base population of 35 full-sib families (Hörstgen-Schwark and Langholz, 1998). Gonad weight, gonadosomatic index (GSI) and the stage of gonadal development were used as indicators of sexual maturation. The heaviest individuals within each selected family were used as spawners for the next generation. After two generations, males of the selected line exhibited lower states of maturity (i.e. lower gonad weights) as compared to the control group, whereas the values of the GSI were lower in the females of the selected lines than in the control females. These selection responses indicated the possibility of delaying sexual maturity in Nile tilapia to a later stage without affecting their growth rates. A similar degree of genetic variability for maturation indices was achieved when the reference strain was tested under field conditions (Oldorf *et al.*, 1989).

The selection for early maturation of Nile tilapia was also investigated several years ago (Uraiwan, 1988). Fish selected for early maturation matured 11–14 days earlier after one generation of selection, but inconsistent results were obtained in the second generation. More recently, the response to bidirectional selection for frequency of early maturation in Nile tilapia females was evaluated in 42 full-sib families within 21 randomly chosen half-sib families of Nile tilapia, among the families representing the third generation of selection for improved growth performance in the GIFT project (Longalong *et al.*, 1999). Sex, body weight and sexual maturity in the females were recorded 2, 3 or 4 weeks after the first swim-up fry were observed in the earthen ponds holding the communally families. The broodstock was selected from full-sib families with a high (> 75%) or a low (< 20%) frequency of mature females. The fry of 16 pairs of high early maturity (83%) and nine pairs of low early maturity (0) were collected and each pair was reared separately until they reached a size of 3–5 g. The response to selection was highly significant, and there was a positive correlation between early maturity and fish size (i.e. early-maturing fish were the fastest-growing).

11.3.3. Red body colour

Body colour is an important trait in some cultured tilapia, in respect of market demand. In some countries, fish consumers prefer the red morph to the common black morph. The market price of red fish is expected to be higher than that of black fish in these countries. Some progress has been achieved in the efforts to select for homogeneity of colour in red tilapia.

A number of 'red' tilapia strains/hybrids have been developed and commercially produced, generating high market acceptance for tilapia, especially in the USA, Europe and Asia (Avtalion and Reich, 1989). This indicates that the improvement of 'colour quality' in tilapia is an important approach for the development of tilapia culture. Accordingly, a number of recent studies have considered mass selection for red colour in tilapia, with promising results. Mather *et al.* (2001) evaluated the mass selection of Fijian hybrid tilapia (*Oreochromis niloticus* × *Oreochromis mossambicus*) in order to improve their red body colour by reducing the amount of black spots on the red phenotypes. The responses of two selection intensities (low-selection line, top 50%, and high-selection line, top 30%) were compared with that of a control line (no selection). After three generations, about 100% improvement in colour quality (i.e. reduction in black spots on the red phenotype) was achieved. These authors suggested that the red phenotype can be improved significantly by applying mass selection, without adversely affecting fish growth.

Very recently, mass selection for red colour in Nile tilapia has been investigated in Mexico (Granduño-Lugo *et al.*, 2004). The selection was applied using the Stirling strain of Nile tilapia to obtain a red-coloured homozygous dominant population from a base population containing wild fish and red fish with black blotches. After five selection generations, the red colour proportion increased from 5.6% in the first generation to 100% in the fifth generation. All the fry produced were red-coloured, indicating that the population was homozygous dominant for that trait. The fecundity and fry production of red tilapia were similar to those of wild fish. Again, these results clearly indicate that mass selection for red colour is very appropriate and unlikely to cause any adverse effects on growth and reproduction performances. Red tilapia were introduced in Jamaica because the people were used to marine fish and preferred the silver or red colour of those fish (R. Stickney, Texas, USA, 2004, personal communication). The red tilapia were sold at a premium price, but, once consumers did a side-by-side taste comparison, they realized that the normal-coloured tilapia were just as good and the red ones were no longer able to demand a higher price.

11.3.4. Cold resistance

As mentioned in Chapter 3, tilapia feeding is sharply reduced below 20°C, and the fish stop feeding at about 16°C, while severe mortality occurs at 12°C. Tilapia can also tolerate temperatures as low as 7–10°C, but only for a short time, depending on tilapia species and strains. It is therefore essential to produce tilapia strains that can tolerate cold weather, especially during winter seasons in temperate regions. This problem has been tackled through mass selection for cold tolerance (Behrends *et al.*, 1996). Based on data derived from bidirectional selection for cold tolerance, the heritability estimates of Nile tilapia, blue tilapia and their hybrids were calculated according to the response : selection ratio. After one generation of selection, estimates for up-selected populations (most cold-tolerant) were blue tilapia (0.33), Nile tilapia (−0.05) and tilapia hybrids (0.31). Estimates for two down-selected populations (least cold-tolerant) were blue tilapia (0.04) and Nile tilapia (0.42). In addition, fingerlings from up-selected populations grew faster than those obtained from down-selected populations at temperatures ranging from 17.2 to 21.2°C. Under low temperature regimes, blue tilapia grew significantly better than Nile tilapia, irrespective of selection criteria. These results demonstrated the possibility of increasing cold tolerance in blue tilapia and related hybrids through directed selection.

The genetic basis of cold tolerance of Mozambique and blue tilapia and their F1 and F2 hybrids was also studied by Cnaani *et al.* (2000). Groups of fish of similar age and grown under similar conditions were housed in net enclosures in a controlled water system. Water temperature was gradually decreased and survival time was observed. There was no correlation between cold tolerance and fish size (within the range of 2.3–10.5 cm standard length), and the distribution for the trait was not normal. Mozambique tilapia were the most cold-sensitive fish, followed by F2 and F1 groups, which were similar to blue tilapia. The authors concluded that genetic variation in cold tolerance seems to have a large dominance component, based on the similarity of the F1 hybrid to the blue tilapia parent.

11.3.5. GenoMar Supreme Tilapia and GET EXCEL tilapia

The success achieved with the GIFT programme has encouraged researchers to further develop breeding programmes in order to maximize the genetic gain for the traits of interest. As a result, new tilapia strains characterized with tremendous genetic gains have been developed, and the GenoMar Supreme Tilapia (GST™) strain is an example of such innovation. This strain is a continuation of the GIFT breeding programme, with some changes. The two major changes were as follows:

1. Applying DNA fingerprinting as an identification tool.
2. Changing to a revolving mating scheme in order to complete the generation after nine monthly batches (Gjøen, 2004).

DNA fingerprinting makes it much easier and faster to produce the strains with the desired characteristics. DNA fingerprinting also provides many advantages, such as increased selection intensity and shorter generation intervals, in addition to other operational benefits. Among the genetic traits contained in GenoMar Supreme Tilapia are salt tolerance, growth rate, feed conversion efficiency, disease resistance and fillet yield.

Gjøen (2004) estimated the genetic gain of GST™ by testing generations 10 and 11 (G10/G11) as a base, against G13. The selection was adequately successful, with an average genetic gain of about 20% per generation compared to conventional schemes. The survival rate also increased to about 11% per generation, leading to a total average survival of more than 80% in G13. The average FCR was also less than 1.1.

GET EXCEL tilapia is another new breed, developed by the government of the Philippines by combining an improved breed of Nile tilapia using within-family selection and a rotational mating scheme (Tayamen, 2004). Consequently, the Bureau of Fisheries and Aquatic Resources–National Freshwater Fisheries Technology Centre (BFAR-NFFTC) has sustained the development of a fast-growing fish known as GET EXCEL 2002, through the use of genetically improved tilapia. GET EXCEL 2002 is a product of a selection programme combining strain crosses and within-family selection with rotational mating, using the following four parent lines:

- G8 of the GIFT strain – developed by crossing the best-performing genetic groups from eight Nile tilapia strains and their crosses.
- G13 Freshwater Aquaculture Centre (FAC)-selected tilapia (FaST) – a product of within-family selection of Nile tilapia in a rotational mating scheme.
- Egypt strain.
- Kenya strain.

The term EXCEL stands for 'EXcellent strain that has a Comparative advantage over other tilapia strains for Entrepreneurial Livelihood projects in support of aquaculture for rural development' (Tayamen, 2004). Preliminary evaluation of GET EXCEL tilapia indicated that the strain grew much faster and with better survival than the other improved commercial Nile tilapia strains. This breed has a great potential for increasing the yield and profitability for tilapia farmers. Therefore, the government of the Philippines has initiated a flagship project known as 'Nationwide Dissemination of GET EXCEL Tilapia', with the ultimate goal of replacing the old tilapia breed currently in use with the latest improved breed, in order to approach the targeted incremental production and participate in the global market for value-added products.

11.4. Gynogenesis, Androgenesis and Cloning

Gynogenesis and androgenesis are forms of uniparental inheritance, induced by fertilizing one gamete with another in which the DNA has been denatured by irradiation. Gynogenesis is therefore the production of progeny with only a maternal genome, while androgenesis produces diploid organisms the nuclear DNA of which is wholly of paternal origin. This process produces a haploid zygote that can be diploidized by application of physical shock, such as temperature or pressure (Mair, 2002). The resulting individuals are completely homozygous and 100% inbred. The induction and survival rate of these homozygous fish are very low (Müller-Belecke and Hörstgen-Schwark, 2000), but the lethal or deleterious genes, in the homozygous state, are selected out during the process. Genetically identical individuals or homozygous clones are produced in the second generation of androgenesis or gynogenesis (Mair, 2002). Meanwhile, crosses between these homozygous fish produce heterozygous clones.

Androgenesis is more difficult to accomplish than gynogenesis (Scheerer *et al.*, 1986), since diploidy can only be induced in androgens at first cell division, a difficult time to manipulate the embryo. Androgens are also totally homozygous, and their mortality rates are very high due to the presence of lethal or deleterious genes (Scheerer *et al.*, 1986). The production of diploid androgenic organisms depends on: (i) the denucleation of 'host' eggs using ultraviolet (UV) irradiation; and (ii) the inhibition of the first mitotic division in order to double the haploid sperm chromosome complement following fertilization of host eggs.

Myers *et al.* (1995) succeeded in producing diploid Nile tilapia androgens by denucleation of their eggs by UV irradiation. A total dose of 450–720 J/m^2 for 5–8 min produced an acceptable 22.9% denucleated eggs, as estimated by the survival of haploid androgenetic fish to 48 h post-fertilization. Successful mitotic inhibition was accomplished using a heat shock of 42.5°C for 3–4 min, applied at 2.5 min intervals 22.5–30 min post-fertilization. The mean survival of androgenetic diploid fish to yolk-sac absorption varied from 0.4% to 5.3%, compared to the controls. A survival peak of 1.6% in androgenic diploid Nile tilapia and 0.9% in hybrid Nile tilapia × blue tilapia at a dose of 5940–6930 erg/mm^2 have also been reported by Marengoni and Onoue (1998). These results demonstrated the possibility of producing androgenetic tilapia, and indicated that the application of this technique may be useful in quantitative and conservation genetics.

The external egg fertilization and embryonic development in fish make the development of homozygous clones with the use of induced gynogenetic or androgenetic reproduction techniques much easier and more successful than in farm animals, such as pigs and poultry (Müller-Belecke and Hörstgen-Schwark, 2000). The probability of producing homozygous genotypes free from expression of harmful genes is also higher in fish due to their high fecundity. Therefore, clones have been developed within many commercial fish species, such as carp (Komen *et al.*, 1991), salmon (Kobayashi *et al.*, 1994) and tilapia (Müller-Belecke and Hörstgen-Schwark, 1995; Hussain *et al.*, 1998).

Cloning can be very useful as a pure 'gene pool' for genetic studies such as cell and tissue culture, genetic fingerprinting, immunology, disease resistance, heritability and sex differentiation studies, in addition to the development of breeding schemes based on the exploitation of heterosis. However, little information is available on the production, reproduction and performance of homozygous and heterozygous clones as compared to normal heterozygous tilapia. Hussain *et al.* (1998) produced heterozygous and homozygous clones of Nile tilapia. An outbred clonal line (OCL) was produced by cross-breeding between a viable mitotic gynogenetic female and male (recessive mutation in a sex-determining gene). In addition, an inbred clonal line (ICL) was produced by gynogenetic reproduction (retention of second polar body), using pressure/heat shock techniques. The ICL showed very poor viability and high levels of abnormality, presumably due to inbreeding or the increased sensitivity of the eggs to the pressure/heat shocks. The OCL clone had higher survival (25–40%) to first feeding and low levels of abnormality. The authors related the superior viability of OCL to ICL to the effect of heterosis. They further suggested that hybrid or heterozygous clonal lines can be utilized for fixing superior genes desirable for selective breeding and genetic improvement of tilapia.

It should be emphasized here that the development and reproduction of ICL and OCL are difficult and expensive. It is also too early to conclude whether the development and continuation of clonal lines in tilapia are worthwhile for practical applications in breeding programmes, especially in developing countries, where funds for this type of research may not be available.

11.5. Triploidy

Under aquaculture conditions, tilapia generally mature at small sizes, and are also characterized by multiple annual spawning if environmental conditions are favourable. This reproduction pattern usually leads to overpopulation and stunting of cultured fish. Therefore, controlling reproduction has been a major challenge in tilapia culture for many years. The culture of monosex tilapia is probably the most common method of tackling the problem. Traditionally, monosex tilapia are produced by: (i) manual sorting; (ii) hormonal sex reversal; and (iii) interspecific hybridization. These methods have been addressed in detail in Chapter 7. In addition, triploidy can be used as a useful tool to induce sexually sterile tilapia (Hussain *et al.*, 1991, 1995; El Gamal *et al.*, 1999). The use of sterile fish has several applications, including controlling reproduction, preventing backcrosses of hybrids with their parents and improving growth of farmed species, because more energy is saved for somatic growth and less for reproduction (Rottmann *et al.*, 1991).

The induction of triploidy occurs after ovulation and sperm entrance into the egg. The cells of triploid individuals have three sets of chromosomes (3n), whereas diploid individuals have two sets (2n). Triploidy is produced by preventing the second meiotic division, after the sperm enters the egg. This process results in two sets of chromosomes from the female and one set from the male. Heat, cold and pressure shocks are commonly used for triploid induction in fish.

Successful induction of triploidy in different tilapia species has been reported. Hussain *et al.* (1991) investigated triploidy induction in Nile tilapia using pressure, heat and cold shocks. The best results were obtained at a pressure of 8000 psi, applied for 2 min, 9 min post-fertilization (pf), heat of 41°C applied for 3.5 min, 5 min pf, and cold of 9°C applied for 30 min, 7 min pf. Pressure and heat shocks produced better triploid yields than cold shocks. A similar study was conducted by El Gamal *et al.* (1999), where heat shock, cold shock, hydrostatic pressure and/or a chemical (cytochalasin) were applied to induce triploidy and tetraploidy in Nile tilapia. Heat shock, 10 min after fertilization, induced triploidy at an incubation temperature of 42°C but not at 31°C. Up to 100% triploidy was induced using a heat shock of 40–41°C, 4–6 min pf, with hatchability similar to controls. Cold shock at 13°C for 45 min, 5 min pf, induced 85–100% triploids.

Heat shock and multiple heat shocks were the most effective treatments for the induction of tetraploidy in Nile tilapia (El Gamal *et al.*, 1999). When two heat treatments of 41°C were applied at 65 and 80 min pf, for 5 min, each of them induced about 80% tetraploidy. The use of cytochalasin was ineffective in inducing triploidy or tetraploidy in the fish.

The aquaculture potential of triploid tilapia has been investigated (Bramick *et al.*, 1995). Diploid and triploid tilapia were reared in earthen ponds and fed with rice bran as a supplemental feed, in addition to weekly fertilization, for 285 days. No differences in growth between triploid and diploid fish were observed at the age of maturation. However, the final weights of triploids were significantly greater than those of diploids. Triploid males were 66% heavier whereas triploid females were 95% heavier than diploid fish. As expected, gonad weights were lower and gonadal development was retarded in triploid fish, but about 15% of triploid females had extremely high gonadosomatic indices. Stocking the ponds with triploid tilapia also prevented fish overproduction and stunting. Chang *et al.* (1996) found that the growth of diploid and triploid blue tilapia did not differ significantly, but the triploids were of more uniform size than the diploids. The genital papillae of triploid fish showed no development compared with diploids at 18 weeks of age. Mean gonad weights of females and males were similar for diploids, but different for triploids. Some triploid testes developed well with high numbers of sperm of variable shapes accumulating in the lobular lumina. Yolk accumulation was pronounced in the oocytes of diploids, but the threadlike triploid ovaries contained mainly oogonia with only a few yolky oocytes.

On the other hand, Puckhaber and Hörstgen-Schwak (1996) reported that the growth and gonadal development of triploid Nile tilapia were poorer than those of diploids. Triploids were also incapable of producing viable progenies. The survival and growth of diploid blue tilapia reared in tanks and fed with pelleted feeds or reared in fertilized ponds were also better than those of triploids (Byamungu *et al.*, 2001), but the performance of both lines was better in tanks than in ponds. It is clear that induced triploidy can be an effective tool for controlling excessive reproduction of farmed tilapia. But the contradictory results over survival and growth performance make it difficult to conclude whether triploid tilapia can be used for growth enhancement. Once again, more research is needed on the growth and feed efficiency of triploid tilapia as compared with their diploid counterparts.

As mentioned in Section 11.2.1, transgenesis can enhance tilapia growth. Triploidy may also have a promising potential in tilapia culture (see above). One may ask: would it be possible to induce triploidy in transgenic fish and what would be the expected results? These questions have been addressed by Razak *et al.* (1999), who evaluated growth performance and gonadal development of growth-enhanced transgenic Nile tilapia following heat shock-induced triploidy. The authors found that the red blood cells of triploid fish were 1.5 times larger than those in diploids. Growth performance and gonadal development were compared between diploids and triploids from both transgenic and non-transgenic full siblings. The growth of transgenic diploids was superior, followed by transgenic triploids, non-transgenic diploids and non-transgenic triploids. The testes of transgenic triploids were smaller than those of non-transgenic triploids and non-transgenic diploids. However, some triploid testes contained some spermatozoa, indicating reproductive functionality, while the ovaries of triploid fish were completely non-functional. The authors suggested that triploid transgenic females could be a good option for tilapia culture, due to their sterility and superior growth over the normal wild tilapia. At the same time, they emphasized that careful monitoring of potential gene flow is required prior to commercial use of the technology.

11.6. Production of Genetically Male Nile Tilapia (GMT)

It is well accepted that tilapia males grow much faster than females, and therefore tilapia farmers prefer male tilapia to female fish for culture. The culture of tilapia males also overcomes the problem of massive recruitment resulting from excessive reproduction. The traditional methods of producing all-male tilapia have been addressed in detail in Chapter 7. An alternative is the use of genetically male tilapia (GMT) technology for producing all-male tilapia populations. The production of GMT, which is also known as YY male or supermale, has been developed through the use of YY chromosome male fish. These YY males are the offspring of a breeding between a normal male and a female produced by hormonal sex reversal of a genetic male. Twenty five per cent of the offspring from such a mating will have a YY configuration of their sex chromosomes, instead of the normal XY. When a YY male is crossed with a normal XX female it produces a high percentage of XY (male) offspring (Fig. 11.1).

In fact, the production of YY *O. mossambicus* was reported many years ago, using the endocrine sex reversal technique with selective breeding or gynogenetic technique, but the numbers of YY males produced were relatively low (Varadaraj and Pandian, 1989). The adoption of endocrine and gynogenetic techniques increased the production of YY males from 25 to 50% and reduced the time cost of production from 22 months to only 8 months.

For Nile tilapia, Mair *et al.* (1995) reported that up to 100% GMT can be produced using YY technology. Under controlled conditions, GMT have a sex ratio of over 95% males, whereas, in commercial production systems, this ratio is variable due to genetic contamination (Mair *et al.*, 2002). The performance of GMT was also compared with hormonally sex-reversed Nile tilapia and mixed-sex Nile tilapia, both on station and on farm (Mair *et al.*, 1995). The growth rates, feed utilization efficiency, harvest weight, total yield and market income were superior in GMT to the other two groups. Farm trials in the Philippines also demonstrated that GMT culture was 100% more profitable than the culture of mixed-sex tilapia. As a result, GMT are currently being cultured in many countries worldwide, with reasonable success. The YY male technology is also being disseminated through the provision of YY male and normal female broodstocks to hatcheries in many regions of the world, particularly in Asia.

11.7. Disease Management

Infectious diseases cause considerable loss in aquaculture, especially on shrimp farms, in many parts of the world. Conventional protective and treatment methods (such as antibiotics and disinfection) may be ineffective, particularly with pathogens such as viruses. Also, it may not be possible to export the product to some countries if it contains antibiotic residues. Thus, unconventional methods of disease management have received great attention in recent years. Molecular techniques, such as DNA- or RNA-based probes and *in vitro* tissue culture, have great potential for disease detection and, in turn, for enhancing protective measures. These techniques have been very successful with shrimp aquaculture (Subasinghe *et al.*, 2003). The production of specific pathogen-free (SPF) and specific pathogen-resistant populations

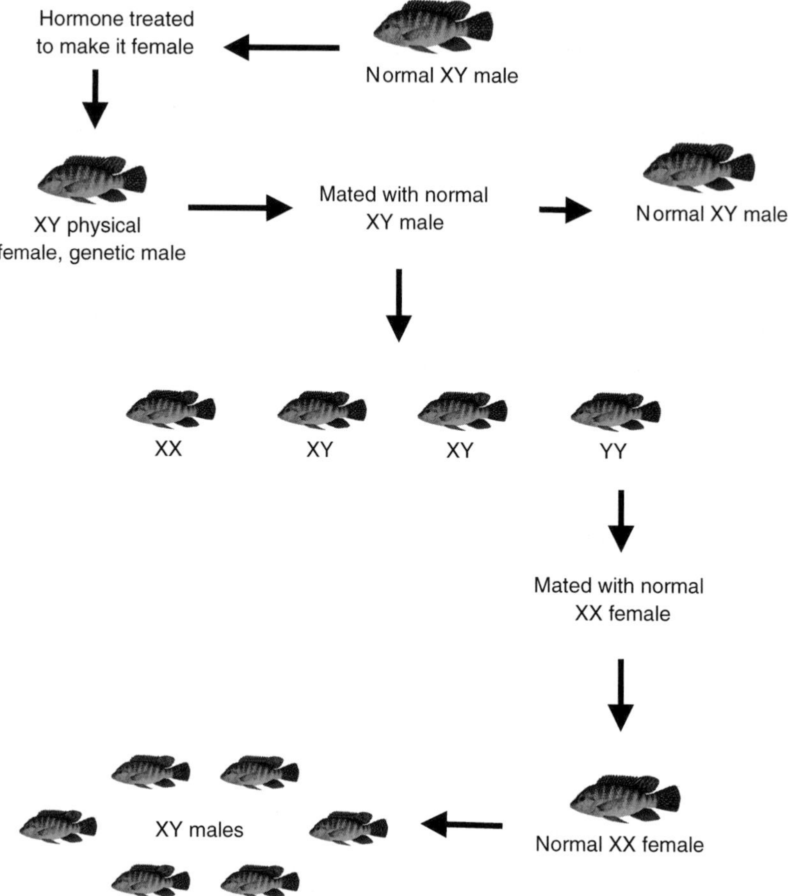

Fig. 11.1. Production of genetically male tilapia via YY chromosome males (from Subasinghe *et al.*, 2003).

has also been successful with shrimp broodstocks. SPF shrimp are produced by selecting animals free of specific pathogens, using them as broodstock and raising their offspring under controlled sanitary conditions.

Despite the significant contribution of tilapia culture to world fish production, little attention has been given to their disease management under culture conditions. Very few studies have considered the development of disease-resistant tilapia through selective breeding programmes. Cai *et al.* (2004) investigated the resistance of Nile tilapia, blue tilapia and their hybrids (♀ Nile tilapia × ♂ blue tilapia) to disease caused by the bacterium *Aeromonas sobria*. Based on non-specific immunity functions, significant differences occurred between the resistance of interspecific hybrids and that of their parental species to *A. sobria*, where the hybrids had the highest resistance and blue tilapia had the lowest resistance. These results suggest that specific variations exist in immunity to a specific pathogen and highlighted the potential for genetic improvement of disease resistance in tilapia.

The injection of tilapia with extracellular products (ECP) of certain bacteria and various adjuvants also seems to induce a non-specific immune response. When Nile tilapia were injected with ECP of various *Mycobacterium* spp. and various adjuvants, the level of lysozyme activity detected in the serum of fish 4 days after being immunized was significantly higher than that found in the serum of the control fish (Chen *et al.*, 1998).

Passive immunization with antibacterial whole serum can also be another protective approach against microbial infection. The immunization can be induced by intraperitoneal injection. This approach has been tested in Nile tilapia to examine the protective effect of anti-*Streptococcus iniae* whole sera (ASI), heat-inactivated anti-*S. iniae* whole sera (HIASI) and normal whole sera (NWS) intraperitoneally injected into the fish (Shelby *et al.*, 2002b). Significantly higher mortality was noted in tilapia passively immunized with NWS than in fish passively immunized with ASI or HIASI, 14 days after *S. iniae* infection. The immunity induced by ASI and HIASI injection against *S. iniae* infection was related to antibodies against *S. iniae*. In other words, *S. iniae* infection stimulates protective antibodies against these bacteria, indicating that ASI antibody plays a primary role in immunity to *S. iniae*.

11.8. Feed and Feeding

Feed represents more than 50% of total operating costs of intensive aquaculture enterprises, while protein alone represents about 50% of feed costs. It is no surprise that a great deal of attention has been paid to tilapia nutrition and feeding management, with the ultimate goal being to develop commercial, cost-effective tilapia feeds, using locally available, cheap and unconventional resources.

As stated in Chapter 6, fishmeal (FM) has been traditionally used as the main protein source in the aquafeed industry. The increased demand for FM, coupled with a significant shortage in global FM production, has created sharp competition for its use by the animal feed industry. As a result, FM has become the most expensive protein commodity in aquaculture feeds. Many developing countries have realized that, in the long run, they will be unable to afford FM as a major protein source in tilapia feeds. Therefore, extensive research has been carried out to partially or totally replace FM with less expensive, locally available protein sources. Remarkable progress has been achieved in this area, with FM-free (or FM-low) tilapia diets now being commercially available in many parts of the world.

A wide variety of unconventional protein sources, including animal proteins, plant proteins, single-cell proteins and industrial and agricultural wastes, have been evaluated with respect to their utility in farmed tilapia feeds. These sources have been reviewed by El-Sayed and Teshima (1991), Jauncey (1998), El-Sayed (1999a) and Shiau (2002). However, the herbivorous/omnivorous nature of tilapia has directed the research on protein nutrition of tilapia towards plant protein sources. Biotechnology plays a great role in enhancing plant protein production, improving processing techniques (to destroy anti-nutritional compounds) and enhancing efficacy of feed storage and delivery and feeding efficiency. The following sections throw light on recent innovations in protein and lipid nutrition of farmed tilapia.

11.8.1. Amino acid versus mineral supplementation

Many of the protein sources used in tilapia feeds are deficient in certain essential amino acids (EAA). The supplementation of these EAA in the diet has been a common practice. However, it was found that the utilization of many protein sources in tilapia feeds may be limited by dietary minerals (such as phosphorus and zinc), rather than the deficient EAA. This means that the inclusion of dietary EAA may not be necessary if the deficient mineral(s) in the protein source is added to the diet. In other words, the incorporation of the deficient mineral(s) in the diet may meet the requirement for deficient EAA. For example, the inclusion of a dietary phosphorus source such as dicalcium phosphate (DCP) in soybean meal (SBM)-based diets may meet the requirement for methionine, which is essential for tilapia but deficient in SBM. Viola *et al.* (1986, 1988) reported that the non-inclusion of the deficient EAA in SBM-based diets did not result in growth retardation, while SBM supplemented with 3% DCP and oil completely replaced FM without any adverse effects on fish growth. The non-necessity of EAA (lysine) supplementation has also been reported with sesame seeds if they are supplemented with zinc (El-Sayed, 1987). The adoption of this approach may therefore improve feed protein quality and reduce the cost of the diet. However, this assumption should not be generalized, at least for now, until it is further supported by laboratory and field experiments on other feed ingredients deficient in various EAA and/or minerals.

11.8.2. Phytase supplementation

Many plant protein sources contain high levels of phytic acid, which binds with divalent minerals such as Ca, P, Zn, Mn, Mg and Fe to form water-insoluble salts, rendering the minerals unavailable. Traditionally, higher supplementary mineral levels are added when those plants are used as the primary source of protein in aquaculture feeds. The inclusion of bacterial phytase in tilapia diets is a new and effective tool used to reduce phytic acid activity and improve the quality of these plant protein sources. Very recent studies indicated that phytase supplementation may also reduce the effect of anti-nutritional factors, protect amino acids from degradation, decrease leaching of water-soluble components and improve feed digestibility and growth rates (Riche *et al.*, 2001; Heindl *et al.*, 2004; Phromkunthong *et al.*, 2004; Liebert and Portz, 2005).

11.8.3. Essential fatty acid (EFA) requirements

Early studies on essential fatty acid (EFA) requirements of tilapia revealed that these fish require n-6 EFA rather than n-3 EFA. However, recent studies demonstrated that they require both n-6 and n-3 highly unsaturated fatty acids (HUFA) (Chou and Shiau, 1999). More recently, El-Sayed *et al.* (2005a) found that Nile tilapia broodstock reared in brackish water required n-3 HUFA for optimum spawning performance, while the reproductive performance of fish reared in fresh water was not affected by dietary oil source. It is evident that the requirements of tilapias for EFA are species-specific, and more work is needed to quantify the requirements of different tilapias for these EFA under different culture conditions.

11.8.4. Feeding regimes

Many other nutritional innovations in respect of diet form, feeding frequency and feeding methods on tilapia farms have been achieved (see Chapter 6 for details). These innovations include the development of low-protein, high-energy supplemental feeds for semi-intensive culture. The adoption of the best timing of supplemental feeding in well-fertilized ponds has also received considerable attention (see Chapter 4), aiming at delaying supplemental feeding as long as the natural food, provided through pond fertilization, supports good fish growth. This approach is very likely to result in a great reduction in operating costs, without significantly affecting fish growth and yield (Diana *et al.*, 1996; Abdelghany *et al.*, 2002). Moreover, recent studies (Lin and Yi, 2003; Yi *et al.*, 2004a) indicated that, when Nile tilapia and red tilapia were fed supplemental diets at only 50% satiation, they produced comparable yields to 75% and 100% satiation levels. At the 50% satiation level, 50% of feed costs were saved, in addition to a significant reduction in nutrient loading.

11.9. Innovations in Farming Systems

A full description of tilapia culture systems is provided in Chapters 4 and 5. These culture systems have undergone notable technology-based developments in many parts of the world. The intensification of tilapia production in indoor closed-recirculation systems is probably one of the most highly developed technologies. These technologies include the use of liquid oxygen, the development and use of mircoscreens and fluidized bead biofilters and the adoption of extremely high stocking densities. Significant expansion of these technology-based systems has occurred in the USA in recent years. For example, Rosati *et al.* (1993) raised Nile tilapia in recirculating systems under commercial conditions at high stocking densities of 263.2 fish/m^3, using vertical screen biofilters and particle filters. The final yield was 65.6 kg/m^3, which is considered extremely high. In another closed system, consisting of six raceways with a total capacity of 160 Mt, the annual yield of Nile tilapia was 11.33 Mt (70.8 kg/m^3), with an average weight of 560 g/fish. However, it should be emphasized that these systems are complicated, difficult to manage and costly. It is hoped that recent innovations will lead to a reduction in costs and enable small-scale tilapia farmers, especially in developing countries, to adopt these systems.

11.9.1. Effluent treatment and management

There has been a switch from the traditional, semi-intensive tilapia culture systems to more intensive systems. This jump was accompanied by a substantial development in the inputs to intensive culture systems. The developments in treatment and management of culture effluents in recirculating systems focused mainly on: (i) waste settlement and removal; and (ii) water discharge and replacement (see Chapter 5 for details).

WASTE SETTLEMENT AND REMOVAL. The removal of solids from culture water is an essential process in closed-recirculating systems. This procedure maintains the quality of water and improves fish performance and yield. As indicated in Chapter 5, traditional biofiltration, such as water exchange, back-flushing, fixed film media, etc., is inconvenient for small-scale tilapia culture enterprises because of occasional system failure and high costs. A number of unconventional biofiltration methods have been developed to tackle this problem. Most of these methods are simple, cheap and efficient and can be easily applied using locally available inputs. For example, carbon-rich agricultural by-products, such as rice hulls, added to fish tanks have been successfully used as primary media for solids removal from recirculating tilapia culture systems (Scott *et al.*, 1998).

Different biofiltration media, such as polypropylene plastic chips and polyethylene blocks, have been developed for ammonia removal in recirculating tilapia culture systems. Different types of bead filters (such as the bubble-washed bead filter) are also widely used for nitrification and clarification of culture water and capturing solid wastes. Slow-sand filters have also been developed for the treatment of effluents within recycling aquaculture systems. These filters are simple, cheap, easy to construct and operate and effective in removing suspended solids.

WATER DISCHARGE. Recirculating systems are generally adopted in areas having limited freshwater sources and harsh weather conditions. Economic use of these water sources becomes a necessity, and waste water discharge may not be a good idea. Instead, water recycling and reuse are highly recommended under these conditions. Extensive research has been carried out to minimize, or even eliminate, water discharge from tilapia recycling systems. This approach has been tried in greenhouses (Shnel *et al.*, 2002), where zero water discharge from a tilapia recirculating system was applied. Neither water nor organic matter was discharged, while sludge was biologically digested and nitrate was reduced to nitrogen gas. This system was very successful in maintaining water quality within acceptable levels for tilapia and removing metabolic wastes. Total tilapia yield was also reasonably high, reaching 81.1 kg/m^3 at maximum stocking densities of 61.8 kg/m^3.

Water exchange may also be sharply reduced without any adverse effect on the performance of tilapia reared intensively in earthen ponds if sufficient aeration is provided. The addition of a carbon source (wheat bran, rice bran, etc.) on the water surface can also be used as a surface for the growth of bacteria that reduce ammonia accumulation, while the tilapia consume the bacteria produced in the system as natural food. Applying this technology can reduce dietary protein by about 10% and the overall cost can be reduced by up to 50% (Chamberlain and Hopkins, 1994).

Active sedimentation and re-suspension can also be applied to reduce water exchange in intensive pond culture. This process is based on the re-suspension of solid particles from pond sediments in the water column by continuous aeration, mixing, stirring and turbulence. Re-suspension favours the aerobic decomposition of organic matter. Suspended particles and bacterial blooms can be used as natural food for tilapia, leading to additional savings in operational and feed costs.

11.9.2. Tilapia production in aquaponic systems

Aquaponics is the integration of hydroponics with aquaculture in recirculating systems, where the wastes and metabolites produced by cultured fish are removed by nitrification and taken up by plants. The bacteria living in the gravel and in association with the plant roots play a critical role in nutrient removal. Aquaponics is a simple, promising and sustainable food production technology, especially in developing countries and arid zones, where resources are limited, fresh water is scarce and populations are increasing.

Tilapia have been successfully cultured in aquaponic systems in many parts of the world (see Chapter 5, Section 5.9, for details). However, the University of the Virgin Islands (USA) is considered among the best centres for aquaponic research, going back over two decades, when Rakocy and Alison (1981) published the first of the papers on the subject. The production of tilapia and vegetables in aquaponics systems appears very promising and profitable. The break-even prices of tilapia and vegetables could be even lower if production costs are reduced.

11.10. Future Prospects

The role of modern technology in tilapia culture is expected to be more pronounced in the future. Biotechnology, especially transgenesis, is also expected to receive considerable attention as an important tool for genetic enhancement and growth improvement of farmed tilapia. Future developments in this regard are likely to include the following:

1. The application of sex-specific markers to increase the efficiency of sex control breeding programmes (such as the YY male technology), aiming at increasing the sex ratio to 100% males.
2. The incorporation of cloned tilapia genes into the production of new strains of transgenic tilapia. The production of tilapia with desirable characteristics such as salinity and cold tolerance, disease resistance, colour and enhanced growth will be of prime priority.
3. A widespread application of genetically improved farmed tilapia (GIFT), with particular emphasis on the developing countries.
4. The improvement of the quality of fishmeal replacers in tilapia feeds, especially plant proteins. More attention will be paid to processing these sources and to anti-nutrient removal.
5. The inclusion of phytase and deficient minerals (not deficient essential amino acid(s)) may open a new gate for improving the quality of feed ingredients and reducing feed costs.
6. A shortage of fresh water, concomitant with competition for it with other agricultural and urban activities, dictates the intensification of tilapia production in closed-recirculation systems in many parts of the world. The development in production technologies in these systems is very likely to include improvements in aeration (e.g. liquid oxygen), biological and mechanical filtration techniques, waste settlement and removal, water reuse, increased fish density and feeding management.
7. Widespread application of tilapia production in aquaponic systems, especially in arid areas characterized by shortages of fresh water and high evaporation rates.

12
Environmental Impacts

12.1. Introduction

The aquaculture industry is expanding worldwide at an outstanding rate. During the 1990s, aquaculture production sharply increased more than threefold, from 16,831,543 Mt in 1990, representing 16.35% of global fish production (102,943,474 Mt), to 51,385,912 Mt in 2002, comprising 35.21% of the fish production (145,942,278 Mt) (FAO, 2004). This rapid expansion of aquaculture has had several ecological, human health and socio-economic impacts.

Traditionally, small-scale, semi-intensive aquaculture has been sustainable in many countries, especially in rural areas, with minimum adverse effects on surrounding environments. In recent years, however, there has been a growing trend towards tilapia intensification, which is usually driven by market forces and competitive use of resources. The use of artificial culture inputs, such as prepared feed, drugs, hormones, fuels, etc., will become inevitable in intensive culture practices. This may pose serious environmental and socio-economic threats. It is also very likely that the high intensity of farmed species will lead to enrichment of aquatic environments. According to the Joint Group of Experts on the Scientific Aspects of Marine Pollution (GESAMP) (1991), the environmental impacts of coastal aquaculture are listed as follows:

1. *Ecological impacts*: These include enrichment, interaction with the food web, oxygen consumption, disturbance of wildlife, habitat destruction, interaction between escaped farmed stock and wild species, introductions and transfers and bioactive compounds (including pesticides, antibiotics, chemicals introduced via construction materials, hormones and growth promoters).

2. *Implications for human health*: These include outbreaks of diseases associated with the consumption of shellfish, typhoid fever, infectious hepatitis and other viral diseases, survival of enteric viruses in the marine environment, cholera, the influence of fish pathogens on human health, phycotoxins and depuration.

3. *Socio-economic considerations*: As mentioned in Chapter 1, tilapia culture is expanding at a very high rate, so that it is now ranked third in terms of global production, only after carps and salmonids. Tilapia are currently cultured in more than 100 countries around the world. The traditional, small-scale, semi-intensive system of tilapia culture, especially in Asia, is now being replaced with more intensive, large-scale farming systems. Intensive tilapia culture can thus pose a possible threat to the environment. This chapter discusses the major environmental impacts of tilapia culture, with emphasis on the following impacts: (i) transfers and introductions; (ii) genetic pollution; and (iii) nutrient enrichment.

12.2. Transfers and Introductions

Transfers of aquatic species are movements of that species within its geographical range. Transfers generally take place to support stressed populations, enhance genetic characteristics or re-establish a species that has failed locally (GESAMP, 1991). Introductions are the movements of a species beyond its present geographical range. Introductions are intended to establish new taxa in the flora and fauna of an environment. Transfers and introductions can pose a wide variety of risks to the integrity of ecosystems, existing species, human health, agriculture, aquaculture and related primary industries. In other words, transfers and introductions may change the biodiversity of the receiving ecosystem through interbreeding, predation, competition for food and space and habitat destruction.

Tilapia have been introduced as alien species into about 90 countries worldwide. They are known to thrive and reproduce in most of their new habitats. The introductions of tilapia within and outside Africa are discussed in Chapter 2. Despite the impacts that they may have on the environment, most introductions have not been preceded by any environmental impact assessment (EIA). Instead, in most cases, evaluation and appraisal started after the introductions took place. At that point it becomes very difficult, or even impossible, to mitigate the impacts of introduced tilapia in their new environments. The message is clear; careful and thorough appraisal must be carried out prior to any introductions or transfers of tilapia and proper management plans must also be adopted.

As mentioned in Chapter 2, the introductions of tilapia can be unsuccessful, successful, or successful but with high environmental costs, ranging from habitat destruction, hybridization with endemic species, to disappearance of native species. In the next sections, some successful examples of tilapia introductions are highlighted. The environmental impacts of tilapia introductions are also discussed.

12.2.1. Successful introductions

One of the most successful examples of tilapia introductions is the introduction of Nile tilapia into Lake Victoria (Uganda) in the early 1900s. After their introduction, Nile tilapia contributed to the disappearance of native tilapia from the Lake Victoria fisheries. Currently, Nile tilapia are the dominant tilapia species in the Lake Victoria region, and second only to another introduced, non-cichlid species, Nile perch (*Lates niloticus*), in commercial value (Fuerst *et al.*, 2000). A few generations after their introduction into Sri Lanka, Mozambique tilapia became a major component in the inland fisheries, contributing over 70% of the landings from reservoir fisheries in the 1980s (De Silva, 1985b). They were found in all inland water bodies in the country (De Silva and Senaratne, 1988). In addition, a field survey of exotic tilapia in Bangladesh found very limited evidence of breeding in the wild, suggesting that the introduced fish have a limited impact on natural fisheries (Ireland *et al.*, 1996). However, because these fish are the most commonly cultured species in cages, there is a chance of their escaping and competing with native species for food and space.

12.2.2. Impacts on native aquatic species

The impacts of tilapia introductions on native populations vary with species and geographical location. Generally, introduced tilapia are known to adapt easily to their new habitats, while native species are forced to contend with environmental changes and competition from exotic species. These impacts range from little or slight to devastating. However, there is still a wide debate over the environmental impacts of tilapia introductions in many parts of the world. The case of the Philippines is an example of such arguments. Eighty-four per cent of fish farmers in the Philippines claimed that tilapia introductions did not cause any displacement of native species in natural waters. Yet 16% claimed species displacement and/or reduction in the landings of other species after the introduction of tilapia (Pullin *et al.*, 1997). However, these claims were based mostly on perception and not supported by solid evidence.

The introduction of tilapia to India has led to a significant reduction in the yield of the native species, such as Indian carp, pearl spot and mirror carp, from many reservoirs (Natarajan and Aravindan, 2002). The decrease in the yield of these fish has been attributed to the prolific breeding nature of introduced tilapia.

The introduction of *Oreochromis spilurus niger* and *Oreochromis esculentus* into Lake Bunyoni (Uganda) has failed. On the other hand, the introduction of Nile tilapia into Lake Bunyoni was successful but had considerable impacts, including hybridization with the other two introduced species (Lowe-McConnell, 1958). Nile tilapia populations have also exhibited retarded growth and are subject to infestation by parasites and poor fishery yields. It has also been suggested that the disappearance of *O. spilurus* from Lake Naivasha and *O. esculentus* and *Oreochromis variabilis* from Lake Victoria was related, partially, to their hybridization with introduced species (Ogutu-Ohwayo and Hecky, 1991). In Australia, the competition between *Oreochromis mossambicus* and indigenous species for food and breeding sites in the Darling River could have devastating impacts on indigenous fish populations (Arthington and Bluhdorn, 1996). In contrast, *O. mossambicus* supported the fisheries on the Sepik River flood plain (Papua, New Guinea) without affecting the indigenous species (Coates, 1987).

Habitat alterations/destruction by tilapia have been reported mainly for macrophyte feeders, *Oreochromis aureus*, *Tilapia rendalli* and *Tilapia zillii*, which are highly destructive to native indigenous vegetation (Philippart and Ruwet, 1982). Introduced *T. zillii* have also been reported to compete with the endemic *O. variabilis* in Lake Victoria, leading to severe harm to that species (Welcomme, 1988). The introduction of *T. rendalli* into Madagascar and Mauritius has also caused serious disturbance to the indigenous fauna and flora in the lakes, rivers and reservoirs (George, 1976). Tilapia introductions also led to changes in the biology of native species. Moreau (1983) reported that the littoral native species in Lake Kyoga (Africa) began to migrate to open water after the introduction of tilapia into the lake.

Tilapia have been found to migrate into the lower Colorado River mainstream in North America from irrigation canals in which they had been stocked for aquatic weed control (Fitzsimmons, 2001b). Tilapia, along with other exotic species introduced as sport fish, completely changed the native fish communities in the river, to the extent that the native species are endangered. In some parts of the river, tilapia represent about 90% of the biomass.

It is clear from the above examples that the introductions and transfers of tilapia into new habitats can be supportive or destructive to the environment. This means that any introductions of tilapia should be preceded by a complete EIA (see below).

12.3. Genetic Pollution

12.3.1. Hybridization

One of the major risks of tilapia introductions is their ability to interbreed with closely related domestic species. Such uncontrolled hybridization is likely to cause a loss of genetic variability. As a result, the loss of pure tilapia species is gradually increasing. Unlimited numbers of hybridizations have been reported in both cultured and wild tilapia. In most cases, hybrids have morphometrically and biologically different characteristics from those of their parents. Natural hybridization between related tilapia species (either introduced or native) is also possible. For example, hybridization between *O. spilurus niger* and *Oreochromis leucosticte* (Lake Naivasha), between *O. spilurus niger* and *Oreochromis niloticus* (Lake Bunyoni) and between *Oreochromis macrochir* and *O. niloticus* (Lake Itasy) has been reported (Moreau, 1983).

12.3.2. Inbreeding

Inbreeding is defined as 'the mating or crossing of individuals more closely related than average pairs in the population'. The smaller the founder stock, the higher the chance of inbreeding. Inbreeding of tilapia has been a major problem both in wild populations established in new environments and in farmed populations around the world. In most cases, the founder stocks used for tilapia aquaculture were small. For example, it has been reported that the founder stock of Mozambique tilapia in Asia was only five individuals (three males and two females) introduced into Indonesia in the 1950s (Agustin, 1999) (Atz, 1954, reported that Mozambique tilapia were introduced into Indonesia prior to 1939 and not in the 1950s). These individuals were bred and their progenies formed the basis of tilapia culture and the establishment of feral stocks throughout much of Asia, except at higher latitudes, where they are not present. As expected, inbreeding of these fish was inevitable, leading to stunting, early sexual maturation, body deformation (Fig. 12.1), low survival and poor growth. As a result, Mozambique tilapia are no longer considered to be a desirable aquaculture candidate in many locations, and the species has been replaced by Nile tilapia or tilapia hybrids in many Asian countries. Inbreeding may also lead to the reduction of heterozygosity, as has been suggested by Kocher (1997), who reported that some strains of farmed tilapia showed less than 10% of the heterozygosity of their wild counterparts.

Fig. 12.1. Normal and healthy Nile tilapia (top) and Nile tilapia showing body deformation because of inbreeding (bottom) (top photo provided by M. Rizk).

12.3.3. Transgenesis

Despite the fact that transgenesis offers several advantages for tilapia culture (see Chapter 11 for details), the rate of genetic change in transgenic tilapia is such that their phenotypic and behavioural properties cannot be easily predicted (Mair, 2002). When transgenic tilapia are introduced into new environments they may have adverse effects both on the environment and on native species. Transgenic tilapia could also escape and the transgene become a part of the gene pool, inducing artificial genetic diversity in native populations. This may increase or decrease fitness or have no phenotypic or ecological effects (Dunham, 1999). Negative impacts include over-dominance and/or replacing the native populations. Guillen *et al.* (1999) found that transgenic tilapia had lower feeding motivation and dominance status than wild tilapia in Cuba. If large numbers of transgenic fish escape or are introduced, they could also reduce reproduction in natural populations through infertile matings (Dunham, 1999).

12.4. Farm Effluents and Organic Enrichments

Considerable amounts of uneaten feed, along with wastes, including organic matter, nutrients such as phosphorus and nitrogen and suspended solids, are released from intensive, land-based, tilapia farms into the water. These compounds are generally toxic to fish and other aquatic animals. They may also have varying degrees of environmental impacts, depending on the intensity of the culture operations. In fish ponds, part of these wastes settle to the pond bottom, while the remainder is discharged with farm effluents into the environment. If not removed or treated, they may be subject to oxidation, leading to oxygen depletion and, in turn, producing anaerobic conditions, which may stress and/or kill the fish, both in the ponds and in receiving waters.

The global expansion of tilapia culture in floating cages is also very likely to have severe environmental and socio-economic impacts. The assimilation of wastes and nutrients accumulated underneath the cages requires a very large area of the aquatic system. Berg *et al.* (1996) reported that the production of *T. rendalli*, *O. mossambicus* and *O. niloticus* in 1 m^2 of intensively managed lacustrine cages required an ecosystem area of 115 m^2 and 160 m^2 for phosphorus assimilation and oxygen production. This simply means that the production of tilapia in large cages, which are currently widely used, would require a tremendous ecosystem area for the assimilation of wastes resulting from the cages.

Tilapia culture in floating cages in Sampaloc Lake (Philippines) is a good example of the catastrophic environmental and socio-economic impacts of unmanaged culture practices. Tilapia cage culture in the lake was introduced in 1976 and has been expanding since then, with overuse of commercial feeds. It was reported that about 6000 Mt of feed are wasted through the cages and settle on the sediments annually, in addition to the accumulation of fish faeces and other organic wastes underneath the cages. This situation has created severe anoxic conditions and toxic waters, resulting in the progressive depletion of dissolved oxygen, a high biological oxygen demand (BOD) load and extremely high concentrations of ammonia and total sulphides. In 1990, the operators were warned about the possibility of a fish kill due to the huge volume of anoxic and toxic waters underneath the cages, and were also advised to harvest their stocks. The warning was ignored, and, as expected, a massive fish kill of market-size tilapia occurred, causing the farmers to lose their investment (A. Santiago, undated, www.iirr.org/aquatic_resources/p5c02.htm).

In Egypt, cage culture of tilapia in the Nile River, particularly in the Damietta and Rosetta branches, has expanded at an outstanding rate during the past few years (Fig. 12.2). This has increased public awareness and created a debate among government authorities, academics and environmental protection bodies in respect of the environmental impacts of these cages. The authority (in Damietta Governorate) decided to remove the cages from the Nile and has warned the farmers to remove their cages within a limited time; otherwise they will be removed by force. This situation has divided public opinion into two broad streams. Those in favour of tilapia cage culture suspect that the cages pollute the Nile, and claim that this culture system is practised all over the globe. They suggest that removing the cages is not the solution, whereas proper management, such as limiting the number of cages/km^2, the use of high-quality feed with optimum feeding strategies, proper selection of cage sites, farming other fish species such as silver carp with tilapia and continuous governmental monitoring and inspections, would certainly minimize the effects of the cages. Researchers, academics and farmers are in favour of this opinion.

Fig. 12.2. Several long arrays of tilapia cages in the Nile River (Egypt) may pose a threat to the environment.

On the other side, the environmental protection authority and governorate administration claim that cage culture is causing the quality of the Nile to deteriorate and is very likely to become a health hazard, and therefore removing these cages is the only solution. Unfortunately, removing the cages will cause a catastrophic loss to the farmers, who will be unable to pay back the loans they got from the banks. The farmers took their case to the court, and are waiting for the decision.

Troell and Berg (1997) suggested that intensive tilapia farming in cages in the tropics can generate severe eutrophication. They found that the average flux of particulate nutrients (ammonium and phosphates) under tilapia cages in Lake Kariba (Zimbabwe) was up to 22 times greater than in control areas. The release of the nutrients usually stimulates phytoplankton production and other algal blooms, which may further increase oxygen consumption during the decomposition of uneaten food and organic wastes. These algal blooms may also be toxic. Moreover, when the algae die, they settle on the bottom, decompose and further deplete the oxygen in the water, making the already eutrophic conditions worse. Not only that, but farm wastes can be a major cause of changes in the structure of benthic communities, because uneaten feed may favour certain aquatic species over others. Sedentary animals may also suffer from mass mortality when the water is depleted of oxygen, while the mobile population may migrate to other areas.

12.5. Bioactive Compounds

Bioactive compounds, including pesticides, antibiotics and other therapeutic drugs, are used in aquaculture for various reasons, including disease control, fertilization, liming, disinfection, oxidation, coagulation, pesticides and adsorption. They are generally added to feed and/or culture water. Some of these compounds, especially drugs, pesticides and antibiotics, may be toxic to aquatic animals and/or may accumulate in the environment. Aquaculture effluents containing these compounds may therefore have adverse ecological impacts, in addition to the possible contamination of the flesh of aquatic animals, which could pose a hazard to consumers.

12.6. Hormones

As mentioned in Chapter 7 (Section 7.8.2), steroid hormones are widely used for producing monosex tilapia. Hormones are generally incorporated either with larval feeds or through the immersion of fertilized eggs or sac fry. However, the use of hormones for sex reversal has been under increasing public criticism due to their potential health and environmental impacts. The hormone residues and metabolites can be a potential environmental contaminant. For example, feeding tilapia with 17α-methyltestosterone (MT)-treated feed has resulted in considerable 'leakage' of MT into pond water and sediments (Contreras-Sánchez *et al.*, 2001, 2002). MT was detected in the water during MT treatment, and it accumulated and remained in the sediments of the ponds for up to 8 weeks. Hormone traces pose a risk to hatchery workers and also to other non-target aquatic organisms.

The exposure of untargeted organisms to steroid hormones can result in biased sex ratios. Thus significant masculinization occurred in common carp exposed to water used in MT-impregnated feeding trials (Gomelsky *et al.*, 1994). Abucay *et al.* (1997) also reported high rates of masculinization and feminization in sexually undifferentiated Nile tilapia fry reared in aquarium water previously used for oral application of hormones used for sex reversal. Also, the sex ratio of fry fed with normal feed in cages adjacent to empty cages containing hormone-incorporated feed changed significantly. These findings clearly demonstrate that hormone (or hormone metabolites) leaching from uneaten food can induce sex reversal in exposed untargeted organisms.

12.7. Reducing Environmental Impacts

12.7.1. Management of introductions and transfers

It is obvious from the previous discussion that unmanaged introductions and/or transfers of tilapia can cause catastrophic environmental impacts. If tilapia are established in their new environment, it would be almost impossible to control and mitigate their destructive impacts. It is therefore essential that firm regulations be put in place to control the introductions of these fishes. Necessary precautions should also be taken before tilapia are introduced into new habitats, with emphasis on the following aspects (FAO, 1995b):

1. Purpose and objectives of the introduction, the stage(s) in the life cycle proposed for introduction, the area of origin and the target area(s) of release.
2. Physical, chemical and biological requirements for reproduction and growth, and natural and human-mediated dispersal mechanisms.
3. Ecological studies on the water and native fish fauna.
4. Comprehensive studies on the biology and ecology of the introduced species, including taxonomy, food and feeding habits, reproduction, behaviour, diseases, environmental adaptability, etc.
5. Detailed analysis of the possible impacts on the aquatic ecosystem of the proposed introduction. This analysis should include a thorough review of the ecological, genetic and disease impacts.

These aspects should include, but are not necessarily limited to: (i) potential habitat breadth; (ii) prey/predator relationships (including the potential for altered diets and feeding strategies); (iii) competition and competitors; (iv) hybridization potential between introduced species and native species; and (v) the role played by disease agents and associated organisms and epibiota.

If it is decided to proceed with the introduction, the following action is recommended (FAO, 1995b):

- A broodstock should be established in a quarantine situation approved by the receiving country in sufficient time to allow adequate evaluation of the health status of the stock.
- The broodstock should be developed from stocks imported as eggs or juveniles, to allow sufficient time for observation in quarantine.
- The first-generation progeny should be placed on a limited scale into open waters to assess their ecological interactions with native species.
- All effluents from hatcheries or establishments used for quarantine purposes should be sterilized in an approved manner.
- A continuing study should be conducted of the introduced species in its new environment.

12.7.2. Effluent management

The treatment and management of tilapia culture effluents have been described in detail in Chapter 5

(Section 5.8). Boyd (2003) listed the following best management practices (BMPs) for minimizing nutrient loads in aquaculture effluents:

1. Use fertilizers only as needed to maintain phytoplankton blooms.
2. Select stocking and feeding rates that do not exceed the assimilation capacity of ponds.
3. Feeds should be of high quality, be water-stable and contain no more nitrogen and phosphorus than necessary.
4. Apply feeds conservatively to avoid overfeeding and to ensure that as much of the feed is consumed as possible.
5. Do not use water exchange, or reduce water exchange rates as much as possible.
6. In intensive aquaculture, apply enough mechanical aeration to prevent chronically low dissolved oxygen concentration and to promote nitrification and other aerobic, natural water purification processes.
7. Provide storage volume for heavy rainfall to minimize storm overflow.
8. Deep-water release structures should not be installed in ponds, for they discharge lower-quality water from near pond bottoms.
9. Where possible, seine-harvest fish without partially or completely draining ponds.
10. Where possible, discharge pond draining effluent through a settling basin or a vegetated ditch.
11. Reuse water where possible.

While these practices are valid for reducing nutrient loads in tilapia culture, there are certain practices that have been exceptionally successful in effluent management at the farm level. Extensive research was conducted at the Asian Institute of Technology in Thailand (AIT) to mitigate the environmental impact of tilapia culture through reducing nutrient loading in pond culture of Nile tilapia. Special attention was given to: (i) optimization of feeding regimes; (ii) maximizing nutrient utilization through integration and recycling systems; and (iii) minimizing waste loading from effluents through appropriate draining techniques during harvest. Lin and Yi (2003) described the results of these practices. It would be useful here to throw light on these results.

OPTIMIZATION OF FEEDING REGIMES. When Nile tilapia reared in fertilized ponds were fed supplemental diets at 50%, 75% and 100% satiation, they produced comparable yields (Lin and Yi, 2003). However, considerable reduction in nutrient loading was achieved at the 50% level. This means that about 50% of the feed can be saved, since the fish are able to supplement their diets with the natural food available in the pond. This also indicates that an optimal feeding regime may reduce both feed costs and nutrient loading in the ponds.

REDUCING EFFLUENT IMPACT WITH A CAGE-CUM-POND INTEGRATED SYSTEM. Tilapia culture in a cage-cum-pond integrated system has been described in detail in Chapter 4 (Section 4.6.2) and Chapter 5 (Section 5.4.5). The system involves rearing hybrid catfish (*Clarias macrocephalus* × *Clarias gariepinus*) in cages in ponds and feeding them with formulated feeds, while their wastes are used as a natural food for tilapia reared in the open pond. This system yields considerable production of catfish and tilapia and also reduces nutrient loading through the use of fish wastes for pond fertilization and natural food production.

EFFLUENT MANAGEMENT DURING HARVEST. Seining fish ponds with nets is a very common harvesting method, but it stirs bottom sediments, leading to deterioration of water quality and its effluents. Appropriate harvest methods and draining treatments can significantly reduce pollutants from Nile tilapia ponds. In a novel study, Lin *et al.* (2001) evaluated four treatments to determine the efficiency of fish harvest and effectiveness in reducing nutrient loading in the effluents. The treatments were: (i) treating the ponds with tea seed cake (10 ppm) to partially anaesthetize Nile tilapia 1.5 h prior to harvest by seining; (ii) liming the ponds (75 g/m^3 of calcium hydroxide) 24 h prior to harvest to precipitate phosphorus and organic matter, followed by sequential complete draining with a pump, and Nile tilapia collected from a harvesting pit; (iii) draining the ponds by sequential complete draining with a pump and Nile tilapia collected from a harvesting pit; and (iv) ponds drawn down from 100 to 50 cm with a pump and Nile tilapia harvested by seining twice, followed by complete draining and collection of the remaining Nile tilapia from a harvesting pit. The results of this study revealed that liming the pond 24 h before draining and gradually draining ponds to 25 cm depth during harvest were most effective.

12.7.3. Reducing nutrient loading through green-water tank culture

The green-water tank culture system can be an efficient method for reducing nutrient loading in fish tanks. In this system, fish wastes and other metabolites are treated in such a way as to increase natural food production in the tanks. The process involves the oxidation of toxic ammonia (NH_3) and nitrite (NO_2) into relatively non-toxic nitrate (NO_3) through nitrifying bacteria grown on suspended organic matter. The bacteria remove the organic matter from the tanks and use it as food, while the bacteria are used as natural food for filter-feeding fishes such as tilapia and carp. However, in green-water systems, solid wastes should be removed continuously and both biofiltration and aeration should be provided. This system has been successfully used for tilapia culture in various parts of the world, particularly in the Virgin Islands. The advantages of green-water culture are summarized in Chapter 5 (Section 5.6).

12.7.4. Removing nutrients through aquatic plants

In semi-intensive tilapia culture, about 80% of the nitrogen and phosphorus added in fertilizers settles to the pond mud (Lin and Yi, 2003). The removal (de-sludging) of pond mud is tedious and labour-demanding. Planting rooted aquatic plants in fish ponds can be an effective way of extracting nutrients from the mud. This approach has been tested with lotus (*Nelumbo mucifera*) seedlings planted in Nile tilapia ponds fertilized weekly with urea and triple superphosphate at a rate of 4 kg N and 1 kg P/ha/day (Yi *et al.*, 2002). Lotus effectively removed nutrients from pond mud. The annual nutrient losses from the mud were 2.4 Mt N and 1 Mt P/ha, about 300 kg N and 43 kg P of which were taken by lotus.

In temperate zones, water temperature drops during wintertime and becomes unsuitable for tilapia culture. Generally, tilapia farmers in these regions harvest their fish during September–November. During the winter season, these ponds are dried and become useless. Growing land winter crops, such as wheat and lucerne, in these ponds can be an effective method of removing nutrients from pond muds, in addition to generating an additional income for the farmer. This approach has been tested in some areas in Egypt, where lucerne and wheat have been grown in tilapia ponds during the winter, with promising results (Fig. 12.3).

12.7.5. Removing nutrients through aquaponic systems

Aquaponics is the integration of hydroponics with aquaculture in a recirculating system. The wastes and metabolites produced by cultured fish are

Fig. 12.3. Wheat crop grown in earthen tilapia ponds at Hamool (Egypt) during the winter season. This technique can be an effective way of removing nutrients from pond mud.

Fig. 12.4. Growing plants (lettuce) with tilapia in an aquaponic system is an effective way of removing nutrients from culture water (photo provided by J. Rakocy and D. Bailey).

removed by nitrification and taken up by the plants, while the bacteria living in the gravel and in association with the plant roots play a critical role in nutrient removal (Fig. 12.4). This means that the plants act as a biological filter by removing fish wastes and improving the quality of the culture water. Many studies have been conducted on aquaponic systems at the University of the Virgin Islands Agricultural Experimental Station and also in other geographical regions, using tilapia and various vegetables with considerable success. The results of these studies are discussed in Chapter 5 (Section 5.9).

12.8. Closing Remarks

1. Introductions and/or transfers of tilapia can be unsuccessful, successful or successful but with environmental impacts ranging from habitat destruction, hybridization with endemic species, to disappearance of native species.
2. Uncontrolled, mismanaged introduced tilapia may interbreed with wild tilapia. This hybridization is likely to cause a loss of genetic variability. Hybrids have morphometrically and biologically different characteristics from their parents.
3. If the founder stock of introduced tilapia is small, crossing of closely related individuals may occur, leading to inbreeding. Inbreeding may lead to stunting, early sexual maturation, low survival and poor growth.
4. Firm regulations should be put in place and necessary precautions taken before tilapia introductions into new habitats. The introductions should also be preceded by preparation of a detailed environmental impact assessment.
5. Varying amounts of uneaten feed and fish wastes are released from land-based and cage culture of tilapia. If the wastes are not removed or treated, they will produce anaerobic conditions that stress and/or kill the fish.
6. If transgenic tilapia are introduced into new environments, deliberately or accidentally, the transgene may become a part of the gene pool, inducing artificial genetic diversity in native populations, which may have negative effects on both the environment and the native species.
7. Public criticism concerning the use of steroid hormones for sex reversal in tilapia is increasing because hormone residues and metabolites can have health and environmental impacts.
8. BMPs should be adopted for minimizing nutrient loads in tilapia effluents, in order to reduce their environmental impacts.

References

Aban, S.M., Ba-ad, E.G.V., Parreno, S.C. and Garcia, A.C. (1999) Bacterial diseases and salinity tolerance of red tilapia *Oreochromis aureus* and their hybrids (red tilapia × *Oreochromis aureus*). Paper presented at the Fourth Symposium on Diseases in Asian Aquaculture: Aquatic Animal Health for Sustainability, Cebu City, Philippines, 22–26 November 1999.

Abdalla, A. (1995) Age and growth and population dynamics of tilapia species in the Egyptian inland water, Edku Lake. PhD thesis, Assiut University, Egypt.

Abdelghany, A.E. (1997) Optimum ratio between anchovy fish meal and soy protein concentrate in formulated diets for Nile tilapia (*Oreochromis niloticus* L.). In: Fitzimmons, K. (ed.) *Proceedings from the Fourth International Symposium on Tilapia in Aquaculture*. Northeast Regional Agriculture Engineering Service, Ithaca, New York, pp. 31–39.

Abdelghany, A.E. (2000a) Optimum dietary protein requirements for *Oreochromis niloticus* L. fry using formulated semi-purified diets. In: Fitzsimmons, K. and Filho, J.C. (eds) *Tilapia Culture in the 21st Century, Proceedings from the Fifth International Symposium on Tilapia Aquaculture, Rio de Janeiro, Brazil.* American Tilapia Association, Charles Town, West Virginia, and ICLARM, Penang, Malaysia, pp. 101–108.

Abdelghany, A.E. (2000b) Replacement value of cystine for methionine in semi-purified diets supplemented with free amino acids for the Nile tilapia *Oreochromis niloticus* L. fry. In: Fitzsimmons, K. and Filho, J.C. (eds) *Tilapia Culture in the 21st Century. Proceedings from the Fifth International Symposium on Tilapia Aquaculture, Rio de Janeiro, Brazil.* American Tilapia Association, Charles Town, West Virginia, and ICLARM, Penang, Malaysia, pp. 109–119.

Abdelghany, A.E. (2003) Partial and complete replacement of fish meal with gambusia meal in diets for red tilapia *'Oreochromis niloticus* × *O. mossambicus'*. *Aquaculture Nutrition* 9, 145–154.

Abdelghany, A.E. and Ahmad, M.H. (2002) Effects of feeding rates on growth and production of Nile tilapia, common carp and silver carp polycultured in fertilized ponds. *Aquaculture Research* 33, 415–423.

Abdelghany, A.E., Ayyat, M.-S. and Ahmad, M.H. (2002) Appropriate timing of supplemental feeding for production of Nile tilapia, silver carp, and common carp in fertilized polyculture ponds. *Journal of the World Aquaculture Society* 33, 307–315.

Abdel-Malek, S.A. (1972) Food and feeding habits of some Egyptian fishes in Lake Quarun. 1. *Tilapia zillii* (Gerv.) B. According to different length groups. *Bulletin of the Institute of Oceanography and Fisheries* 2, 204–213.

Absalom, K.V. and Omenaihe, O. (2000) Effects of water replacement rate on growth and survival of the Nile tilapia *Oreochromis niloticus* fry. *Journal of Aquatic Sciences* 15, 19–22.

Abucay, J.S., Mair, G.C., Skibinski, D.O. and Beardmore, J.A. (1997) The occurrence of incidental sex reversal in *Oreochromis niloticus* L. In: Fitzsimmons, K. (ed.) *Proceedings from the Fourth International Symposium on Tilapia in Aquaculture*. Northeast Regional Agriculture Engineering Service, Ithaca, New York, pp. 729–738.

Adams, A.M., Johnson, P.B. and Hong-qui, Z. (1988) Chemical enhancement of feeding for herbivorous fish *Tilapia zillii*. *Aquaculture* 72, 95–107.

Adeparusi, E.O. and Olute, B.W. (2000) Effects of methionine-supplemented toasted lima bean (*Phaseolus lunatus*) diets on growth of *Oreochromis niloticus*. In: Fitzsimmons, K. and

Filho, J.C. (eds) *Tilapia Culture in the 21st Century. Proceedings from the Fifth International Symposium on Tilapia Aquaculture, Rio de Janeiro, Brazil.* American Tilapia Association, Charles Town, West Virginia, and ICLARM, Penang, Malaysia, pp. 125–130.

Afolabi, J.A., Imoudu, P.B. and Fagbenro, O.A. (2000) Economic and technical viability of tilapia culture in homestead concrete tanks. In: Fitzsimmons, K. and Filho, J.C. (eds) *Tilapia Culture in the 21st Century, Proceedings from the Fifth International Symposium on Tilapia Aquaculture, Rio de Janeiro, Brazil.* American Tilapia Association, Charles Town, West Virginia, and ICLARM, Penang, Malaysia, pp. 575–581.

Afonso, L.O.B. and Leboute, E.M. (2003) Sex reversal in Nile tilapia: is it possible to produce all male stocks through immersion in androgens? *World Aquaculture* 34 (3), 16–19.

Afonso, L.O.B, Wassermann, G.J. and De Oliveira, R.T. (2001) Sex reversal in Nile tilapia (*Oreochromis niloticus*) using a nonsteroidal aromatase inhibitor. *Journal of Experimental Zoology* 290, 177–181.

Agius, C. (2001) Commercial culture of *Oreochromis spilurus* in open sea cages. In: Subasinghe, S. and Singh, T. (eds) *Tilapia: Production, Marketing and Technical Developments. Proceedings of the Tilapia 2001 International Technical and Trade Conference on Tilapia.* Infofish, Kuala Lumpur, Malaysia, pp. 132–135.

Aguilar-Manjarrez, J. and Nath, S.S. (1998) *A Strategic Reassessment of Fish Farming Potential in Africa.* FAO Technical Paper No. 32, FAO, Rome, 170 pp.

Agustin, L. (1999) Effects of genetic bottlenecks on levels of genetic diversity and differentiation in feral populations of *Oreochromis mossambicus.* PhD thesis, Queensland University of Technology, Australia.

Ahmed, M. and Lorica, M.H. (2002) Improving developing country food security through aquaculture development – lessons from Asia. *Food Policy* 27, 125–141.

Ahmed, M., Abdur Rab, M. and Bimbao, M.A.P. (1993) *Household Socioeconomics, Resource Use and Fish Marketing in Two Thanas of Bangladesh.* ICLARM Technical Report No. 40, ICLARM, Manila, Philippines, 82 pp.

Ahmed, M., Rab, M.A. and Gupta, M.V. (1995) Impact of improved aquaculture technologies: results of an extension program on the farming systems of Bangladesh. *Asian Fisheries Science* 8, 27–39.

Ahmed, M., Bimbao, M.P. and Gupta, M.V. (1996) Economics of tilapia aquaculture in small water bodies in Bangladesh. In: Pullin, R.S.V., Lazard, J., Legendre, M., Amon Kottias, J.B. and Pauly, D. (eds) *Proceedings of the Third International Symposium on Tilapia in Aquaculture.* ICLARM Conference Proceedings No. 41, ICLARM, Penang, Malaysia, pp. 471–475.

Ahmed, N.A., El-Serafy, S.S., El-Shafey, A.A.M. and Abdel-Hamid, N.H. (1992) Effect of ammonia on some haematological parameters of *Oreochromis niloticus. Proceedings of Zoological Society of Arab Republic of Egypt* 23, 155–160.

Akel, E.H.K. (1989) Effect of water pollution on tilapia populations in Lake Mariout. MSc thesis, Alexandria University, Alexandria, Egypt.

Al-Ahmad, T.A., Ridha, M. and Al-Ahmad, A.A. (1988) Reproductive performance of the tilapia *Oreochromis spilurus* in seawater and brackish ground-water. *Aquaculture* 73, 323–332.

Al-Ahmed, A.A. (2001) A review of tilapia culture in Kuwait. *World Aquaculture* 32 (2), 47–48.

Al-Ahmed, A.A. (2002) Tilapia culture in Kuwait. *Global Aquaculture Advocate* 5 (6), 31.

Al-Amoudi, M.M. (1987a) Acclimation of commercially cultured *Oreochromis* species to sea water – an experimental study. *Aquaculture* 65, 333–342.

Al-Amoudi, M.M. (1987b) The effect of high salt diet on the direct transfer of *Oreochromis mossambicus, O. spilurus* and *O. aureus/O. niloticus* hybrids to seawater. *Aquaculture* 64, 333–338.

Al-Amoudi, M.M., El-Nakkadi, A.M.N. and El-Nouman, B.M. (1992) Evaluation of optimum dietary requirement of vitamin C for the growth of *Oreochromis spilurus* fingerlings in water from the Red Sea. *Aquaculture* 105, 165–173.

Al-Amoudi, M.M., El-Sayed, A.-F.M. and El-Ghobashy, A. (1996) Effects of thermal and thermo-haline shocks on survival and osmotic concentration of the tilapias *Oreochromis mossambicus* and *Oreochromis aureus* × *Oreochromis niloticus* hybrids. *Journal of the World Aquaculture Society* 27, 456–461.

Alceste, C.C., Illingworth, J.A. and Jory, D.E. (2001) Tilapia farming industry in Ecuador. *Aquaculture Magazine* 27, 77–82.

Alfredo, M.H. and Hector, S.L. (2002) Blood gasometric trends in hybrid red tilapia *Oreochromis niloticus* (Linnaeus) × *O. mossambicus* (Peters) while adapting to increasing salinity. *Journal of Aquaculture in the Tropics* 17, 101–112.

Al Hafedh, Y.S. (1999) Effects of dietary protein on growth and body composition of Nile tilapia, *Oreochromis niloticus* L. *Aquaculture Research* 30, 385–393.

Al-Harbi, A.H. (1994) First isolation of *Streptococcus* sp. from hybrid tilapia (*Oreochromis niloticus* × *O. aureus*) in Saudi Arabia. *Aquaculture* 128, 195–201.

Al-Harbi, A.H. and Naim Uddin, M. (2004) Seasonal variations in the intestinal bacterial flora of hybrid tilapia (*Oreochromis niloticus* × *Oreochromis aureus*) cultured in earthen ponds in Saudi Arabia. *Aquaculture* 229, 37–44.

Allanson, B.R., Bok, A. and VanWyk, N.I. (1971) The influence of exposure to low temperature on *Tilapia mossambica* Peters (Cichlidae). 2. Changes in serum osmolarity, sodium and chloride ion concentrations. *Journal of Fish Biology* 3, 181–185.

Allison, R., Smitherman, R.O. and Gabrero, J. (1979) Effects of high density culture and form of feed on reproduction and yield of *Tilapia aurea*. In: Pillay, T.V.R. and Dill, W.A. (eds) *Advances in Aquaculture*. Fishing News Books, Farnham, UK, pp. 168–170.

Almazan, G.J., Pullin, R.S.V., Angeles, A.F., Manalo, T.A., Agbayani, R.A. and Trono, M.T.B. (1986) *Azolla pinnata* as a dietary component for Nile tilapia *Oreochromis niloticus*. In: Maclean, J.L., Dizon, L.B. and Hosillos, L.V. (eds) *Proceedings of the First Asian Fisheries Forum*. Asian Fisheries Society, Manila, Philippines, pp. 523–528.

Aloo, P., Okelo, R.O. and Ntiba, M.J. (1995) Helminth parasites of *Oreochromis leucostictus* (Trewavas) in Lake Naivasha, Kenya. In: *Pan African Fisheries Congress on Sustainable Development of Fisheries in Africa, Nairobi, Kenya, 31 July–4 August*. Fisheries Society of Africa, Nairobi, Kenya, Abstract No. 172.

Alsagoof, S.A.K., Clonts, H.A. and Jolly, C.M. (1990) An integrated poultry, multi-species aquaculture for Malaysian rice farmers: a mixed integer programming approach. *Agricultural Systems* 32, 207–223.

Amin, N.E., Abdallah, I.S., Faisal, M., Easa, E.M., Alaway, T. and Alyan, S.A. (1988) *Columnaris* infection among cultured Nile tilapia *Oreochromis niloticus*. *Antonie Van Leeuwenhoek* 54, 509–520.

Anadu, D.I. and Barho, L. (2002) The production of tilapia and vegetables in aquaponics system. In: *Proceedings of the Fourth International Conference on Recirculating Aquaculture*. US Department of Agriculture, Virginia Polytechnic Institute and State University, Blacksburg, Virginia, pp. 551–559.

Anderson, J., Jackson, A.J., Matty, A.J. and Capper, B.S. (1984) Effects of dietary carbohydrate and fibre on the tilapia *Oreochromis niloticus* (Linn.). *Aquaculture* 37, 303–314.

Anene, A. (1999) Morphometric and meristic description of *Tilapia mariae* (Boulenger, 1901) and *Tilapia zillii* (Gervais, 1848) from the Umuoseriche Lake in the freshwater reaches of the Niger delta floodplains. *Acta Hydrobiologica* 41, 211–218.

Anon. (2002) Cleaner technique to produce all-male tilapia. *Asian Aquaculture Magazine*, July/August, p. 6.

Anuta, J.D. (1995) Effects of chicken manure application on water quality and production of *Tilapia guineensis* in freshwater concrete tanks. *Journal of Aquaculture in the Tropics* 10, 167–176.

Appler, H.N. (1985) Evaluation of *Hydrodictyon reticulatum* as a protein source in feeds for *Oreochromis* (*Tilapia*) *niloticus* and *Tilapia zillii*. *Journal of Fish Biology* 27, 327–334.

Ardjosoediro, I. and Ramnarine, I.W. (2002) The influence of turbidity on growth, feed conversion and survivorship of the Jamaica red tilapia strain. *Aquaculture* 212, 159–165.

Areerat, S. (1987) Clarias culture in Thailand. *Aquaculture* 63, 355–362.

Ariel, E. and Owens, L. (1997) Epizootic mortalities in tilapia *Oreochromis mossambicus*. *Diseases of Aquatic Organisms* 29, 1–6.

Armstrong, R.D. and Ferguson, H.W. (1989) Systematic viral disease of the chromide cichlid *Etroplus maculatus*. *Diseases of Aquatic Organisms* 7, 155–157.

Arrumm, A. (1987) *Effects of parasite* Contracaecum *on tilapia (Oreochromis niloticus)*. Annual Report. 1986/1987, Kenya Marine Fisheries Research Institute, Mombassa, 21 pp.

Arthington, A.H. and Bluhdorn, D.R. (1996) The effects of species introductions resulting from aquaculture opertations. In: Baird, D.J., Beveridge, M.C.M., Kelly, L.A. and Muir, J.F. (eds) *Aquaculture and Water Resources Management*. Blackwell, Oxford, pp. 114–139.

Arumugam, P.T. (1997) Suitability of a continuous-flow chamber for investigating fish larvae/fry growth responses. *Aquaculture* 151, 365–370.

Atwood, H.L., Fontenot, Q.C., Tomasso, J.R. and Isely, J.J. (2001) Toxicity of nitrite to Nile tilapia: effect of fish size and environmental chloride. *North American Journal of Aquaculture* 63, 49–51.

Atz, J.M. (1954) The peregrinating tilapia. *Animal Kingdom* 57, 148–155.

Avarindan, C.M. and Padmanabhan, K.G. (1972) Source of a new crop of oocytes in *Tilapia mossambica*. *Helgolander Wissenschaftliche Meeresuntersuchungen* 23, 184–192.

Avnimelech, Y. (1998) Minimal discharge from intensive fish ponds. *World Aquaculture* 29 (1), 32–37.

Avnimelech, Y. (1999) Carbon/nitrogen ratio as a control element in aquaculture systems. *Aquaculture* 176, 227–235.

Avnimelech, Y. (2003) Control of microbial activity in aquaculture systems: active suspension ponds. *World Aquaculture* 34 (4), 19–21.

Avnimelech, Y., Kochva, M. and Hargreaves, J.A. (1999) Sedimentation and re-suspension in earthen fish ponds. *Journal of the World Aquaculture Society* 30, 401–409.

Avtalion, R.R. and Reich, L. (1989) Chromatophore inheritance in red tilapias. *Israeli Journal of Aquaculture/Bamidgeh* 41, 98–104.

Avtalion, R.R. and Shlapobersky, M. (1994) A whirling viral disease of tilapia larvae. *Israeli Journal of Aquaculture/Bamidgeh* 46, 102–104.

Ayinla, A.A., Oladosu, G.A., Ajiboye, M.O. and Ansa, E.J. (1994) Pollution and hazards of integrated livestock-cum-fish farming systems in Nigeria. Paper presented at the CIFA Seminar on African Inland Fisheries, Aquaculture and Environment, Harare, Zimbabwe, 5–7 December 1994.

Azad, I.S., Rajendran, K.V., Rajan, J.J.S., Vijayan, K.K. and Santiago, T.C. (2001) Virulence and histopathology of *Aeromonas hydrophila* (Sah 93) in experimentally infected tilapia, *Oreochromis mossambicus* (L.). *Journal of Aquaculture in the Tropics* 16, 265–275.

Azim, M.E., Verdegem, M.C.J., Wahab, M.A., van Dam, A.A. and Beveridge, M.C.M. (2001) Periphyton boosts production in pond aquaculture systems. *World Aquaculture* 32 (4), 57–61.

Azim, M.E., Verdegem, M.C.J., Mantingh, I., van Dam, A.A. and Beveridge, M.C.M. (2003) Ingestion and utilisation of periphyton grown on artificial substrates by Nile tilapia *Oreochromis niloticus* L. *Aquaculture Research* 34, 85–92.

Azim, M.E., Wahab, M.A., Biswas, P.K., Asaeda T., Fujino, T. and Verdegem, M.C.J. (2004a) The effect of periphyton substrate density on production in freshwater polyculture ponds. *Aquaculture* 232, 441–453.

Azim, M.E., Rahaman, M.M., Wahab, M.A., Asaeda, T., Little, D.C. and Verdegem, M.C.J. (2004b) Periphyton-based pond polyculture system: a bioeconomic comparison of on-farm and on-station trials. *Aquaculture* 242, 381–396.

Bailey, D.S., Rakocy, J.E., Martin, J.M. and Shultz, M.R. (2000) Intensive production of tilapia fingerlings in a recirculating system. In: Fitzsimmons, K. and Filho, J.C. (eds) *Tilapia Culture in the 21st Century. Proceedings from the Fifth International Symposium on Tilapia Aquaculture, Rio de Janeiro, Brazil.* American Tilapia Association, Charles Town, West Virginia, and ICLARM, Penang, Malaysia, pp. 328–333.

Bakhoum, S.A. (2002) Occurrence of natural hybrids between Nile tilapia, *Oreochromis niloticus* (L.) and blue tilapia, *O. aureus* (Steind.) in Lake Edku, Egypt. *Egyptian Journal of Aquatic Biology and Fisheries* 6, 143–162.

Balarin, J.D. and Haller, R.D. (1982) The intensive culture of tilapia in tanks, raceways and cages. In: Muir, J.F. and Roberts, R.J. (eds) *Recent Advances in Aquaculture.* Croom Helm, London and Canberra, and Westview Press, Boulder, Colorado, pp. 267–355.

Balarin, J.D. and Haller, R.D. (1983) Commercial tank culture of tilapia. In: Fishelson, L. and Yaron, Z. (eds) *International Symposium on Tilapia in Aquaculture.* Tel Aviv University, Tel Aviv, Israel, pp. 473–483.

Balarin, J.D. and Hatton, J.P. (1979) *Tilapia: A Guide to their Biology and Culture in Africa.* University of Stirling, Stirling, UK.

Balfry, S.K., Shariff, M. and Iwama, G.K. (1997) Strain differences in non-specific immunity of tilapia (*Oreochromis niloticus*) following challenge with *Vibrio parahaemolyticus. Diseases of Aquatic Organisms* 30, 77–80.

Balogun, A.M. and Fagbenro, O.A. (1995) Use of macadamia presscake as a protein feedstuff in practical diets for tilapia, *Oreochromis niloticus* (L.). *Aquaculture Research* 26, 371–377.

Bao-tong, H. and Hua-zhu, Y. (1984) *Integrated Management of Fish-cum-Duck Farming and its Economic Efficiency and Revenue.* NACA/WP/84/14, Network of Aquaculture Centres in Asia (NACA), Bangkok, Thailand.

Baras, E., Jacobs, B. and Melard, C. (2001) Effect of water temperature on survival, growth and phenotypic sex of mixed (XX–XY) progenies of Nile tilapia *Oreochromis niloticus. Aquaculture* 192, 187–199.

Barcellos, L.J.G., Nicolaiewsky, S., de Souza, S.M.G. and Lulhier, F. (1999a) Plasmatic levels of cortisol in the response to acute stress in Nile tilapia, *Oreochromis niloticus* (L.), previously exposed to chronic stress. *Aquaculture Research* 30, 437–444.

Barcellos, L.J.G., Nicolaiewsky, S., de Souza, S.M.G. and Lulhier, F. (1999b) The effects of stocking density and social interaction on acute stress response in Nile tilapia (*Oreochromis niloticus*) fingerlings. *Aquaculture Research* 30, 887–892.

Bardach, J.E., Ryther, J.H. and McLarney, W.O. (1972) *Aquaculture: The Farming and Husbandry of Freshwater and Marine Organisms.* Wiley Interscience, New York, 868 pp.

Barlow, C.G., Pearce, M.G., Rodgers, L.J. and Clayton, P. (1995) Effects of photoperiod on growth, survival and feeding periodicity of larval and juvenile barramundi *Lates calcarifer* (Bloch). *Aquaculture* 138, 159–168.

Barman, B.K., Little, D. and Edwards, P. (2002) Small-scale fish culture in Northwest Bangladesh: a participatory appraisal focussing on the role of tilapia. In: Edwards, P., Little, D. and Demaine, H. (eds) *Rural Aquaculture.* CAB International, Wallingford, UK, pp. 227–244.

Baroiller, J.F., Clota, F. and Geraz, E. (1995) Temperature sex determination in two tilapia, *Oreochromis niloticus* and the red tilapia (red Florida strain): effect of high or low temperature. In: Goetz, F.W. and Thomas, P. (eds) *Proceedings of the Fifth International Symposium on the Reproductive Physiology of Fish*. University of Texas Press, Austin, Texas, pp. 158–160.

Baroiller, J.F., Desprez, D., Carteret, Y., Tacon, P., Borel, F., Hoaraeu, M.C., Mélard, C. and Jalabert, B. (1997) Influence of environmental and social factors on the reproductive efficiency in three tilapia species, *Oreochromis niloticus, O. aureus,* and the red tilapia (Red Florida Strain). In: Fitzsimmons, K. (ed.) *Tilapia Aquaculture. Proceedings from the Fourth International Symposium on Tilapia in Aquaculture*. Northeast Regional Agriculture Engineering Service, Ithaca, New York, pp. 238–252.

Bart, A. (2002) The application of ultrasound to produce all-male tilapia using immersion protocol. In: McElwee, K., Lewis, K., Nidiffer, M. and Buitrago, P. (eds) *Nineteenth Annual Technical Report*. Pond Dynamics/Aquaculture CRSP, Oregon State University, Corvallis, Oregon, pp. 53–57.

Barton, B.A. and Iwama, G.K. (1991) Physiological changes in fish from stress in aquaculture with emphasis on the response and effects of corticosteroids. *Annual Reviews of Fish Diseases* 10, 3–26.

Bauer, J. (1968) Vergleichende Untersuchungen zum Kontaktverhalten verschiedener Arten der Gattung *Tilapia* (Cichlidae, Pisces) und ihrer Bastarde. *Zeitschrift für Tierpsychologie* 25, 22–70.

Beadle, L.C. (1981) *The Inland Waters of Tropical Africa. An Introduction to Tropical Limnology*, 2nd edn. Longman, London.

Beamish, F.W.H. (1970) Influence of temperature and salinity acclimation on temperature preferenda of the euryhaline fish *T. nilotica*. *Journal of the Fisheries Research Board of Canada* 27, 1209–1214.

Beardmore, J.A., Mair, G.C. and Lewis, R.I. (2001) Monosex male production in finfish as exemplified by tilapias: applications, problems and prospects. *Aquaculture* 197, 283–301.

Becker, K. and Fishelson, L. (1986) Standard and routine metabolic rate, critical oxygen tension and spontaneous scope for activity of tilapias. In: Maclean, J.L., Dizon, L.B. and Hosillos, L.V. (eds) *Proceedings of the First Asian Fisheries Forum*. Asian Fisheries Society, Manila, Philippines, pp. 623–628.

Behrends, L.L. and Smitherman, R.O. (1983) Use of warm water effluents to induce winter spawning of tilapia in a temperate climate. In: Fishelson, L. and Yaron, Z. (eds) *Proceedings of the International Symposium on Tilapia in Aquaculture*. Tel Aviv University, Tel Aviv, Israel, pp. 446–454.

Behrends, L.L. and Smitherman, R.O. (1984) Development of a cold-tolerant population of red tilapia through introgressive hybridization. *Journal of the World Mariculture Society* 14, 172–178.

Behrends, L.L., Kingsley, J.B. and Price, A.H. III (1993) Hatchery production of blue tilapia, *Oreochromis aureus* (Steindachner), in small suspended hapa nets. *Aquaculture and Fisheries Management* 24, 237–243.

Behrends, L.L., Kingsley, J.B. and Bulls, M.J. (1996) Cold tolerance in maternal mouthbrooding tilapias: heritability estimates and correlated growth responses at suboptimal temperatures. In: Pullin, R.S.V., Lazard, J., Legendre, M., Amon Kottias, J.B. and Pauly, D. (eds) *Proceedings of the Third International Symposium on Tilapia in Aquaculture*. ICLARM Conference Proceedings No. 41, ICLARM, Penang, Malaysia, pp. 257–265.

Belal, I.E.H. (1999) Replacing dietary corn with barley seeds in Nile tilapia, *Oreochromis niloticus* (L.), feed. *Aquaculture Research* 30, 265–269.

Belal, I.E.H., Al-Owaifeir, A. and Al-Dosari, M. (1995) Replacing fish meal with chicken offal silage in commercial *Oreochromis niloticus* (L.) feed. *Aquaculture Research* 26, 855–858.

Bello-Olusoji, O.A., Aderiye, B.K., Borede, A.A. and Oyekanmi, F.B. (2000) Ectoparasitic studies of pond cultured and wild tilapia in major cocoa producing area of Nigeria. In: Fitzsimmons, K. and Filho, J.C. (eds) *Tilapia Culture in the 21st Century. Proceedings from the Fifth International Symposium on Tilapia Aquaculture, Rio de Janeiro, Brazil.* American Tilapia Association, Charles Town, West Virginia, and ICLARM, Penang, Malaysia, pp. 496–498.

Berg, H., Michélsen, P., Troell, M., Folke, C. and Kautsky, N. (1996) Managing aquaculture for sustainability in tropical Lake Kariba, Zimbabwe. *Ecological Economics* 18, 141–159.

Berridge, B.R., Gonzalez, M. and Frelier, P.F. (1998) Outbreak of *S. difficile* in hybrid tilapia (*Oreochromis aura* × *O. nilotica*) persistently infected with *S. iniae*. In: *Book of Abstracts – Aquaculture '98*. World Aquaculture Society, Baton Rouge, Louisiana, pp. 186–187.

Beveridge, M.C.M. and McAndrew, B.J. (eds) (2000) *Tilapias: Biology and Exploitation*. Kluwer Academic Publishers, Dordrecht/Boston/London, 505 pp.

Bevis, R. (1994) The effect of artificial nests on reproductive performance in Nile tilapia (*Oreochromis niloticus*, L.) spawned in net hapas. MSc thesis, Asian Institute of Technology, Bangkok, Thailand.

Bhujel, R.C. (1999) Management of Nile tilapia (*Oreochromis niloticus*) broodfish for commercial seed production in Thailand. PhD thesis, Asian Institute of Technology, Bangkok, Thailand.

Bhujel, R.C. (2000) A review of strategies for the management of Nile tilapia broodfish in seed production systems, especially hapa-based systems. *Aquaculture* 181, 37–59.

Bhujel, R.C., Yakupitiyage, A., Turner, W.A. and Little, D.C. (2001a) Selection of a commercial feed for Nile tilapia (*Oreochromis niloticus*) broodfish breeding in a hapa-in-pond system. *Aquaculture* 194, 303–314.

Bhujel, R.C., Turner, W.A., Yakupitiyage, A. and Little, D.C. (2001b) Impacts of environmental manipulation on the reproductive performance of Nile tilapia (*Oreochromis niloticus*). *Journal of Aquaculture in the Tropics* 3, 197–209.

Bimbao, G.B., Paraguas, F.J., Dey, M.M. and Eknath, A.E. (2000) Socioeconomics and production efficiency of tilapia hatchery operations in the Philippines. *Aquaculture Economics and Management* 4, 49–63.

Bimbao, M.P. and Smith, I.R. (1988) Philippine tilapia economics: industry growth and potential. In: Pullin, R.S.V., Bhukaswan, T., Tonguthai, K. and Maclean, J.L. (eds) *Proceedings of the Second International Symposium on Tilapia in Aquaculture*. ICLARM Conference Proceedings No. 15, Department of Fisheries, Bangkok, Thailand, and ICLARM, Manila, Philippines, pp. 539–551.

Bishop, C.D., Angus, R.A. and Watts, A. (1995) The use of feather meal as a replacement for fishmeal in the diet of *Oreochromis niloticus* fry. *Bioresource Technology* 54, 291–295.

Biswas, A.K. and Takeuchi, T. (2002) Effect of different photoperiod cycles on metabolic rate and energy loss of both fed and unfed adult tilapia *Oreochromis niloticus*: part II. *Fisheries Science* 68, 543–553.

Biswas, A.K., Endo, M. and Takeuchi, T. (2002) Effect of different photoperiod cycles on metabolic rate and energy loss of both fed and unfed young tilapia *Oreochromis niloticus*: part I. *Fisheries Science* 68, 465–477.

Biswas, A.K., Morita, T., Yoshizaki, G., Maita, M. and Takeuchi, T. (2005) Control of reproduction in Nile tilapia *Oreochromis niloticus* (L.) by photoperiod manipulation. *Aquaculture* 243, 229–239.

Boeuf, G. and Le Bail, P. (1999) Does light have an influence on fish growth? *Aquaculture* 177, 129–152.

Bolivar, R.B. and Newkirk, G.F. (2002) Response to within family selection for body weight in Nile tilapia (*Oreochromis niloticus*) using a single-trait animal model. *Aquaculture* 204, 371–381.

Bolivar, R.B., Eknath, A.E., Bolivar, H.L. and Abella, T.A. (1993) Growth and reproduction of individually tagged Nile tilapia (*Oreochromis niloticus*) of different strains. *Aquaculture* 111, 159–169.

Bondad-Reantaso, M.G. and Arthur, J.R. (1989) Trichodinids (Protozoa: Ciliophora: Peritrichida) of Nile tilapia (*Oreochromis niloticus*) in the Philippines. *Asian Fisheries Science* 3, 27–44.

Bonga, S.E.W., Flike, G. and Balm, P.H.M. (1987) Physiological adaptation to acid stress in fish. In: Witters, H. and Vanderborght, O. (eds) *Symposium on Ecophysiology of Acid Stress in Aquatic Organisms*. Annales de Société Royale Zoologique de Belgique, Brussels, Belgium, pp. 243–254.

Bowen, S.H. (1976) Feeding ecology of the cichlid fish *Sarotherodon mossambicus* in Lake Sibaya, KwaZulu. PhD thesis, Rhodes University, Grahamstown, South Africa.

Bowen, S.H. (1979) A nutritional constraint in detritivory by fishes: the stunted population of *Sarotherodon mossambicus* in Lake Sibaya, South Africa. *Ecological Moneograph* 49, 17–31.

Bowen, S.H. (1982) Feeding, digestion and growth – qualitative considerations. In: Pullin, R.V.S. and Lowe-McConnell, R.H. (eds) *The Biology and Culture of Tilapias*. ICLARM Conference Proceedings No. 7, ICLARM, Manila, Philippines, pp. 141–156.

Bowser, P.R., Wooster, G.A., Getchell, R.G. and Timmons, M.B. (1998) *Streptococcus iniae* infection of tilapia *Oreochromis niloticus* in a recirculation production facility. *Journal of the World Aquaculture Society* 29, 335–339.

Boyd, C.E. (1974) *Lime Requirements of Alabama Fish Ponds*. Alabama Agricultural Experimental Station, Bulletin 459, Auburn University, Auburn, Alabama.

Boyd, C.E. (1976) Nitrogen fertilizer effects on production of *Tilapia* in ponds fertilized with phosphorus and potassium. *Aquaculture* 7, 385–390.

Boyd, C.E. (1990) *Water Quality in Ponds for Aquaculture*. Auburn University, Auburn, Alabama, 482 pp.

Boyd, C.E. (2003) Guidelines for aquaculture effluent management at the farm level. *Aquaculture* 226, 101–112.

Bramick, U., Puckhaber, B., Langholz, H.-J. and Hörstgen-Schwark, G. (1995) Testing of triploid tilapia (*Oreochromis niloticus*) under tropical pond conditions. *Aquaculture* 137, 343–353.

Brem, G., Brenig, B., Hörstgen-Schwark, G. and Winnacker, E.-L. (1988) Gene transfer in tilapia (*Oreochromis niloticus*). *Aquaculture* 68, 209–219.

Brock, J.A., LeaMaster, B.R. and Lee, C.-S. (1993) An overview of pathogens and diseases in marine finfish hatcheries in Hawaii with comments on strategies for health management and disease prevention. In: Lee, C.-S., Su, M.S. and Liao, I.C. (eds) *Proceedings of Finfish Hatchery in Asia '91*. Tungkang Marine Laboratory, Keelung, Taiwan, pp. 221–238.

Bromage, N., Porter, M. and Randall, C. (2001) The environmental regulation of maturation in farmed finfish with special reference to the role of photoperiod and melatonin. *Aquaculture* 197, 63–98.

Broussard, M.C., Reyes, R. and Raguindin, F. (1983) Evaluation of hatchery management schemes for large-scale production of *Oreochromis niloticus* fingerlings in Central Luzon, Philippines. In: Fishelson, L. and Yaron, Z. (eds) *Proceedings of the International Symposium on Tilapia in Aquaculture*. Tel Aviv University, Tel Aviv, Israel, pp. 414–424.

Brown, C.L., Bolivar, R.B., Jimenez, E.T. and Szyper, J. (2000) Timing of the onset of supplemental feeding of Nile tilapia (*Oreochromis niloticus*) in ponds. In: Fitzsimmons, K. and Filho, J.C. (eds) *Tilapia Culture in the 21st Century. Proceedings from the Fifth International Symposium on Tilapia Aquaculture, Rio de Janeiro, Brazil.* American Tilapia Association, Charles Town, West Virginia, and ICLARM, Penang, Malaysia, pp. 237–240.

Brown, C.L., Bolivar, R.B., Jimenez, E.B.T. and Szyper, J.P. (2002a) Reduction of feed rations below satiation levels in tilapia pond production. In: McElwee, K., Lewis, K., Nidiffer, M. and Buitrago, P. (eds) *Nineteenth Annual Technical Report*. Pond Dynamics/Aquaculture CRSP, Oregon State University, Corvallis, Oregon, pp. 21–23.

Brown, C.L., Bolivar, R.B., Jimenez, E.B.T. and Szyper, J.P. (2002b) Educational development activities in support of tilapia aquaculture in the Philippines. In: McElwee, K., Lewis, K., Nidiffer, M. and Buitrago, P. (eds) *Ninteenth Annual Technical Report*. Pond Dynamics/Aquaculture CRSP, Oregon State University, Corvallis, Oregon, pp. 25–26.

Brummett, R.E. (2000) Factors influencing fish prices in Southern Malawi. *Aquaculture* 186, 243–251.

Brummett, R.E. (2002a) Comparison of African tilapia partial harvesting systems. *Aquaculture* 214, 103–114.

Brummett, R.E. (2002b) Seasonality, labour and integration of aquaculture into southern African smallhold farming systems. *Naga: The ICLARM Quarterly* 25, 23–27.

Brummett, R.E. and Noble, R.P. (1995) *Aquaculture for African Smallholders*. ICLARM Technical Report No. 46. ICLARM, Manila, the Philippines, 69 pp.

Brummett, R.E. and Williams M.J. (2000) The evolution of aquaculture in African rural and economic development. *Ecological Economics* 33, 193–203.

Buddington, R.K. (1979) Digestion of an aquatic macrophyte by *Tilapia zillii* (Gervais). *Journal of Fish Biology* 15, 449–455.

Bunch, E.C. and Bejerano, I. (1997) The effect of environmental factors on the susceptibility of hybrid tilapia, *Oreochromis niloticus* × *Oreochromis aureus*, to streptococcosis. *Israeli Journal of Aquaculture/Bamidgeh* 49, 67–76.

Byamungu, N., Darras, V.M. and Kühn, E.R. (2001) Growth of heat-shock induced triploids of blue tilapia, *Oreochromis aureus*, reared in tanks and in ponds in eastern Congo: feeding regimes and compensatory growth response of triploid females. *Aquaculture* 198, 109–122.

Cagauan, A.G., Baleta, F.N. and Abucay, J.S. (2004) Sex reversal of Nile tilapia *Oreochromis niloticus* L. by egg immersion technique: the effect of hormone concentration and immersion time. In: Bolivar, R., Mair, G. and Fitzsimmons, K. (eds) *Proceedings of the Sixth International Symposium on Tilapia in Aquaculture*. Bureau of Fisheries and Aquatic Resources, Manila, Philippines, and American Tilapia Association, Charles Town, West Virginia, pp. 127–136.

Cai, W.-Q. and Sun, P.F. (1994) Within species variation of resistance of common carp in the case of outbreak fish disease. *Journal of Fisheries of China* 18, 290–301.

Cai, W.-Q. and Sun, P.F. (1995) Among species variation of resistance of grass carp, bighead carp and silver carp to the case of outbreak fish disease. *Journal of Fisheries Sciences of China* 2, 71–75.

Cai, W.-Q., Li, S.-F. and Ma, J.-Y. (2004) Disease resistance of Nile tilapia (*Oreochromis niloticus*), blue tilapia (*Oreochromis aureus*) and their hybrid (female Nile tilapia × male blue tilapia) to *Aeromonas sobria*. *Aquaculture* 229, 79–87.

Campbell, D. (1985) Large scale cage farming of *Sarotherodon niloticus*. *Aquaculture* 48, 57–69.

Campos-Mendoza, A., McAndrew, B.J., Coward, K. and Bromage, N. (2004) Reproductive response of Nile tilapia (*Oreochromis niloticus*) to photoperiodic manipulation: effects on spawning periodicity, fecundity and egg size. *Aquaculture* 231, 299–314.

Canagaratnam, P. (1966) Growth of *Tilapia mossambica* Peters in different salinities. *Bulletin of Fisheries Research Station, Ceylon* 19, 1–2.

Cao, T.B. (1998) Development of pond culture of Thai Nile tilapia (*Oreochromis niloticus* L.) and its marketability in Hanoi, Vietnam. PhD thesis, Asian Institute of Technology, Bangkok, Thailand.

Carmelo, A. (2002) Commercial culture of *Oreochromis spilurus* in open seawater cages and onshore tanks. *Israeli Journal of Aquaculture/Bamidgeh* 54, 27–33.

Caulton, M.S. (1976) The importance of pre-digestive food preparation to *Tilapia rendalli* Boulenger when feeding on aquatic macrophytes. *Transactions of Rhodesian Scientific Association* 57, 22–28.

Chamberlain, G.W. and Hopkins, J.S. (1994) Reducing water use and feed cost in intensive ponds. *World Aquaculture* 25 (3), 29–32.

Chang, P.H. and Plumb, J.A. (1996a) Effects of salinity on *Streptococcus* infection of Nile tilapia, *Oreochromis niloticus*. *Journal of Applied Aquaculture* 6, 39–46.

Chang, P.H. and Plumb, J.A. (1996b) Histopathology of experimental *Streptococcus* sp. infection in tilapia, *Oreochromis niloticus* (L.) and Channel catfish *Ictalurus punctatus* (Rafinesque). *Journal of Fish diseases* 19, 235–241.

Chang, S.L., Chang, C.F. and Liao, I.C. (1996) Studies on the growth performance and gonadal development of triploid tilapia *Oreochromis aureus*. In: Pullin, R.S.V., Lazard, J., Legendre, M., Amon Kottias, J.B. and Pauly, D. (eds) *Proceedings of the Third International Symposium on Tilapia in Aquaculture*. ICLARM Conference Proceedings No. 41, ICLARM, Penang, Malaysia, p. 539.

Chapman, G. and Fernando, C.H. (1994) The diets and related aspects of feeding Nile tilapia (*Oreochromis niloticus* L.) and common carp (*Cyprinus carpio*, L.) in lowland rice fields in northeast Thailand. *Aquaculture* 123, 281–307.

Charles, A.T., Agbayani, R.F., Agbayani, E.C., Agüero, M., Belleza, E.T., González, E., Stomal, B. and Weigel, J.-Y. (1997) *Aquaculture Economics in Developing Countries: Regional Assessments and an Annotated Bibliography*. FAO Fisheries Circular No. 932, FAO, Rome, 401 pp.

Chatakondi, N. (1995) Evaluation of transgenic common carp *Cyprinus carpio*, containing rainbow trout growth hormone in ponds. PhD thesis, Auburn University, Auburn, Alabama.

Chen, S.-C., Yoshida, T., Adams, A., Thompson, K.D. and Richards, R.H. (1998) Non-specific immune response of Nile tilapia, *Oreochromis nilotica*, to the extracellular products of *Mycobacterium* spp. and to various adjuvants. *Journal of Fish Diseases* 21, 39–46.

Chen, S.-M., Chen, P.-C., Hsiao, I.-C. and Chen, J.-C. (2001) Effects of pH on the nitrogenous excretion and lethal DO of tilapia *Oreochromis mossambica*. In: *Sixth Asian Fisheries Forum, Book of Abstracts*. Asian Fisheries Society, Manila, Philippines, p. 51.

Chen, W., Sun, L., Tsai, C., Song, Y. and Chang, C. (2002) Cold-stress induced the modulation of catecholamines, cortisol, immunoglobulin M and leukocyte phagocytosis in tilapia. *General and Comparative Endocrinology* 126, 90–100.

Cheong, L., Chan, F.K., Wong, J. and Chou, R. (1987) Observations on the culture of red tilapia (*Oreochromis niloticus* hybrid) in seawater under intensive tank condition using a biodrum. *Singapore Journal of Primary Industries* 15, 42–56.

Chervinski, J. (1982) Environmental physiology of tilapias. In: Pullin, R.V.S. and Lowe-McConnell, R.H. (eds) *The Biology and Culture of Tilapias*. ICLARM Conference Proceedings No. 7, ICLARM, Manila, Philippines, pp. 119–128.

Chervinski, J. and Lahav, M. (1976) The effect of exposure to low temperature on fingerlings of local tilapia (*Tilapia aurea*) (Steindachner) and imported tilapia (*Tilapia vulcani*) (Trewavas) and *Tilapia nilotica* (Linne) in Israel. *Bamidgeh* 28, 25–29.

Chervinski, J. and Yashouv, A. (1971) Preliminary experiments on the growth of *Tilapia aurea* in seawater ponds. *Bamidgeh* 23, 125–129.

Chervinski, J. and Zorn, M. (1974) Note on the growth of *Tilapia aurea* and *Tilapia zillii* in seawater ponds. *Aquaculture* 4, 249–255.

Chiang, F.-S. (2001) Marketing Taiwanese tilapia to the US market. In: Subasinghe, S. and Singh, T. (eds) *Tilapia: Production, Marketing and Technical Developments. Proceedings of the Tilapia 2001 International Technical and Trade Conference on Tilapia*. Infofish, Kuala Lumpur, Malaysia, pp. 75–80.

Chiayvareesajja, J., Roeed, K.H., Eknath, A.E., Danting, J.C. and De Vera, M.P. (1999) Genetic variation in lytic activities of blood serum from Nile tilapia and genetic associations with survival and body weight. *Aquaculture* 175, 49–62.

Chiayvareesajja, S., Wongwit, C. and Tansakul, R. (1990) Cage culture of tilapia (*Oreochromis niloticus*) using aquatic weed-based pellets. In: Hirano, R. and Hanyu, I. (eds) *Proceedings of the Second Asian Fisheries Forum*. Asian Fisheries Society, Manila, Philippines, pp. 287–290.

Chikafumbwa, F.J.K. (1998) Use of different fishpond inputs on pond mud nutrient levels in integrated crop fish culture. In: Coetzee, L., Gon, J. and Kulongowski, C. (eds) *International Conference for the Paradi Association and the Fisheries Society of Africa*. Paradi Association and Fisheries Society of Africa, Grahamstown, South Africa, p. 201.

Chimatiro, S.K. and Janke, A. (1994) Socioeconomic assessment of small-holder aquaculture: a case study of smallholder farmers in Mwanza and

Zomba Districts. In: Brummett, R.E. (ed.) *Aquaculture Policy Options for Integrated Resource Management in Sub Saharan Africa*. ICLARM Conference Proceedings (Extended Abstracts) No. 46 GTZ/ICLARM, Manila, Philippines, 38 pp.

Chinabut, S. (2002) A case study of isopod infestation in tilapia cage culture in Thailand. In: Arthur, J.R., Phillips, M.J., Subasinghe, R.P., Reantaso, M.B. and MacRae, I.H. (eds) *Primary Aquatic Animal Health Care in Rural, Small-scale, Aquaculture Development*. FAO Fisheries Technical Paper No. 406, FAO, Rome, pp. 201–202.

Chitminat, C. (1996) Predator avoidance of transgenic channel catfish containing salmon growth hormone genes. MSc thesis, Auburn University, Auburn, Alabama.

Chou, B.S. and Shiau, S.-Y. (1996) Optimal dietary lipid level for growth of juvenile hybrid tilapia, *Oreochromis niloticus* × *Oreochromis aureus*. *Aquaculture* 143, 185–195.

Chou, B.S. and Shiau, S.-Y. (1999) Both n-6 and n-3 fatty acids are required for maximum growth of juvenile hybrid tilapia, *Oreochromis niloticus* × *Oreochromis aureus*. *North American Journal of Aquaculture* 61, 13–20.

Chow, F., Macchiavello, J., Santa, S.S., Fonck, E. and Olivares, J. (2001) Utilization of *Gracilaria chilensis* (Rhodophyta: Gracilariaceae) as a biofilter in the depuration of effluents from tank cultures of fish, oysters and sea urchins. *Journal of the World Aquaculture Society* 32, 215–220.

Chun, S.-K. and Sohn, S.-G. (1985) Characteristics of *Flexibacter columnaris* isolated from tilapia (*Tilapia* sp.). *Bulletin of the Korean Fisheries Society* 18, 369–373.

Cissé, A. (1988) Effects of varying protein levels on spawning frequency and growth of *Sarotherodon melanotheron*. In: Pullin, R.S.V., Bhukaswan, T., Tonguthai, K. and Maclean, J.L. (eds) *Proceedings of the Second International Symposium on Tilapia in Aquaculture*. ICLARM Proceedings No. 15, Department of Fisheries, Bangkok, Thailand, and ICLARM, Manila, Philippines, pp. 329–333.

Clark, J.H., Watanabe, W.O. and Ernst, D.H. (1990) Effect of feeding rate on growth and diet conversion of Florida red tilapia reared in floating marine cages. *Journal of the World Aquaculture Society* 21, 16–24.

Cnaani, A., Gall, G.A.E. and Hulata, G. (2000) Cold tolerance of tilapia species and hybrids. *Aquaculture International* 8, 289–298.

Coates, D. (1987) Considerations of fish introductions into the Sepik River, Papua New Guinea. *Aquaculture and Fisheries Management* 18, 231–241.

Coche, A.G. (1982) Cage culture of tilapia. In: Pullin, R.S.V. and Lowe-McConnell, R.H. (eds) *The Biology and Culture of Tilapias*. ICLARM Conference Proceedings No. 7, ICLARM, Manila, Philippines, pp. 205–246.

Cole, W.M., Rakocy, J.E., Shultz, K.A. and Bailey, D.S. (1997) Effects of solids removal on tilapia production and water quality in continuously aerated, outdoor tanks. In: Fitzsimmons, K. (ed.) *Proceedings from the Fourth International Symposium on Tilapia in Aquaculture*. Northeast Regional Agriculture Engineering Service, Ithaca, New York, pp. 373–384.

Constantino Casas, F., Armijo Ortiz, A., Osorio Sarabia, D. and Chavez Soriano, L.A. (1997) Clinical and pathological descriptions of an infection due to *Aeromonas hydrophila* and *Ichthyophthirius multifiliis* in rainbow trout (*Oncorhynchus mykiss*, Walbaum) and tilapia (*Oreochromis aureus*, L). *Veterinaria Mexico* 28, 59–62.

Contreras-Sánchez, W.M., Fitzpatrick, M.S. and Schreck, C.B. (2001) Fate of methyltestosterone in the pond environment: detection of MT in pond soil from a CRSP site. In: Gupta, A., McElwee, K., Burke, D., Burright, J., Cummings, X. and Egna, H. (eds) *Eighteenth Annual Technical Report*. Pond Dynamics/Aquaculture CRSP, Oregon State University, Corvallis, Oregon, pp. 79–82.

Contreras-Sánchez, W.M., Couturrier, G.M. and Schreck, C.B. (2002) Fate of methyltestosterone in the pond environment: use of MT in earthen ponds with no record of hormone usage. In: McElwee, K., Lewis, K., Nidiffer, M. and Buitrago, P. (eds) *Nineteenth Annual Technical Report*. Pond Dynamics/Aquaculture CRSP, Oregon State University, Corvallis, Oregon, pp. 103–106.

Copin, Y. and Oswald, M. (1993) The orientation of local fishfarming techniques in the west of the Ivory Coast. In: *Production, Environment and Quality*. Special Publication No. 18, European Aquaculture Society, Ghent, Belgium, pp. 407–420.

Corbet, S.A., Green, J., Griffith, J. and Betney, E. (1973) Ecological studies on crater lakes in West Cameroon: Lakes Kotto and Mboandong. *Journal of Zoology* 170, 309–324.

Cornish, D.A., Smit, G.L. and Campbell, I.M. (1997) Seasonal fluctuations in gonadotropin levels in the plasma and gonads of male and female tilapia, *Oreochromis mossambicus*. *Water South Africa* 23, 263–270.

Corre, V.L. (1981) Culture of Nile tilapia in brackishwater ponds using supplemental feeds containing various levels of protein. MSc thesis,

University of the Philippines, Holio City, Philippines.

Costa, F.H.F., Sampaio, A.H., Sake-Sampaio, S., Lima, F.M., Matias, J.F.N., Rocha, I.R.C.B., Santos, J.A.R. and Rocha, P.J.C. (2000) Tilapia cage culture in reservoirs in Cear State. In: Fitzsimmons, K. and Filho, J.C. (eds) *Tilapia Culture in the 21st Century. Proceedings from the Fifth International Symposium on Tilapia Aquaculture, Rio de Janeiro, Brazil.* American Tilapia Association, Charles Town, West Virginia, and ICLARM, Penang, Malaysia, pp. 393–399.

Costa-Pierce, B.A. and Hadikusumah, H. (1995) Production management of double-net tilapia *Oreochromis* spp. hatcheries in a eutrophic tropical reservoir. *Journal of the World Aquaculture Society* 26, 453–459.

Coulibaly, N.D. and Salembere, S. (1998) La biocénose terrestre des zones humides et ses incidences sur l'infestion des poissons cichlides du lac de Kompienga *Clinostomum* sp. (Trematode: Clinostomatidae). In: Coetzee, L. and Kulongowski, C. (eds) *African Fisheries Diversity and Utilization.* Paradi Association and the Fisheries Society of Africa, Grahamstown, South Africa, p. 352.

Coward, K. and Bromage, N.R. (1998) Histological classification of ovarian growth and the dynamics of ovarian recrudescence in *Tilapia zillii* (Gervais), a substrate-spawning cichlid. *Journal of Fish Biology* 53, 285–302.

Coward, K. and Bromage, N.R. (1999) Spawning periodicity, fecundity and egg size in laboratory-held stocks of a substrate-spawning tilapiine, *Tilapia zillii* (Gervais). *Aquaculture* 171, 251–267.

Coward, K. and Bromage, N.R. (2000) Reproductive physiology of female tilapia broodstock. *Reviews in Fish Biology and Fisheries* 10, 1–25.

Cozzo-Siqueira, A., Oetterer, M. and Gallo, C.R. (2003) Effects of irradiation and refrigeration on the nutrients and shelf-life of tilapia (*Oreochromis niloticus*). *Journal of Aquatic Food Product Technology* 12, 85–102.

Crosby, D. (2001) Fish health issues and problems in Virginia's aquaculture industry. In: *Book of Abstracts – Aquaculture 2001.* World Aquaculture Society, Baton Rouge, Louisiana, p. 145.

Cruz, E.M. and Ridha, M. (1991) Production of tilapia *Oreochromis spilurus* Gunther stocked at different densities in sea cages. *Aquaculture* 99, 95–103.

Dabrowska, H., Meyer-Burgdorff, K. and Gunther, K.D. (1989) Interaction between dietary protein and magnesium level in tilapia (*Oreochromis niloticus*). *Aquaculture* 76, 277–291.

Dambo W.B. and Rana, K.J. (1992) Effects of stocking density on growth and survival of Nile tilapia *Oreochromis niloticus* (L.) fry in the hatchery. *Aquaculture and Fisheries Management* 23, 71–80.

Dan, N.C. and Little, D.C. (2000a) The culture performance of monosex and mixed-sex new season and over-wintered fry in two strains of Nile tilapia (*Oreochromis niloticus*) in Northern Vietnam. *Aquaculture* 184, 221–231.

Dan, N.C. and Little, D. (2000b) Overwintering performance of Nile tilapia *Oreochromis niloticus* (L.) broodfish and seed at ambient temperatures in northern Vietnam. *Aquaculture Research* 31, 485–493.

Darwish, A.M. and Griffin, B.R. (2002) Study shows oxytetracycline controls *Streptococcus* in tilapia. *Global Aquaculture Advocate* 5 (6), 34–35.

Daud, S.K., Hasbollah, A. and Law, A.T. (1988) Effect of unionized ammonia on red tilapia (*Oreochromis mossambicus/O. niloticus* hybrids) fry. In: Pullin, R.S.V., Bhukaswan, T., Tonguthai, K. and Maclean, J.L. (eds) *Proceedings of the Second International Symposium on Tilapia in Aquaculture.* ICLARM Conference Proceedings No. 15, Department of Fisheries, Bangkok, Thailand, and ICLARM, Manila, Philippines, pp. 411–414.

Davies, S.J., Williamson, J., Robinson, M. and Bateson, R.I. (1989) Practical inclusion levels of common animal by-products in complete diets for tilapia (*Oreochromis mossambicus*, Peters). In: Takeda, M. and Watanabe, T. (eds) *Proceedings of the Third International Symposium on Feeding and Nutrition of Fish.* Tokyo University of Fisheries, Tokyo, Japan, pp. 325–332.

Davies, S.J., McConnell, S. and Bateson, R.I. (1990) Potential of rapeseed meal as an alternative protein source in complete diets for tilapia (*Oreochromis mossambicus*, Peters). *Aquaculture* 87, 145–154.

Degani, G., Viola, S. and Yehuda, Y. (1997) Apparent digestibility of protein and carbohydrate in feed ingredients for adult tilapia (*Oreochromis aureus* × *O. niloticus*). *Israeli Journal of Aquaculture/Bamidgeh* 49, 115–123.

De Graaf, G.J., Galemoni, F. and Huisman, E.A. (1999) Reproductive biology of pond-reared Nile tilapia, *Oreochromis niloticus* L. *Aquaculture Research* 30, 25–33.

Dela Cruz, C.R. and Lopez, E.A. (1980) Rotational farming of rice and fish in paddies. *Fishery Research Philippines* 5, 39–52.

de la Fuente, J., Hernandez, O., Martinez, R., Guillen, I., Estrada, M.P. and Lleonart, R. (1996) Generation, characterization and risk assessment of transgenic tilapia with accelerated growth. *Biotecnologia* 13, 221–230.

de la Fuente, J., Guillen, I., Martinez, R. and Estrada, M.P. (1999) Growth regulation and enhancement in tilapia: basic research findings and their applications. *Genetic Analysis: Biomolecular Engineering* 15, 85–90.

Delmendo, M.N. (1980) A review of integrated livestock–fowl–fish farming systems. In: Pullin, R.S.V. and Shehadeh, Z.H. (eds) *Proceedings of Integrated Agriculture – Aquaculture Farming Systems.* ICLARM Conference Proceedings No. 4, ICLARM, Manila, Philippines, pp. 59–71.

Delvin, R.H., Yesaki, T.Y., Biagi, C.A., Donaldson, E.M., Swanson, P. and Chan, W.-K. (1994) Extraordinary salmon growth. *Nature* 371, 209–210.

de Moor, F.C., Wilkinson, R.C. and Herbst, H.M. (1986) Food and feeding habits of *Oreochromis mossambicus* (Peters) in hypertrophic Hartbeespoort Dam, South Africa. *South African Journal of Zoology* 21, 170–176.

Denzer, H.W. (1968) Studies on the physiology of young *Tilapia. FAO Fisheries Report* 44 (4), 356–366.

de Ocampo, A.A. and Camberos, L.O. (1998) Histopatologica de la respuesta de la tilapia (*Oreochromis* sp.) a una infeccion mixta por myxosporidios. Estudio en un caso natural. *Veterinaria Mexico* 29, 213–216.

De Silva, S.S. (1985a) Body condition and nutritional ecology of *Oreochromis mossambicus* (Pisces, Cichlidae) populations of man-made lakes in Sri Lanka. *Journal of Fish Biology* 27, 621–633.

De Silva, S.S. (1985b) Status of the introduced cichlid *Sarotherodon mossambicus* (Peters) in the reservoir fishery of Sri Lanka: a management strategy and ecological implications. *Aquaculture and Fisheries Management* 16, 91–102.

De Silva, S.S. (1995) Supplementary feeding in semi-intensive aquaculture systems. In: New, M.B., Tacon, A.G.J. and Csavas, I. (eds) *Farm-made Aquafeeds.* FAO Fisheries Technical Paper No. 434, FAO, Rome, pp. 24–60.

De Silva, S.S. and Gunasekera, R.M. (1989) Effect of dietary protein level and amount of plant ingredient (*Phaseolus aureus*) incorporated into the diets on consumption, growth performance and carcass composition in *Oreochromis niloticus* fry. *Aquaculture* 80, 121–133.

De Silva, S.S. and Radampola, K. (1990) Effect of dietary protein level on the reproductive performance of *Oreochromis niloticus.* In: Hirano, R. and Hanyu, I. (eds) *Proceedings of the Second Asian Fisheries Forum.* Asian Fisheries Society, Manila, Philippines, pp. 559–563.

De Silva, S.S. and Senaratne, K.A.D.U. (1988) *Oreochromis mossambicus* is not universally a nuisance species: the Sri Lankan experience. In: Pullin, R.S.V., Bhukaswan, T., Tonguthai, K. and Maclean, J.L. (eds) *Proceedings of the Second International Symposium on Tilapia in Aquaculture.* Conference Proceedings No. 15, Department of Fisheries, Bangkok, Thailand, and ICLARM, Manila, Philippines, pp. 445–450.

De Silva, S.S., Perera, M.K. and Maitipe, P. (1984) The composition, nutritional status and digestibility of the diets of *Sarotherodon mossambicus* from nine man-made lakes in Sri Lanka. *Environmental Biology of Fishes* 11, 205–219.

De Silva, S.S., Gunasekera, R.M. and Keembiyahetty, C. (1986) Optimum ration and feeding frequency in *Oreochromis niloticus* young. In: Mclean, J.L., Dizon, L.B. and Hosillos, L.V. (eds) *Proceedings of the First Asian Fisheries Forum.* Asian Fisheries Society, Manila, Philippines, pp. 559–564.

De Silva, S.S., Gunasekera, R.M. and Shim, K.F. (1991) Interactions of varying protein and lipid levels in young red tilapia: evidence of protein sparing. *Aquaculture* 95, 305–318.

Desprez, D., Géraz, E., Hoareau, M.C., Mélard, C., Bosc, P. and Baroiller, J.F. (2003) Production of a high percentage of male offspring with a natural androgen, 11β-hydroxyandrostenedione (11βOHA4), in Florida red tilapia. *Aquaculture* 216, 55–65.

Dewan, S. and Saha, S.N. (1979) Food and feeding habits of *Tilapia nilotica* (L.) (Perciformes: Cichlidae). 2. Diel and seasonal patterns of feeding. *Bangladesh Journal of Zoology* 7, 75–80.

Dey, M.M. (2000) The impact of genetically improved farmed Nile tilapia in Asia. *Aquaculture Economics and Management* 4, 107–124.

Dey, M.M. (2001) Tilapia production in South Asia and the Far East. In: Subasinghe, S. and Singh, T. (eds) *Tilapia: Production, Marketing and Technical Developments. Proceedings of the Tilapia 2001 International Technical and Trade Conference on Tilapia.* Infofish, Kuala Lumpur, Malaysia, pp. 17–27.

Dey, M.M., Bimbao, G.B., Kohinoor, A.H.M. and Hussain, M.G. (1997) Socio-economic status of small aquaculture farmers, consumption patterns and fish marketing in Bangladesh. Paper presented in the Final Workshop of the DEGITA Project, 18–19 February 1997, Manila, Philippines.

Dey, M.M, Bimbao, G.B., Yong, L., Regaspi, P., Kohinoor, A.H.M., Chung, D., Pongthana, N. and Paraguas, F.J. (2000a) Current status of production and consumption of tilapia in selected Asian countries. *Aquaculture Economics and Management* 4, 13–31.

Dey, M.M., Paraguas, F.J., Bimbao, G.B. and Regaspi, P.B. (2000b) Technical efficiency of tilapia growout pond operations in the Philippines.

Aquaculture Economics and Management 4, 33–47.

Dey, M.M., Eknath, A.E., Sifa, L., Hussain, M.G., Thien, T.M., Van Hao, N., Aypa, S. and Pongthana, N. (2000c) Performance and nature of genetically improved farmed tilapia: a bioeconomic analysis. *Aquaculture Economics and Management* 4, 83–106.

Diana, J.S. and Lin, C.K. (1998) The effects of fertilization and water management on growth and production of Nile tilapia in deep ponds during the dry season. *Journal of the World Aquaculture Society* 29, 405–413.

Diana, J.S., Lin, C.K. and Schneeberger, P.J. (1991) Relationships among nutrient inputs, primary production, and yield of *Oreochromis niloticus* in ponds. *Aquaculture* 92, 323–342.

Diana, J.S., Lin, C.K. and Jaiyen, K. (1994) Supplemental feeding of tilapia in fertilized ponds. *Journal of the World Aquaculture Society* 25, 497–506.

Diana, J.S., Lin, C.K. and Yi, Y. (1996) Timing of supplemental feeding for tilapia production. *Journal of the World Aquaculture Society* 27, 410–419.

do Carmo e Sá, M.V., Pezzato, L.E., Lima, M.M.B. and Padilha, P.M. (2004) Optimum zinc supplementation level in Nile tilapia *Oreochromis niloticus* juveniles diets. *Aquaculture* 238, 385–401.

Dos Santos, M.J.M. and Valenti, W.C. (2002) Production of Nile tilapia *Oreochromis niloticus* and freshwater prawn *Macrobrachium rosenbergii* stocked at different densities in polyculture systems in Brazil. *Journal of the World Aquaculture Society* 33, 369–376.

Doueellou, L. and Erlwanger, K.H. (1994) Crustacean parasites of fishes in Lake Kariba, Zimbabwe, preliminary results. *Hydrobiologia* 287, 233–242.

Drennan, D.G. II., Rahman, M.M. and Malone, R.F. (1993) Slow sand filtration technologies for the containment of non-indigenous species in recirculating aquaculture effluents. In: Wang, J.-K. (ed.) *Techniques for Modern Aquaculture*. American Society of Agricultural Engineers, St Joseph, Michigan pp. 335–344.

Drenner, R.W., Vinyard, G.L., Hambright, K.D. and Gophen, M. (1987) Ingestion by *Tilapia galilaea* is not affected by removal of gill rakers and microbranchiospines. *Transactions of the American Fisheries Society* 116, 272–276.

D'Silva, A.M. and Maughan, O.E. (1995) Effects of density and water quality on red tilapia (*Oreochromis mossambicus* × *O. urolepis hornorum*) in pulsed-flow culture systems. *Journal of Applied Aquaculture* 5, 69–76.

Duarte, S.A. (1989) Effects of water depth and water fluctuation on the breeding performance of *Oreochromis niloticus* in concrete tanks. MSc thesis, Auburn University, Auburn, Alabama.

Dunham, R.A. (1999) Utilization of transgenic fish in developing countries: potential benefits and risks. *Journal of the World Aquaculture Society* 30, 1–11.

Duponchelle, F. and Legendre, M. (1997) Influence of space structure on reproductive traits of *Oreochromis niloticus* females. In: Fitzsimmons, K. (ed.) *Proceedings from the Fourth International Symposium on Tilapia in Aquaculture*. Northeast Regional Agriculture Engineering Service, Ithaca, New York, pp. 305–314.

Duponchelle, F. and Legendre, M. (2000) *Oreochromis niloticus* (Cichlidae) in Lake Ayame, Côte d'Ivoire: life history traits of a strongly diminished population. *Cybium* 24, 161–172.

Duponchelle, F. and Panfili, J. (1998) Variations in size and age at maturity of female Nile tilapia, *Oreochromis niloticus*, populations from man-made lakes of Côte d'Ivoire. *Environmental Biology of Fishes* 52, 453–465.

Duponchelle, F., Pouyaud, L. and Legendre, M. (1997) Variation in reproductive characteristics of *Oreochromis niloticus* populations: genetic or environmental effects? In: Fitzsimmons, K. (ed.) *Proceedings from the Fourth International Symposium on Tilapia in Aquaculture*. Northeast Regional Agricultural Engineering Services, Ithaca, New York, pp. 253–261.

Edirisinghe, U. (1990) Suitability of common carp for inclusion in bighead carp–Nile tilapia polyculture systems in ponds fertilized with duck litter. In: Hirano, R. and Hanyu, I. (eds) *The Second Asian Fisheries Forum*. Asian Fisheries Society, Manila, Philippines, pp. 201–204.

Edwards, P., Demaine H., Komolmarl, S., Little, D.C., Innes-Taylor, N., Turongruang, D., Yakupitiyage, A. and Warren, T.J. (1991) Towards the improvement of fish culture by small-scale farmers in Northeast Thailand. *Journal of the Asian Farming System Association* 1, 278–302.

Edwards, P., Pacharaprakiti, C. and Yomjinda, M. (1994a) An assessment of the role of buffalo manure for pond culture of tilapia. I. On-station experiment. *Aquaculture* 126, 83–95.

Edwards, P., Kaewpaitoon, K., Little, D.C. and Siripandh, N. (1994b) An assessment of the role of buffalo manure for pond culture of tilapia. II. Field trial. *Aquaculture* 126, 97–106.

Edwards, P., Demaine, H., Innes-Taylor, N. and Turongruang, D. (1996) Sustainable aquaculture for small-scale farmers: need for a balanced model. *Outlook on Agriculture* 25, 19–26.

Edwards, P., Lin, C.K. and Yakupitiyage, A. (2000) Semi-intensive pond aquaculture. In: Beveridge, M.C.M. and McAndrew, B.J. (eds) *Tilapias: Biology and Exploitation*. Kluwer Academic Publishers, Dordrecht/Boston/London, pp. 377–403.

Edwards, P., Little, D. and Demaine, H. (2002) Issues in rural aquaculture. In: Edwards, P., Little, D. and Demaine, H. (eds) *Rural Aquaculture*. CAB International, Wallingford, UK, pp. 323–340.

Eguia, R.V. and Romana-Eguia, M.R. (2004) *Tilapia Farming in Cages and Ponds*. SEAFDEC Aquaculture Department, Tigbauan, Iloilo, Philippines, 40 pp.

Eid, A.E. and Ghonim, S. (1994) Dietary zinc requirement of fingerling *Oreochromis niloticus*. *Aquaculture* 119, 259–264.

Eknath, A.E. and Acosta, B.O. (1998) *Genetic Improvement of Farmed Tilapias (GIFT) Project: Final Report, March 1988 to December 1997*. ICLARM, Manila, Philippines, 75 pp.

Eknath, A.E., Tayamen, M.M., Palada-de Vera, M.S., Danting, J.C., Reyes, R.A., Dionisio, E.E., Capili, J.B., Bolivar, H.L., Abella, T.A., Circa, A.V., Bentsen, H.B., Gjerde, B., Gjedrem, T. and Pullin, R.S.V. (1993) Genetic improvement of farmed tilapias: the growth performance of eight strains of *Oreochromis niloticus* tested in different farm environments. *Aquaculture* 111, 171–188.

El-Ebiary, E.H., Zaki, M.A. and Mourad, M.H. (1997) Effect of salinity on growth, feed utilization and haematological parameters of Florida red tilapia fingerlings. *Bulletin of the National Institute of Oceanography and Fisheries (Egypt)* 23, 203–216.

El Gamal, A.-R. (1988) Reproductive performance, sex ratios, gonadal development, cold tolerance, viability and growth of red and normally pigmented hybrids of *Tilapia aurea* and *T. nilotica*. PhD thesis, Auburn University, Auburn, Alabama.

El Gamal, A.-R., Davis, K.B., Jenkins, J.A. and Torrans, E.L. (1999) Induction of triploidy and tetraploidy in Nile tilapia, *Oreochromis niloticus* (L.). *Journal of the World Aquaculture Society* 30, 269–275.

El-Gohary, M.S.A. (2004) Effects of some organophosphorus compounds on some cultured freshwater fishes. PhD thesis, University of Alexandria, Alexandria, Egypt.

El-Kateib, T. and Afifi, S.H. (1993) Inspection of (*Tilapia nilotica*) Nile fish exposed to organophosphorus compound (Diazinon). *Assiut Veterinary Medical Journal* 29, 119–124.

El-Naggar, G.O., El Nady, M.A., Kamar, M.G. and Al-Kobabay, A.I. (2000) Effect of photoperiod, dietary protein and temperature on reproduction in Nile tilapia (*Oreochromis niloticus*) In: Fitzsimmons, K. and Filho, J.C. (eds) *Tilapia Culture in the 21st Century. Proceedings from the Fifth International Symposium on Tilapia Aquaculture, Rio de Janeiro, Brazil*. American Tilapia Association, Charles Town, West Virginia, and ICLARM, Penang, Malaysia, pp. 352–358.

El-Sayed, A.-F.M. (1987) Protein and energy requirements of *Tilapia zillii*. PhD thesis, Michigan State University, East Lansing, Michigan.

El-Sayed, A.-F.M. (1989) Evaluation of semipurified test diets for *Tilapia zilli* fingerlings. *Journal of the World Aquaculture Society* 20, 240–244.

El-Sayed, A.-F.M. (1990) Long-term evaluation of cotton seed meal as a protein source for Nile tilapia, *Oreochromis niloticus*. *Aquaculture* 84, 315–320.

El-Sayed, A.-F.M. (1991) Evaluation of sugarcane bagasse as a feed ingredient for the tilapias *Oreochromis niloticus* and *Tilapia zillii*. *Asian Fisheries Science* 4, 53–60.

El-Sayed, A.-F.M. (1992) Effects of substituting fish meal with *Azolla pinnata* in practical diets for fingerling and adult Nile tilapia *Oreochromis niloticus* L. *Aquaculture and Fisheries Management* 23, 167–173.

El-Sayed, A.-F.M. (1998) Total replacement of fishmeal with animal protein sources in Nile tilapia, *Oreochromis niloticus* (L.), feeds. *Aquaculture Research* 29, 275–280.

El-Sayed, A.-F.M. (1999a) Alternative dietary protein sources for farmed tilapia spp. *Aquaculture* 179, 149–168.

El-Sayed, A.-F.M. (1999b) *Aquaculture Feed and Fertilizer Resource Atlas of Egypt*. FAO, Regional Office for the Near East, Cairo, Egypt, FIRI/670, 105 pp.

El-Sayed, A.-F.M. (2002) Effects of stocking density and feeding levels on growth and feed efficiency of Nile tilapia, *Oreochromis niloticus* (L.) fry. *Aquaculture Research* 33, 621–626.

El-Sayed, A.-F.M. (2003) Effects of fermentation methods on the nutritive value of water hyacinth for Nile tilapia, *Oreochromis niloticus* (L.) fingerlings. *Aquaculture* 218, 471–478.

El-Sayed, A.-F.M. (2004) Feed colour affects growth, feed utilization of Nile tilapia. *Global Aquaculture Advocate* 7 (2), 76.

El-Sayed, A.-F.M. and Garling, D.L. Jr (1988) Carbohydrate-to-lipid ratio in diets for *Tilapia zillii* fingerlings. *Aquaculture* 73, 157–163.

El-Sayed, A.-F.M. and Kawanna, M. (2004) Effects of photoperiod on the performance of farmed Nile tilapia *Oreochromis niloticus*: I. Growth, feed utilization efficiency and survival of fry and fingerlings. *Aquaculture* 231, 393–402.

El-Sayed, A.-F.M. and Teshima, S. (1991) Tilapia nutrition in aquaculture. *Reviews in Aquatic Sciences* 5, 247–265.

El-Sayed, A.-F.M. and Teshima, S. (1992) Protein and energy requirement of Nile tilapia, *Oreochromis niloticus*, fry. *Aquaculture* 103, 55–63.

El-Sayed, A.-F.M., El-Ghobashy, A. and Al-Amoudi, M.M. (1996) Effects of pond depth and water temperature on the growth, mortality and body composition of Nile tilapia, *Oreochromis niloticus* (L.). *Aquaculture Research* 27, 681–687.

El-Sayed, A.-F.M., Moyano, F.J. and Martinez, I. (2000) Assessment of the effect of plant inhibitors on digestive protease of Nile tilapia using *in vitro* assays. *Aquaculture International* 8, 403–415.

El-Sayed, A.-F.M., Mansour, C.R. and Ezzat, A.A. (2003) Effects of dietary protein levels on spawning performance of Nile tilapia (*Oreochromis niloticus*) broodstock reared at different water salinities. *Aquaculture* 220, 619–632.

El-Sayed, A.-F.M., Mansour, C.R. and Ezzat, A.A. (2005a) Effects of dietary lipid source on spawning performance of Nile tilapia (*Oreochromis niloticus*) broodstock reared at different water salinities. *Aquaculture* 248, 187–196.

El-Sayed, A.-F.M., Kawanna, M. and Mudar, M. (2005b) Effects of water flow rates on growth and survival of Nile tilapia fry. *World Aquaculture* 36(1), 5–6.

El-Shafai, S.A., El-Gohary, F.A., Nasr, F.A., van der Steen, N.P. and Gijzen, H.J. (2004) Chronic ammonia toxicity to duckweed-fed tilapia (*Oreochromis niloticus*). *Aquaculture* 232, 117–127.

El-Shafey, A.A.M. (1998) Effect of ammonia on respiratory functions of blood of *Tilapia zillii*. *Comparative Biochemistry and Physiology* 121A, 305–313.

El-Sharouny, H.M. and Badran, R.A.M. (1995) Experimental transmission and pathogenicity of some zoosporic fungi to tilapia fish. *Mycopathologia* 132, 95–103.

El-Tantawy, S.A.M. and Kazaubski, S.L. (1986) The trichodinid ciliates from fish, *Tilapia nilotica* from the Nile Delta (Egypt). *Acta Protozoologica* 25, 439–444.

Endo, M., Kumahara, C., Yoshida, T. and Tabata, M. (2002) Reduced stress and increased immune responses in Nile tilapia kept under self-feeding conditions. *Fisheries Science* 68, 253–257.

Enujiugha, V.N. and Nwanna, L.C. (1998) The impacts of post harvest losses on supply and demand for *Clarias gariepinus* (Clariidae) and *Oreochromis niloticus* (Cichlidae) in Nigeria. In: Coetzee, L., Gon, J. and Kulongowski, C. (eds) *African Fishes and Fisheries Diversity and Utilisation. International Conference for the Paradi Association and the Fisheries Society of Africa.* Paradi Association and Fisheries Society of Africa, Grahamstown, South Africa, p. 111.

Ernst, D.H., Watanabe, W.O. and Ellingson, L.J. (1991) Commercial-scale production of Florida red tilapia seed in low- and brackish-salinity tanks. *Journal of the World Aquaculture Society* 22, 36–44.

Eskelinen, P. (1989) Effects of different diets on egg production and egg quality of Atlantic salmon (*Salmo salar* L.). *Aquaculture* 79, 275–281.

Essa, M.A. (1997) Utilization of some aquatic plants in diets for Nile tilapia, *Oreochromis niloticus*, fingerlings. *Egyptian Journal of Aquatic Biology and Fisheries* 1, 19–34.

Evans, J.J., Shoemaker, C.A. and Klesius, P.H. (2000) Experimental *Streptococcus iniae* infection of hybrid striped bass (*Morone chrysops* × *Morone saxatilis*) and tilapia (*Oreochromis niloticus*) by nares inoculation. *Aquaculture* 189, 197–210.

Everitt, S. and Leung, C.-F.A. (1999) Fish disease in Hong Kong. Paper presented at the Fourth Symposium on Diseases in Asian Aquaculture: Aquatic Animal Health for Sustainability, 22–26 November, Cebu City, Philippines.

Ewart, K.V., Lin, Q. and Hew, C.L. (1999) Structure, function and evolution of antifreeze proteins. *Cellular and Molecular Life Sciences* 55, 271–283.

Ewing, M.S. and Kocan, K.M. (1992) Invasion and development strategies of *Ichthyophthirius multifiliis*, a parasitic ciliate of fish. *Parasitology Today* 8, 204–208.

Fagade, S.O. (1971) The food and feeding habits of *Tilapia* species in the Lagos Lagoon. *Journal of Fish Biology* 3, 151–156.

Fagbenro, O.A. (1988) Evaluation of defatted cocoa cake as a direct feed in the monosex culture of *Tilapia guineensis* (Pisces: Cichlidae). *Aquaculture* 73, 201–206.

Fagbenro, O.A. (1993) Observations on macadamia presscake as supplemental feed for monosex *Tilapia guineensis* (Pisces: Cichlidae). *Journal of Aquaculture in the Tropics* 7, 91–94.

Fagbenro, O.A. (1994) Dried fermented fish silage in diets for *Oreochromis niloticus*. *Israeli Journal of Aquaculture/Bamidgeh* 46, 140–147.

Fagbenro, O.A. (2000) Validation of the essential amino acid requirements of Nile tilapia *Oreochromis niloticus* (Linne. 1758), assessed by the ideal protein concept. In: Fitzsimmons, K. and Filho, J.C. (eds) *Tilapia Culture in the 21st Century. Proceedings from the Fifth International Symposium on Tilapia Aquaculture, Rio de Janeiro, Brazil.* American Tilapia Association, Charles Town, West Virginia, and ICLARM, Penang, Malaysia, pp. 154–156.

Fagbenro, O.A., Jauncey, K. and Haylor, G. (1994) Nutritive value of diets containing dried lactic acid-fermented fish silage and soybean meal for juvenile *Oreochromis niloticus* and *Clarias gariepinus*. *Aquatic Living Resources* 7, 79–85.

Fagbenro, O.A., Adeparusi, E.O. and Jimoh, W.A. (2004) Nutrient quality of detoxified jackbean (*Canavalia ensiformis* L. DC) seeds coked in distilled water or trona solution and evaluation of the meal as a substitute for soybean meal in practical diets for Nile tilapia. In: Bolivar, R., Mair, G. and Fitzsimmons, K. (eds) *Proceedings of the Sixth International Symposium on Tilapia in Aquaculture*. Bureau of Fisheries and Aquatic Resources, Manila, Philippines, and American Tilapia Association, Charles Town, West Virginia, pp. 289–300.

FAO (Food and Agriculture Organization of the United Nations) (1995a) *Precautionary Approach to Fisheries. Part 1: Guidelines on the Precautionary Approach to Capture Fisheries and Species Introductions*. FAO Fisheries Technical Paper No. 350, Part 1, FAO, Rome, 52 pp.

FAO (Food and Agriculture Organization of the United Nations) (1995b) *Review of the State of World Fishery Resources: Aquaculture*. FAO, Rome.

FAO (Food and Agriculture Organization of the United Nations) (2002) *The Role of Aquaculture in Rural Development*. Committee on Fisheries, Sub-committee on Aquaculture, Beijing, China, 18–22 April, COFI:AQ/1/2002/3, Available at: www.fao.org/docrep/meeting/004/y3018e.htm

FAO (Food and Agriculture Organization of the United Nations) (2004) *Fishstat Plus*. FAO, Rome.

Fasakin, E.A., Balogun, A.M. and Fasuru, B.E. (1999) Use of duckweed, *Spirodela polyrrhiza* L. Schleiden, as a protein feedstuff in practical diets for tilapia, *Oreochromis niloticus* L. *Aquaculture Research* 30, 313–318.

Faunce, C.H. (2000) Reproduction of black tilapia, *Sarotherodon melanotheron*, within an impounded mangrove ecosystem in east-central Florida. *Environmental Biology of Fishes* 57, 353–361.

Ferdouse, F. (2001) Tilapia in Asian markets. . . Can we sell more? In: Subasinghe, S. and Singh, T. (eds) *Tilapia: Production, Marketing and Technical Developments. Proceedings of the Tilapia 2001 International Technical and Trade Conference on Tilapia*. Infofish, Kuala Lumpur, Malaysia, pp. 71–74.

Fineman-Kalio, A.S. (1988) Preliminary observations on the effect of salinity on the reproduction and growth of freshwater Nile tilapia, *Oreochromis niloticus* (L.), cultured in brackishwater ponds. *Aquaculture and Fisheries Management* 19, 313–320.

Finn, R.N. (1994) Physiological energetics of developing marine fish embryos and larvae. PhD thesis, University of Bergen, Norway.

Fish, G.R. (1956) Some aspects of the respiration of six species of fish from Uganda. *Journal of the Experimental Biology* 33, 186–195.

Fishelson, L. and Yaron, Z. (eds) (1983) *Proceedings of the International Symposium on Tilapia in Aquaculture*. Tel Aviv University, Tel Aviv, Israel, p. XI.

Fitzsimmons, K. (2000) Tilapia aquaculture in Mexico. In: Costa-Pierce, B.A. and Rakocy, J.E. (eds) *Tilapia Aquaculture in the Americas*, Vol. 2. World Aquaculture Society, Baton Rouge, Louisiana, pp. 171–183.

Fitzsimmons, K. (2001a) Tilapia production in the Americas. In: Subasinghe, S. and Singh, T. (eds) *Tilapia: Production, Marketing and Technical Developments. Proceedings of the Tilapia 2001 International Technical and Trade Conference on Tilapia*. Infofish, Kuala Lumpur, Malaysia, pp. 7–16.

Fitzsimmons, K. (2001b) Environmental and conservation issues in tilapia aquaculture. In: Subasinghe, S. and Singh, T. (eds) *Tilapia: Production, Marketing and Technical Developments. Proceedings of the Tilapia 2001 International Technical and Trade Conference on Tilapia*. Infofish, Kuala Lumpur, Malaysia, pp. 128–131.

Fitzsimmons, K. (2003) Tilapia aquaculture in recirculating systems. *Aquaculture Magazine* 29, 73–76.

Fitzsimmons, K., Circa, A., Jimenez, E.B. and Pereda, D. (1999) Development of low-cost supplemental feeds for tilapia in pond and cage culture. In: McElwee, E., Bruke, D., Niles, M. and Egna, H. (eds) *Sixteenth Annual Technical Report*. Pond Dynamics/Aquaculture CRSP, Oregon State University, Corvallis, Oregon, pp. 57–63.

Foo, J.T.W. and Lam, T.J. (1993) Serum cortisol response to handling stress and the effect of cortisol implantation on testosterone level in the tilapia, *Oreochromis mossambicus*. *Aquaculture* 115, 145–158.

Fortes, R.D. (1985) Tarpon as biological control in milkfish–tilapia polyculture. *Fisheries Journal, University of Philippines in the Visayas* 1, 47–55.

Fouz, B., Alcaide, E., Barrera, R. and Amaro, C. (2002) Susceptibility of Nile tilapia (*Oreochromis niloticus*) to vibriosis due to *Vibrio vulnificus* biotype 2 (serovar E). *Aquaculture* 212, 21–30.

Franklin, C.E., Johnston, I.A., Crockford, T. and Kamunde, C. (1995) Scaling of oxygen consumption of Lake Magadi tilapia, a fish living at 37°C. *Journal of Fish Biology* 46, 829–834.

FRRC (Fisheries Resources Research Centre at Qatif) (2001) *The Most Common Diseases in Fish Farms and Some Marine Fishes in the Eastern Region, Saudi Arabia*. Extension News Bulletin No. 225/226, Ministry of Agriculture and Waters, Qatif, Saudi Arabia, 15 pp. (in Arabic).

Fryer, G. and Iles, T.D. (1972a) Alternative routes to evolutionary success as exhibited by African cichlid fishes of the genus *Tilapia* and the species flocks of the great lakes. *Evolution* 23, 359–369.

Fryer, G. and Iles, T.D. (1972b) *The Cichlid Fishes of the Great Lakes of Africa: Their Biology and Evolution*. T.F.H. Publishing, Neptune City, New Jersey. Also published by Oliver & Boyd, Edinburgh.

Fuerst, P.A., Mwanja, W.W. and Kaufman, L. (2000) The genetic history of the introduced Nile tilapia of Lake Victoria (Uganda – E. Africa): the population structure of *Oreochromis niloticus* (Pisces: Cichlidae) revealed by DNA microsatellite markers. In: Fitzsimmons, K. and Filho, J.C. (eds) *Tilapia Culture in the 21st Century. Proceedings from the Fifth International Symposium on Tilapia Aquaculture. Rio de Janeiro, Brazil.* American Tilapia Association, Charles Town, West Virginia, and ICLARM, Penang, Malaysia, pp. 30–40.

Fukusho, K. (1968) The specific difference of temperature response among cichlid fishes of genus *Tilapia*. *Bulletin of the Japanese Society for Scientific Fisheries* 34, 103–111.

Fúnez, O., Neira, I. and Engle, C. (2001) Honduras survey: 50% of supermarkets to sell tilapia. *Global Aquaculture Advocate* 4 (2), 89.

Gale, W.L., Fitzpatrick, M.S., Lucero, M., Contreras-Sánchez, W.M. and Schreck, C.B. (1999) Masculinization of Nile tilapia (*Oreochromis niloticus*) by immersion in androgens. *Aquaculture* 178, 349–357.

Gall, G.A.E. and Bakar, Y. (1999) Stocking density and tank size in the design of breed improvement programs for body size of tilapia. *Aquaculture* 173, 197–205.

Galman, O.R., Moreau, J. and Avtalion, R. (1988) Breeding characteristics and growth performance of Philippines red tilapia. In: Pullin, R.S.V., Bhukaswan, T., Tonguthai, K. and Maclean, J.L. (eds) *Proceedings of the Second International Symposium on Tilapia in Aquaculture*. ICLARM Conference Proceedings No. 15, Department of Fisheries, Bangkok, Thailand, and ICLARM, Manila, Philippines, pp. 169–175.

Galvez, J.I., Morrison, J.R. and Phelps, R.P. (1996) Efficacy of trenbolone acetate in sex inversion of the blue tilapia *Oreochromis aureus*. *Journal of the World Aquaculture Society* 27, 483–486.

Garcia, C.L., Clavijo, A.M. and Santander, J. (1999) Incidence of *Aeromonas* spp. in some cultured fish species in the states of Aragua, Carabobo, Monagas and Cojedes in Venezuela. In: Cabrera, T., Jory, D. and Silva, M. (eds) *Aquaculture 99. Congreso sur American de Acuicultura,* World Aquaculture Society, Latin American Chapter, Puerto la Cruz, Venezuela, pp. 218–222.

Gbankoto, A., Pampoulie, C., Marques, A. and Sakiti, G.N. (2001) Occurrence of myxosporean parasites in the gills of two tilapia species from Lake Nokoué (Bénin, West Africa): effect of host size and sex, and seasonal patterns of infection. *Diseases of Aquatic Organisms* 44, 217–222.

George, T.T. (1976) Introduction and transplantation of cultivable species into Africa. *FAO/CIFA Technical Paper* 4, Suppl. 1, pp. 407–432.

GESAMP (IMO/FAO/Unesco/WMO/WHO/IAEA/UN/UNEP (Joint Group of Experts on the Scientific Aspects of Marine Pollution) (1991) *Reducing Environmental Impacts of Coastal Aquaculture.* Reports and Studies, No. 47, GESAMP, Rome, 35 pp.

Getchell, R. (1998) *S. iniae* causes tilapia infection. *Fish Farming News* 6, 16.

Gilling, C.J., Skibinski, D.O.F. and Beardmore, J.A. (1996) Sex reversal of tilapia fry by immersion in water containing estrogens. In: Pullin, R.S.V., Lazard, J., Legendre, M., Amon Kottias, J.B. and Pauly, D. (eds) *Proceedings of the Third International Symposium on Tilapia in Aquaculture.* ICLARM Conference Proceedings No. 41, ICLARM, Penang, Malaysia, pp. 314–319.

Gjøen, H.M. (2004) A new era: the merging of quantitative and molecular genetics – prospects for tilapia breeding programs. In: Bolivar, R.B., Mair, G.C. and Fitzsimmons, K. (eds) *Proceedings of the Sixth International Symposium on Tilapia in Aquaculture.* Bureau of Fisheries and Aquatic Resources, Manila, Philippines, and American Tilapia Association, Charles Town, West Virginia, pp. 53–59.

Gomelsky, B., Cherfas, N.B., Peretz, Y., Ben-Dom, N. and Hulata, G. (1994) Hormonal sex inversion in the common carp (*Cyprinus carpio* L.). *Aquaculture* 126, 265–270.

Goncalves-de-Freitas, E. and Nishida, S.M. (1998) Sneaking behaviour of the Nile tilapia. *Boletim Tecnico do CEPTA* 11, 71–79.

Gonzales-Corre, K. (1988) Polyculture of tiger shrimp (*Penaeus monodon*) with Nile tilapia (*Oreochromis niloticus*) in brackishwater fishponds. In: Pullin, R.S.V., Bhukaswan, T., Tonguthai, K. and Maclean, J.L. (eds) *Proceedings of the Second International Symposium on Tilapia in Aquaculture.* ICLARM Conference Proceedings No. 15, Department of Fisheries, Bangkok, Thailand, and ICLARM, Manila, Philippines, pp. 15–20.

Gopalakrishnan, V. (1988) Role of tilapia (*Oreochromis andersonii*) in integrated farming systems in Zambia. In: Pullin, R.S.V., Bhukaswan, T., Tonguthai, K. and Maclean, J.L. (eds) *Proceedings of the Second International Symposium on Tilapia in Aquaculture.* ICLARM Conference Proceedings No. 15, Department of Fisheries, Bangkok, Thailand, and ICLARM, Manila, Philippines, pp. 21–28.

Granduño-Lugo, M., Muñoz-Córdova, G. and Olvera-Novoa, M.Ă. (2004) Mass selection for red colour in *Oreochromis niloticus* (Linnaeus 1758). *Aquaculture Research* 35, 340–344.

Green, B.W. (1992) Substitution of organic manure for pelleted feed in tilapia production. *Aquaculture* 101, 213–222.

Green, B.W., Phelps, R.P. and Alvarenga, H.R. (1989) The effect of manures and chemical fertilizers on the production of *Oreochromis niloticus* in earthen ponds. *Aquaculture* 76, 37–42.

Green, B.W., Teichert-Coddington, D.R. and Phelps, R.P. (1990) Response of tilapia yield and economics to varying rates of organic fertilization and season in two Central American countries. *Aquaculture* 90, 279–290.

Green, B.W., Verrica, K.L. and Fitzpatrick, M.S. (1997) Fry and fingerling production. In: Egna, H.S. and Boyd, C.E. (eds) *Dynamics of Pond Aquaculture*. CRC Press, Boca Raton, Florida, pp. 215–243.

Green, B.W., El Nagdy, Z. and Hebicha, H. (2002) Evaluation of Nile tilapia pond management strategies in Egypt. *Aquaculture Research* 33, 1037–1048.

Grover, J.J., Olla, B.L., O'Brien, M. and Wicklund, R.I. (1989) Food habits of Florida red tilapia fry in manured seawater pools in the Bahamas. *Progressive Fish Culturist* 51, 152–156.

Guerrero, R.D. III (1975) Use of androgens for the production of all-male tilapia *Tilapia aurea* (Steindachner). *Transactions of the American Fisheries Society* 104, 342–348.

Guerrero, R.D. III (1980) Studies on the feeding of *Tilapia nilotica* in floating cages. *Aquaculture* 20, 169–175.

Guerrero, R.D. III (1982) Control of tilapia reproduction. In: Pullin, R.S.V. and Lowe-McConnell, R.H. (eds) *The Biology and Culture of Tilapias*. ICLARM Conference Proceedings No. 7, ICLARM, Manila, Philippines, pp. 309–316.

Guerrero, R.D. III (2001) Tilapia culture in Southeast Asia. In: Subasinghe, S. and Singh, T. (eds) *Tilapia: Production, Marketing and Technical Developments. Proceedings of the Tilapia 2001 International Technical and Trade Conference on Tilapia*. Infofish, Kuala Lumpur, Malaysia, pp. 97–103.

Guerrero, R.D. III and Garcia, A.M. (1983) Studies on the fry production of *Sarotherodon niloticus* in a lake-based hatchery. In: Fishelson, L. and Yaron, Z. (eds) *Proceedings of the International Symposium on Tilapia in Aquaculture*. Tel Aviv University, Tel Aviv, Israel, pp. 388–393.

Guerrero, R.D. III and Guerrero, L.A. (1988) Feasibility of commercial production of sex-reversed Nile tilapia fingerlings in the Philippines. In: Pullin, R.S.V., Bhukaswan, T., Tonguthai, K. and Maclean, J.L. (eds) *Proceedings of the Second International Symposium on Tilapia in Aquaculture*. ICLARM Conference Proceedings No. 15, Department of Fisheries, Bangkok, Thailand, and ICLARM, Manila, Philippines, pp. 183–186.

Guerrero, R.D. III and Guerrero, L.A. (2001) Domestic marketing of tilapia in Southeast Asia. In: Subasinghe, S. and Singh, T. (eds) *Tilapia: Production, Marketing and Technical Developments. Proceedings of the Tilapia 2001 International Technical and Trade Conference on Tilapia*. Infofish, Kuala Lumpur, Malaysia, pp. 81–83.

Guillen, I., Berlanga, J., Valenzuela, C.M., Morales, A., Toledo, J., Estrada, M.P., Puentes, P., Hayes, O. and de la Fuente, J. (1999) Safety evaluation of transgenic tilapia with accelerated growth. *Marine Biotechnology* 1, 2–14.

Gunasekera, R.M. and Lam, T.J. (1997) Influence of dietary protein level on ovarian recrudescence in Nile tilapia, *Oreochromis niloticus* (L.). *Aquaculture* 149, 57–69.

Gunasekera, R.M., Shim, K.F. and Lam, T.J. (1995) Effect of dietary protein level on puberty, oocyte growth and egg chemical composition in the tilapia, *Oreochromis niloticus* (L.). *Aquaculture* 134, 169–183.

Gunasekera, R.M., Shim, K.F. and Lam, T.J. (1996a) Effect of dietary protein level on spawning performance and amino acid composition of eggs of Nile tilapia, *Oreochromis niloticus* (L.). *Aquaculture* 146, 121–134.

Gunasekera, R.M., Shim, K.F. and Lam, T.J. (1996b) Influence of protein content of broodstock diets on larval quality and performance in Nile tilapia, *Oreochromis niloticus* (L.). *Aquaculture* 146, 245–259.

Gunasekera, R.M., Shim, K.F. and Lam, T.J. (1997) Influence of dietary protein content on the distribution of amino acids in oocytes, serum and muscle of Nile tilapia, *Oreochromis niloticus* (L.). *Aquaculture* 152, 205–221.

Gupta, M.V. (2001) Culture of short-cycle species in seasonal ponds and ditches in Bangladesh. In: *Integrated Agriculture–Aquaculture: A Primer*. FAO Fisheries Technical Paper 407, FAO, Rome, pp. 40–42.

Gupta, M.V., Ahmed, M., Bimbao, M.A.P. and Lightfoot, C. (1992) *Socioeconomic Impact and Farmers' Assessment of Nile Tilapia (Oreochromis niloticus) Culture in Bangladesh*. ICLARM Technical Reports, No. 35, Manila, Philpppines, 50 pp.

Gupta, M.V., Mazid, M.A., Islam, M.S., Rahman, M. and Hussain, M.G. (1999) *Integration of Aquaculture into the Farming Systems of the Flood-prone Ecosystems of Bangladesh: an Evaluation of Adoption and Impact*. ICLARM Technical Report, No. 56, ICLARM, Manila, Philippines, 32 pp.

Gwahaba, J.J. (1973) Effects of overfishing on *Tilapia nilotica* populations of Lake George, Uganda, over the past 20 years. *East African Wildlife Journal* 11, 317–328.

Halwart, M. (1998) Trends in rice-fish farming. *FAO Aquaculture Newsletter* 18, 3–11.

Halwart, M., Funge-Smith, S. and Moehl, J. (2003) The role of aquaculture in rural development. In: *Review of the State of World Aquaculture*. FAO Fisheries Circular No. 886, Revision 2, FAO, Rome, pp. 47–58.

Hanley, F. (2000) Digestibility coefficients of feed ingredients for tilapia. In: Fitzsimmons, K. and Filho, J.C. (eds) *Tilapia Culture in the 21st Century. Proceedings from the Fifth International Symposium on Tilapia Aquaculture, Rio de Janeiro, Brazil*. American Tilapia Association, Charles Town, West Virginia, and ICLARM, Penang, Malaysia, pp. 163–172.

Harbott, B.J. (1975) Preliminary observations on the feeding of *Tilapia nilotica* Linn. in Lake Rudolf. *African Journal of Tropical Hydrobiology and Fisheries* 4, 27–37.

Harbott, B.J. (1982) Studies on the feeding activity of *Sarotherodon niloticus* (L.) in Lake Turkana. In: *A Report on the Findings of the Lake Turkana Project, 1972–1975*, Vol. 5. Government of Kenya and the Ministry of Overseas Development, London, pp. 1357–1368.

Hardjamulia, I.A. and Rukyani, A. (2000) Improved Nile tilapia (GIFT strain) cultured in Indonesia. *NAGA: The ICLARM Quarterly* 23, 38–39.

Hardy, R.W., Shearer, K.D., Stone, F.E. and Weig, D.H. (1983) Fish silage in aquaculture diets. *Journal of the World Mariculture Society* 14, 695–703.

Harel, M., Tandler, A. and Kissil, G.W. (1994) The kinetics of nutrient incorporation into body tissues of gilthead sea bream (*Sparus aurata*) females and the subsequent effects on egg composition and egg quality. *British Journal of Nutrition* 72, 45–58.

Hargreaves, J.A. and Behrends, L.L. (1997) Improving the economics of intensive tilapia aquaculture in the southeastern United States through integration: a conceptual assessment. In: Fitzsimmons, K. (ed.) *Proceedings from the Fourth International Symposium on Tilapia in Aquaculture*. Northeast Regional Agriculture Engineering Service, Ithaca, New York, pp. 642–649.

Hargreaves, J.A. and Kucuk, S. (2001) Effects of diel un-ionized ammonia fluctuation on juvenile hybrid striped bass, channel catfish, and blue tilapia. *Aquaculture* 195, 163–181.

Hargreaves, J.A., Rakocy, J.E. and Nair, A. (1986) An evaluation of fixed and demand feeding regimes for cage culture of *Oreochromis aureus*. In: Mclean, J.L., Dizon, L.B. and Hosillos, L.V. (eds) *The First Asian Fisheries Forum*. Asian Fisheries Society, Manila, Philippines, pp. 335–340.

Haroon, A.K.Y. and Alam, M. (1992) *Integrated Paddy-cum-fish/shrimp Farming: Final Report*. Fisheries Research Institute, Mymensingh, 41 pp.

Haroon, A.K.Y. and Pittman, K.A. (1997) Rice–fish culture: feeding, growth and yield of two size classes of *Puntius gonionotus* Bleeker and *Oreochromis* spp. in Bangladesh. *Aquaculture* 154, 261–281.

Haroon, A.K.Y., Pittman, K.A. and Blom, G. (1998) Diel feeding pattern and ration of two sizes of tilapia, *Oreochromis* spp. in pond and paddy field. *Asian Fisheries Science* 10, 281–301.

Hart, L.J., Smith, S.A., Smith, B.J., Robertson, J., Besteman, E.G. and Holladay, S.D. (1998) Subacute immunotoxic effects of the polycyclic aromatic hydrocarbon (PAH), 7,12-dimethylbenzanthracene (DMBA) on spleen and pronephros leukocytic cell counts and phagocytic cell activity in tilapia *Oreochromis niloticus*. *Aquatic Toxicology* 41, 17–29.

Hart, P.R., Hutchinson, W.G. and Purser, G.J. (1996) Effects of photoperiod, temperature and salinity on hatchery-reared larvae of the greenback flounder (*Rhombosolea tapirina* Gunther, 1862). *Aquaculture* 144, 303–311.

Harvey D.J. (2004) *Aquaculture Outlook*. USDA (United States Department of Agriculture), Washington, DC, 22 pp. Available at: www.ers.usda.gov

Hatch, U. and Falck, J. (2001) Assessing economic risk in tilapia production in Honduras. In: *Aquaculture 2001 – Book of Abstracts*. World Aquaculture Society, Baton Rouge, Louisiana, p. 281.

Hauser, W.J. (1977) Temperature requirements for *Tilapia zillii*. *California Fish and Game* 63, 228–233.

Haylor, G.S., Beveridge, M.C.M. and Jauncey, K. (1988) Phosphorus nutrition of juvenile *Oreochromis niloticus*. In: Pullin, R.S.V., Bhukaswan, T., Tonguthai, K. and Maclean, J.L. (eds) *Proceedings of the Second International Symposium on Tilapia in Aquaculture*. ICLARM Conference Proceedings No. 15, Department of Fisheries, Bangkok, Thailand, and ICLARM, Manila, Philippines, pp. 341–345.

Head, W.D. and Watanabe, W.O. (1995) Economic analysis of a commercial-scale, recirculating, brackishwater hatchery for Florida red tilapia. *Journal of Applied Aquaculture* 5, 1–24.

Head, W.D., Zerbi, A. and Watanabe, W.O. (1996) Economic evaluation of commercial-scale, saltwater pond production of Florida red tilapia in Puerto Rico. *Journal of the World Aquaculture Society* 27, 275–289.

Hedrick, R.P., Fryer, J.L., Chen, S.N. and Kou, G.H. (1983) Characteristics of four birnaviruses isolated from fish in Taiwan. *Fish Pathology* 18, 91–97.

Heindl, U., Kliangpradit, A. and Phromkunthong, W. (2004) The effect of phytase on the utilization of plant phosphorus in sex-reversed tilapia (*Oreochromis niloticus*). In: *The 11th International Symposium on Nutrition and Feeding in Fish, Phuket Island, Thailand, 2–7 May 2004*. Department of Fisheries, Bangkok, Thailand, Abstract No. 158.

Helmy, Z.A. (2004) Production of improved Nile tilapia by using gene transfer techniques. PhD thesis, Alexandria University, Alexandria, Egypt.

Hepher, B. and Pruginin, Y. (1981) *Commercial Fish Farming with Special Reference to Fish Culture in Israel*. Wiley-Interscience, New York, 261 pp.

Hepher, B. and Pruginin, Y. (1982) Tilapia culture in ponds under controlled conditions. In: Pullin, R.S.V. and Lowe-McConnell, R.H. (eds) *The Biology and Culture of Tilapias*. ICLARM Conference Proceedings No. 7, ICLARM, Manila, Philippines, pp. 185–203.

Hernandez, O., Guillen, I., Estrada, M.P., Cabrera, E., Pimentel, R., Pina, J.C., Abad, Z., Sanchez, V., Hidalgo, Y., Martinez, R., Lleonart, R. and de la Fuente, J. (1997) Characterization of transgenic tilapia lines with different ectopic expression of tilapia growth hormone. *Molecular Marine Biology and Biotechnology* 6, 364–375.

Hew, C.L. and Fletcher, G.L. (2001) The role of aquatic biotechnology in aquaculture. *Aquaculture* 197, 191–204.

Hickling, C.E. (1960) The Malacca *Tilapia* hybrids. *Journal of Genetics* 57, 1–10.

Hickling, C.E. (1962) *Fish Culture*. Faber & Faber, London.

Hickling, C.E. (1963) The cultivation of tilapia. *Scientific America* 208, 143–152.

Hines, G.A. and Watts, S.A. (1995) Non-steroidal chemical sex manipulation of tilapia. *Journal of the World Aquaculture Society* 26, 98–102.

Hishamunda, N. and Jolly, C. (1998) Evaluation of small-scale aquaculture with intra-rural household trade as an alternative enterprise for limited-resource farmers: the case of Rwanda. *Food Policy* 23, 143–154.

Hofer, S.C. and Watts, S.A. (2002) Cold tolerance in genetically male tilapia (GMT registered), *Oreochromis niloticus*. *World Aquaculture* 33 (2), 19–21.

Hoffman, G.L. and Meyer, F.P. (1974) *Parasites of Freshwater Fishes*. T.F.H. Publications, Neptune City, New Jersey, 224 pp.

Hopkins, K.D. and Cruz, E.M. (1982) *The ICLARM–CLSU Integrated Animal–Fish Farming Project: Final Report*. ICLARM Technical Report 5, ICLARM, Manila, Philippines.

Hörstgen-Schwark, G. and Langholz, H.-J. (1998) Prospects of selecting for late maturity in tilapia (*Oreochromis niloticus*): III. A selection experiment under laboratory conditions. *Aquaculture* 167, 123–133.

Hossain, M.A. and Little, D.C. (1996) Farmers' attitudes to the production and consumption of commonly cultured fish species in two districts of Bangladesh. Paper presented at the World Aquaculture Symposium '96, 29 January – 2 February 1996, Bangkok, Thailand.

Huang, S.-L., Chen, W.C., Shei, M.C., Liao, I.C. and Chen, S.N. (1999) Studies on epizootiology and pathogenicity of *Staphylococcus epidermidis* in tilapia (*Oreochromis* spp.) cultured in Taiwan. *Zoological Studies* 38, 178–188.

Huang, S.-L., Liao, I.-C. and Chen, S.-N. (2000) Induction of apoptosis in tilapia, *Oreochromis aureus* Steindachner, and in TO- 2 cells by *Staphylococcus epidermidis*. *Journal of Fish Diseases* 23, 363–368.

Huang, W.-B. and Chiu, T.-S. (1997) Effects of stocking density on survival, growth, size variation, and productivity of *Tilapia* fry. *Aquaculture Research* 28, 165–173.

Hughes, D.G. and Behrends, L.L. (1983) Mass Production of *Tilapia nilotica* seed in suspended net enclosures. In: Fishelson, L. and Yaron, Z. (eds) *Proceedings of the International Symposium on Tilapia in Aquaculture*. Tel Aviv University, Tel Aviv, Israel, pp. 394–401.

Hussain, M.G., Chatterji, A., McAndrew, B.J. and Johnstone, R. (1991) Triploidy induction in Nile tilapia, *Oreochromis niloticus* L. using pressure, heat and cold shocks. *Theoretical and Applied Genetics* 81, 6–12.

Hussain, M.G., Rao, G.P.S., Humayun, N.M., Randall, C.F., Penman, D.J., Kime, D., Bromage, N.R., Myers, J.M. and McAndrew, B.J. (1995) Comparative performance of growth, biochemical composition and endocrine profiles in diploid and triploid tilapia *Oreochromis niloticus* L. *Aquaculture* 138, 87–97.

Hussain, M.G., Penman, D.J. and McAndrew, B.J. (1998) Production of heterozygous and homozygous clones in Nile tilapia. *Aquaculture International* 6, 197–205.

ICLARM–GTZ (1991) *The Context of Small-Scale Integrated Agriculture–Aquaculture Systems in Africa: A Case Study of Malawi*. ICLARM, Manila, Philippines.

Imsland, A., Folkvord, A.F. and Steffansson, S.O. (1995) Growth, oxygen consumption and activity of juvenile turbot (*Scophthalmus maximus*) reared under different temperatures and photoperiods. *Netherlands Journal of Sea Research* 34, 149–159.

Ireland, M.J., Tapash, K.R., Nuran Nabi, S.M., Rahman, M.A., Huque, S.M.Z. and Aleem, N.A. (1996) Are tilapia breeding in the open waters of Bangladesh? The results of a preliminary countrywide survey. In: *CARE Bangladesh ANR Workshop, 2–3 February 1996, Jessore.*

Irianto, G. and Irianto, H.E. (1997) Post-harvest technology of Nile tilapia in Indonesia: a review. In: *FAO Fisheries Report R563,* FAO, Rome, pp. 71–83.

Ishak, M. and Dollar, A.M. (1968) Studies on Manganese uptake in *Tilapia mossambica* and *Salmo gairdneri.* 1. Growth and survival of *Tilapia mossambica* in response to manganese. *Hydrobiologia* 31, 572–584.

Jackson, A.J. and Capper, B.S. (1982) Investigations into the requirements of tilapia *Sarotherodon mossambicus,* for dietary methionine, lysine and arginine in semi-synthetic diets. *Aquaculture* 20, 289–297.

Jackson, A.J., Capper, B.S. and Matty, A.J. (1982) Evaluation of some plant proteins in complete diets for the tilapia, *Sarotherodon mossambicus. Aquaculture* 27, 97–109.

Jamu, D. (2001) Tilapia culture in Africa: opportunities and challenges. In: Subasinghe, S. and Singh, T. (eds) *Tilapia: Production, Marketing and Technical Developments. Proceedings of the Tilapia 2001 International Technical and Trade Conference on Tilapia.* Infofish, Kuala Lumpur, Malaysia, pp. 105–112.

Jarding, S., Windmar, L., Paterson, R. and Fjallsbak, J.P. (2000) Quality issues in commercial processing of tilapia (*Oreochromis niloticus*) in Zimbabwe. In: Fitzsimmons, K. and Filho, J.C. (eds) *Tilapia Culture in the 21st Century. Proceedings from the Fifth International Symposium on Tilapia Aquaculture, Rio de Janeiro, Brazil.* American Tilapia Association, Charles Town, West Virginia, and ICLARM, Penang, Malaysia, pp. 588–594.

Jauncey, K. (1982) The effects of varying dietary protein level on the growth, food conversion, protein utilization and body composition of juvenile tilapias (*Sarotherodon mossambicus*). *Aquaculture* 27, 43–54.

Jauncey, K. (1998) *Tilapia Feeds and Feeding.* Pisces Press, Stirling, Scotland, 241 pp.

Jauncey, K. and Ross, B. (1982) *A Guide to Tilapia Feeds and Feeding.* University of Stirling, Stirling, Scotland, 111 pp.

Jauncey, K., Tacon, A.G.J. and Jackson, A.J. (1983) The quantitative essential amino acid requirements of *Oreochromis* (*Sarotherodon*) *mossambicus.* In: Fishelson, L. and Yaron, Z. (eds) *Proceedings of the International Symposium on Tilapia in Aquaculture.* Tel Aviv University, Tel Aviv, Israel, pp. 328–337.

Jennings, D.P. (1991) Behavioural aspects of cold tolerance in blackchin tilapia, *Sarotherodon melanotheron,* at different salinities. *Environmental Biology of Fishes* 31, 185–195.

Jimenez Badillo, M.L. and Nepita Villanueva, M.R. (2000) Trophic spectrum of *Oreochromis aureus* (Perciformes: Cichlidae) in Infiernillo Dam, Michoacan-Guerrero, Mexico. *Revista de Biologia Tropical* 48, 487–494.

Jiménez-Montealegre, R., Verdegem, M., Zamora, J.E. and Verreth, J. (2002) Organic matter sedimentation and re-suspension in tilapia (*Oreochromis niloticus*) ponds during a production cycle. *Aquacultural Engineering* 26, 1–12.

Jirawong, P. (2000) Toxicity of copper sulfate to fishes, phytoplankton, and bacteria, *Aeromonas hydrophila.* MSc thesis, Kasetsart University, Bangkok, Thailand.

Job, S.V. (1969a) The respiratory metabolism of *Tilapia mossambica* (Teleostei). I. The effect of size, temperature and salinity. *Marine Biology* 2, 121–126.

Job, S.V. (1969b) The respiratory metabolism of *Tilapia mossambica* (Teleostei). II. The effect of size, temperature, salinity and partial pressure of oxygen. *Marine Biology* 3, 222–226.

Kanazawa, A., Teshima, S. and Sakamoto, M. (1980a) Requirements of *Tilapia zillii* for essential fatty acids. *Bulletin of the Japanese Society for Scientific Fisheries* 46, 1353–1356.

Kanazawa, A., Teshima, S. and Imai, K. (1980b) Biosynthesis of fatty acids in *Tilapia zillii* and the puffer Fish. *Memoir of Faculty of Fisheries, Kagoshima University* (*Japan*) 29, 313–318.

Kapetsky, J.M. (1994) *A Strategic Assessment of Warm-water Fish Farming Potential in Africa.* CIFA Technical Paper No. 27, FAO, Rome, 67 pp.

Kaufman, L. and Liem, K.F. (1982) Fishes of the suborder Labroidei (Pisces: Perciformes): phylogeny, ecology and evolutionary significance. *Breviora* 472, 1–19.

Kaunda, E. and Costa-Pierce, B.A. (1993) Catchability and selectivity of a low cost, traditional fishing gear in aquaculture ponds. *Aquaculture and Fisheries Management* 24, 783–789.

Kebus, M.J., Collins, M.T., Brownfield, M.S., Amundson, C.H., Kayes, T.B. and Malison, J.A. (1992) Effects of rearing density on stress response and growth of rainbow trout. *Journal of Aquatic Animal Health* 4, 1–6.

Keenleyside, M.H.A. (1991) *Cichlid Fishes.* Chapman & Hall, London, 378 pp.

Keshavanath, P., Gangadhar, B., Ramesh, T.J., van Dam, A.A., Beveridge, M.C.M. and Verdegem, M.C.J. (2004) Effects of bamboo substrate and supplemental feeding on growth and production of hybrid red tilapia fingerlings (*Oreochromis mossambicus* × *Oreochromis niloticus*). *Aquaculture* 235, 303–314.

Khalil, R.H. (1998) Effect of Bayluscide® on some cultured freshwater fish (*Oreochromis niloticus*). PhD thesis, Alexandria University, Egypt.

Khater, A. and Smitherman, R.O. (1988) Cold tolerance and growth of three strains of *Oreochromis niloticus*. In: Pullin, R.S.V., Bhukaswan, T., Tonguthai, K. and Maclean, J.L. (eds) *Proceedings of the Second International Symposium on Tilapia in Aquaculture*. ICLARM Conference Proceedings No. 15, Department of Fisheries, Bangkok, Thailand, and ICLARM, Manila, Philippines, pp. 215–218.

Kirk, R.G. (1972) A review of the recent development in tilapia culture with special reference to fish farming in the heated effluents of power stations. *Aquaculture* 1, 45–60.

Kleemann, G.K., Barros, M.M., Pezzato, L.E., Sampaio, F.G., Ferrari, J.C., Valle, J.B., Freire, E.S. and Zuanon, J.A. (2003) Iron requirement for Nile tilapia *Oreochromis niloticus*. *World Aquaculture 2003 – Book of Abstracts*, Vol. 1. World Aquaculture Society, Baton Rouge, Louisiana, p. 82.

Klesius, P.H., Shoemaker, C.A. and Evans, J.J. (2000) Efficacy of single and combined *Streptococcus iniae* isolate vaccine administered by intraperitoneal and intramuscular routes in tilapia (*Oreochromis niloticus*). *Aquaculture* 188, 237–246.

Klesius, P.H., Shoemaker, C.A. and Evans, J.J. (2001) Efficacy of *Streptococcus iniae* vaccine administered in Nile tilapia *Oreochromis niloticus*. In: *Sixth Asian Fisheries Forum. Book of Abstracts*. Asian Fisheries Society, Quezon City, Philippines, p. 130.

Klinnavee, S., Tansakul, R. and Promkuntong, W. (1990) Growth of Nile tilapia (*Oreochromis niloticus*) fed with aquatic plant mixtures. In: Hirano, R. and Hanyu, I. (eds) *The Second Asian Fisheries Forum*. Asian Fisheries Society, Manila, Philippines, pp. 283–286.

Knud-Hansen, C.F. (1992) Pond history as a source of error in fish culture experiments: a quantitative assessment using covariance analysis. *Aquaculture* 105, 21–36.

Knud-Hansen, C.F. (1998) *Pond Fertilization: Ecological Approach and Practical Applications*. Pond Dynamics/Aquaculture CRSP, Oregon State University, Corvallis, Oregon, 125 pp.

Knud-Hansen, C.F. and Batterson, T.R. (1994) Effect of fertilization frequency on the production of Nile tilapia (*Oreochromis niloticus*). *Aquaculture* 123, 271–280.

Knud-Hansen, C.F. and Lin, C.K. (1996) Startegies for stocking Nile tilapia (*Oreochromis niloticus*) in fertilized ponds. In: Pullin, R.S.V., Lazard, J., Legendre, M., Amon Kottias, J.B. and Pauly, D. (eds) *Proceedings of the Third International Symposium on Tilapia in Aquaculture*. ICLARM Conference Proceedings No. 41, ICLARM, Penang, Malaysia, pp. 70–76.

Knud-Hansen, C.F., Batterson, T.R. and McNab, C.D. (1993) The role of chicken manure in the production of Nile tilapia, *Oreochromis niloticus* (L.). *Aquaculture and Fisheries Management* 24, 483–493.

Kobayashi, T., Ide, A., Hiasa, T., Fushiki, S. and Ueno, K. (1994) Production of cloned amago salmon *Oncorhynchus rhodurus*. *Fisheries Science* 60, 275–281.

Kocher, T. (1997) Introduction to the genetics of tilapia. In: Fitzsimmons, K. (ed.) *Proceedings from the Fourth International Symposium on Tilapia in Aquaculture*. Northeast Regional Agriculture Engineering Service, Ithaca, New York, pp. 61–64.

Koffi, C., Oswald, M. and Lazard, J. (1996) Rural development of tilapia culture in Africa: from myth to reality. In: Pullin, R.S.V., Lazard, J., Legendre, M., Amon Kottias, J.B. and Pauly, D. (eds) *Proceedings of the Third International Symposium on Tilapia in Aquaculture*. ICLARM Conference Proceedings No. 41, ICLARM, Penang, Malaysia, pp. 505–514.

Komen, J., Bongers, A.B.J., Richter, C.J.J., van Muiswinkel, W.B. and Huisman, E.A. (1991) Gynogenesis in common carp (*Cyprinus carpio* L.). II. The production of homozygous gynogenetic clones and F_1 hybrids. *Aquaculture* 92, 127–142.

Kubitza, F. (2004) An overview of tilapia aquaculture in Brazil. In: Bolivar, R., Mair, G. and Fitzsimmons, K. (eds) *Proceedings of Sixth International Symposium on Tilapia in Aquaculture*. Bureau of Fisheries and Aquatic Resources, Manila, Philippines, and American Tilapia Association, Charles Town, West Virginia, pp. 709–710.

Kuperman, B.I. and Matey, V.E. (1999) Massive infestation by *Amyloodinium ocellatum* (Dinoflagellida) of fish in a highly saline lake, Salton Sea, California, USA. *Diseases of Aquatic Organisms* 39, 65–73.

Kuperman, B.I., Matey, V.E. and Barlow, S.B. (2002) Flagellate *Cryptobia branchialis* (Bodonida: Kinetoplastida), ectoparasite of tilapia from the Salton Sea. *Hydrobiologia* 473, 93–102.

Kuroda, T. (1977) Japanese aquaculture with thermal water from power plant – Present considerations and problems. In: *Atoms in Japan*. Japanese Atomic Industrial Forum, Tokyo, p. 6.

Kuwamura, T. (1987) Male mating territory and sneaking in a maternal mouthbrooder, *Pseudosimochromis curvifrons* (Pisces; Cichlidae). *Journal of Ethology* 5, 203–206.

Kuwaye, T.T., Okimoto, D.K., Shimoda, S.K., Howerton, R.D., Lin, H.-R., Pang, P.K.T. and Grau, E.G. (1993) Effect of 17α-methyltestosterone on the growth of the euryhaline tilapia, *Oreochromis mossambicus*, in fresh water and in sea water. *Aquaculture* 113, 137–152.

Kwon, J.Y., Haghpanah, V., Kogson-Hurtado, L.M., McAndrew, B.J. and Penman, D.J. (2000) Masculinization of genetic female Nile tilapia (*Oreochromis niloticus*) by dietary administration of an aromatase inhibitor during sexual differentiation. *Journal of Experimental Zoology* 287, 46–53.

Lahav, M. and Sarig, S. (1972) Control of unicellular parasites using formalin. *Bamidgeh* 24, 3–11.

Lalitha, K.V. and Gopakumar, K. (2001) Sensitivity of tilapia (*Oreochromis mossambicus*) to *Clostridium botulinum* toxins. *Aquaculture Research* 32, 761–764.

Lanzing, W.J.R. and Higginbotham, D.R. (1976) Scanning microscopy of pharyngeal and oral teeth of the teleost *Tilapia mossambica*. *Hydrobiologia* 48, 137–139.

Lautenslager, G.T. (1986) Cichlid infectious agent virus: a characterization of a new rhabdovirus of cichlid fish. PhD thesis, Lehigh University, Bethlehem, Pennsylvania.

Lazard, J. and Oswald, M. (1995) Mixed culture of African catfish–tilapia: polyculture or police fish system. *Aquatic Living Resources* 8, 455–463.

Lazzaro, X. (1991) Feeding convergence in South American and African zooplanktivorous cichlids *Geophagus brasiliensis* and *Tilapia rendalli*. *Environmental Biology of Fishes* 31, 283–293.

Ledgerwood, R.D., Hughes, D.G. and Ortis, V. (1978) Wet and dry season comparison of *Tilapia aurea* production in El Salvador in new fertilized and unfertilized earthen ponds. Paper presented at the 9th Annual Meeting of the World Mariculture Society, 3 January 1978, Atlanta, Georgia.

Lee, J.C. (1979) Reproduction and hybridization of three cichlid fishes, *Tilapia aurea, T. hornorum* and *T. nilotica* in aquaria and in plastic pools. PhD thesis, Auburn University, Auburn, Alabama.

Leng, R.A., Stanbolie, J.H. and Bell, R. (1995) Duckweed – a potential high-protein feed resource for domestic animals and fish. *Livestock Research for Rural Development* 7 (1), 11 pp. Available at: www.cipav.org.co/lrrd/lrrd7/1/3.htm

Lenger, T.A. and Pfeiffer, T.J. (2002) Evaluation of supplemental solids removal effects on nitrification rates of floating bead filters. In: *Proceedings of the Fourth International Conference on Recirculating Aquaculture.* US Department of Agriculture, Virginia Polytechnic Institute and State University, Blacksburg, Virginia, p. 418.

Lester, L.J., Abell, T.A., Palada, M.S. and Keus, H.J. (1988) Genetic variation in size and sexual maturation of *Oreochromis niloticus* under hapa and cage culture conditions. In: Pullin, R.S.V., Bhukaswan, T., Tonguthai, K. and Maclean, J.L. (eds) *Proceedings of the Second International Symposium on Tilapia in Aquaculture.* ICLARM Conference Proceedings No. 15, Department of Fisheries, Bangkok, Thailand, and ICLARM, Manila, Philippines, pp. 223–230.

Li, Y., Bai, J., Jian, Q., Ye, X., Lao, H., Li, X., Luo, J. and Liang, X. (2003) Expression of common carp growth hormone in the yeast *Pichia pastoris* and growth stimulation of juvenile tilapia (*Oreochromis niloticus*). *Aquaculture* 216, 329–341.

Liao, I.C. and Chang, S.L. (1983) Studies on the feasibility of red tilapia culture in saline water. In: Fishelson, L. and Yaron, Z. (eds) *Proceedings of the International Symposium on Tilapia in Aquaculture.* Tel Aviv University, Tel Aviv, Israel, pp. 524–533.

Liebert, F. and Portz, L. (2005) Nutrient utilization of Nile tilapia *Oreochromis niloticus* fed plant-based low phosphorus diets supplemented with graded levels of different sources of microbial phytase. *Aquaculture* 248, 111–119.

Lightfoot, C., Costa-Pierce, B.A., Bimbao, M.A.P. and Dela Cruz, C.R. (1992) Introduction to rice–fish research and development in Asia. In: Dela Cruz, C.R., Lightfoot, C., Costa-Pierce, B.A., Carangal, V.R. and Bimbao, M.A.P. (eds) *Rice–Fish Research and Development in Asia.* ICLARM Conference Proceedings No. 24, ICLARM, Manila, Philippines, pp. 1–10.

Lightner, D., Redman, R., Mohney, L., Dickenson, G. and Fitzsimmons, K. (1988) Major diseases encountered in controlled environment culture of tilapias in fresh- and brackishwater over a three-year period in Arizona. In: Pullin, R.S.V., Bhukaswan, T., Tonguthai, K. and Maclean J.L. (ed) *Proceedings of the Second International Symposium on Tilapia in Aquaculture.* ICLARM Conference Proceeding No. 15, Department of Fisheries, Bangkok, Thailand, and ICLARM, Manila, Philippines, pp. 111–116.

Likongwe, J.S. (2002) Studies on potential use of salinity to increase growth of tilapia in aquaculture in Mali. In: McElwee, K., Lewis, K., Nidiffer, M. and Buitrago, P. (eds) *Nineteenth Annual Technical Report.* Pond Dynamics/Aquaculture CRSP, Oregon State University, Corvallis, Oregon, pp. 167–174.

Likongwe, J.S., Stecko, T.D., Stauffer, J.R. and Carline, R.F. (1996) Combined effects of water temperature and salinity on growth and feed utilisation of juvenile Nile tilapia *Oreochromis niloticus* (Linneaus). *Aquaculture* 146, 37–46.

Lim, C. and Dominy, W.G. (1991) Utilization of plant proteins by warmwater fish. In: Akiyama, D.M and Tan, R.K.H. (eds) *Proceedings of Aquaculture Feed Processing and Nutrition Workshop.* American Soybean Association, Singapore, pp. 163–172.

Lim, C. and Klesius, P.H. (2001) Influence of dietary levels of folic acid on growth response and resistance of Nile tilapia *Oreochromis niloticus* to *Streptococcus iniae*. In: *Sixth Asian Fisheries Forum. Book of Abstracts.* Asian Fisheries Society, Quezon City, Philippines, p. 150.

Lim, C. and Leamaster, B. (1991) Thiamine requirements of red hybrid tilapia grown in seawater. *Journal of the World Aquaculture Society* 22, 36A.

Lim, C., Leamaster, B. and Brock, J.A. (1993) Riboflavin requirements of fingerling red tilapia grown in seawater. *Journal of the World Aquaculture Society* 24, 451–458.

Lim, H.-A., Ng, W.-K., Lim, S.-L. and Ibrahim, C.O. (2001) Contamination of palm kernel meal with *Aspergillus flavus* affects its nutritive value in pelleted feeds for tilapia, *Oreochromis mossambicus. Aquaculture Research* 32, 895–905.

Lin, C.K. (1990) Integrated culture of walking catfish (*Clarias macrocephalus*) and tilapia (*Oreochromis niloticus*). In: Hirano, R. and Hanyu, I. (eds) *The Second Asian Fisheries Forum.* Asian Fisheries Society, Manila, Philippines, pp. 209–212.

Lin, C.K. and Diana, J. (1995) Co-culture of Nile tilapia (*Oreochromis niloticus*) and hybrid catfish (*Clarias macrocephalus* × *C. gariepinus*) in ponds. *Aquatic Living Resources* 8, 449–454.

Lin, C.K. and Kaewpaitoon, K. (2000) An overview of freshwater cage culture in Thailand. In: Lin, C.K. and Liao, I.C. (eds) *Proceedings of the First International Symposium on Cage Aquaculture in Asia.* Asian Fisheries Society, Manila, Philippines, pp. 237–242.

Lin, C.K. and Yi, Y. (1998) Economics of tilapia production in Thailand. Paper presented at the Asian Fisheries Society Conference, Chiang Mai, Thailand, December 1998.

Lin, C.K. and Yi, Y. (2003) Minimizing environmental impacts of freshwater aquaculture and reuse of pond effluents and mud. *Aquaculture* 226, 57–68.

Lin, C.K., Yi, Y. and Diana, J. (1997) The effects of pond management strategies on nutrient budgets: Thailand. In: *Fourteenth Annual Technical Report.* Pond Dynamics/Aquaculture CRSP, Oregon State University, Corvallis, Oregon, pp. 19–24.

Lin, C.K., Shrestha, M.K., Yi, Y. and Diana, J.S. (2001) Management to minimize the environmental impacts of pond effluent: harvest draining techniques and effluent quality. *Aquacultural Engineering* 25, 125–135.

Lin, C.-L., Ho, J.-S. and Chen, S.-N. (1996) Developmental stages of *Caligus epidemicus* Hewitt, a copepod parasite of tilapia culture in brackish water. *Journal of Natural History* 30, 661–684.

Lin, J.H. and Shiau, S.-Y. (1995) Hepatic enzyme adaptation to different dietary carbohydrates in juvenile tilapia *Oreochromis niloticus* × *Oreochromis aureus. Fish Physiology and Biochemistry* 14, 165–170.

Lio-Po, G. and Wakabayashi, H. (1986) Immunoresponse in tilapia *Sarotherodon niloticus* with *Edwardsiella tarda* by hyperosmotic infiltration method. *Veterinary Immunology and Immunopathology* 12, 351–357.

Liti, D.M., McElwee, O.E. and Veverica, K.L. (2002) Growth performance and economic benefits of *Oreochromis niloticus/Clarias gariepinus* polyculture fed on three supplementary feeds in fertilized tropical ponds. In: McElwee, K., Lewis, K., Nidiffer, M. and Buitrago, P. (eds) *Nineteenth Annual Technical Report.* Pond Dynamics/Aquaculture CRSP, Oregon State University, Corvallis, Oregon, pp. 11–16.

Little, D.C. (1989) An evaluation of strategies for production of Nile tilapia (*Oreochromis niloticus* L.) fry suitable for hormonal treatment. PhD thesis, Institute of Aquaculture, University of Stirling, Stirling, UK.

Little, D.C. (1992) *The Development of Fry Production and Nursing Strategies for the Nile Tilapia* Oreochromis niloticus. *Final Report.* Asian Institute of Technology, Bangkok, Thailand, 24 pp.

Little, D.C. (1998) Options in the development of the `aquatic chicken'. *Fish Farmer* (July/August). Available at: www.aquafind.com/articles/opt.html

Little, D.C. and Hulata, G. (2000) Strategies for tilapia seed production. In: Beveridge, M.C.M. and McAndrew, B.J. (eds) *Tilapias: Biology and Exploitation.* Kluwer Academic Publishers, Dordrecht/Boston/London, pp. 267–326.

Little, D.C., Macintosh, D.J. and Edwards, P. (1993) Improving spawning synchrony in the Nile tilapia, *Oreochromis niloticus* (L). *Aquaculture and Fisheries Management* 24, 399–405.

Little, D.C., Sikawa, D. and Juntana, J. (1994) Commercial production and marketing of Nile tilapia (*Oreochromis niloticus*) fry in Chonburi and Chachoengsao provinces, Thailand. *NAGA: ICLARM Quarterly* 17, 2–3.

Little, D.C., Surentaraseree, P. and Innes-Taylor, N.L. (1996) Review: fish culture in rainfed ricefields of Northeast Thailand. *Aquaculture* 140, 295–321.

Little, D.C., Bhujel, R.C. and Turner, W.A. (1997) Commercialization of a hatchery process to produce MT-treated Nile tilapia in Thailand.

In: Alston, D.E., Green, B.W. and Clifford, H.C. (eds) *Proceedings of the IV Symposium on Aquaculture in Central America: Focusing on Shrimp and Tilapia*. National Honduran Aquaculturists and the Latin American Chapter of the World Aquaculture Society, Boca Raton, Florida, pp. 108–118.

Little, D.C., Coward, K., Bhujel, R.C., Pham, T.A. and Bromage, N.R. (2000) Effect of broodfish exchange strategy on the spawning performance and sex steroid hormone levels of *Oreochromis niloticus* broodfish in hapas. *Aquaculture* 186, 77–88.

Liu, C.I., Huang, J.L., Hung, H.H. and Pen, C.F. (1990) The experimental infection of *Streptococcus* and/or *Aeromonas* to induce bacterial septicemia in cultured tilapia. *COA Fisheries Series* 24, 40–45.

Liu, X., Zhou, X. and Mo, X. (1993) Characteristics on pathogen of caudal fin rot disease of tilapia (*Oreochromis niloticus*) in mariculture. *Tropic Oceanography/Redai Haiyang* 12, 100–103.

Lone, K.P. and Ridha, M.T. (1993) Sex reversal and growth of *Oreochromis spilurus* (Guenther) in brackish and seawater by feeding 17α-methyltestosterone. *Aquaculture and Fisheries Management* 24, 593–602.

Longalong, F.M., Eknath, A.E. and Bentsen, H.B. (1999) Response to bi-directional selection for frequency of early maturing females in Nile tilapia (*Oreochromis niloticus*). *Aquaculture* 178, 13–25.

Lorico-Querijero, B. and Chiu, Y.N. (1989) Protein digestibility studies on *Oreochromis niloticus* using chromic oxide indicator. *Asian Fisheries Science* 2, 177–191.

Losordo, T.M. and Westerman, P.W. (1994) An analysis of biological, economic, and engineering factors affecting the cost of fish production in recirculating aquaculture systems. *Journal of the World Aquaculture Society* 25, 193–203.

Losordo, T.M., Masser, M.P. and Rakocy, J.E. (1999) *Recirculating Aquaculture Tank Production Systems: a Review of Component Options.* Publication No. 453, Southern Regional Aquaculture Center, Stoneville, Mississippi, 12 pp.

Lovell, R.T. and Limsuwan, T. (1982) Intestinal synthesis and dietary nonessentiality of vitamin B_{12} for *Tilapia nilotica*. *Transactions of the American Fisheries Society* 11, 485–490.

Lovshin, L.L. (1982) Tilapia hybridization. In: Pullin, R.S.V. and Lowe-McConnell, R.H. (eds) *The Biology and Culture of Tilapias*. ICLARM Conference Proceedings No. 7, ICLARM, Manila, Philippines, pp. 279–308.

Lovshin, L.L. (2000a) Tilapia culture in Brazil. In: Costa-Pierce, B.A. and Rakocy, J.E. (eds) *Tilapia Aquaculture in the Americas*, Vol. 2. The World Aquaculture Society, Baton Rouge, Louisiana, pp. 133–140.

Lovshin, L.L. (2000b) Evaluation of tilapia culture by resource-limited farmers in Panama and Guatemala. In: Fitzsimmons, K. and Filho, J.C. (eds) *Tilapia Culture in the 21st Century. Proceedings from the Fifth International Symposium on Tilapia Aquaculture, Rio de Janeiro, Brazil.* American Tilapia Association, Charles Town, West Virginia, and ICLARM, Penang, Malaysia, pp. 633–638.

Lovshin, L.L. and Ibrahim, H.H. (1988) Effects of broodstock exchange on *Oreochromis niloticus* egg and fry production in net enclosures. In: Pullin, R.S.V., Bhukaswan, T., Tonguthai, K. and Maclean, J.L. (eds) *Proceedings of the Second International Symposium on Tilapia in Aquaculture*. ICLARM Conference Proceedings No. 15, Department of Fisheries, Bangkok, Thailand, and ICLARM, Manila, Philippines, pp. 231–236.

Lowe-McConnell, R.H. (1955) The fecundity of tilapia species. *East African Agricultural Journal* 21, 45–52.

Lowe-McConnell, R.H. (1958) Observations on the biology of *Tilapia nilotica* L. in East African waters. *Revue de Zoologie et de Botanique Africaines* 57, 129–170.

Lowe-McConnell, R.H. (1959) Breeding behaviour patterns and ecological differences between tilapia species and their significance for evolution within the genus *Tilapia* (Pisces: Cichlidae). *Proceedings of the Zoological Society of London* 132, 1–30.

Lowe-McConnell, R.H. (1982) Tilapias in fish communities. In: Pullin, R.S.V. and Lowe-McConnell, R.H. (eds) *The Biology and Culture of Tilapias.* ICLARM Conference Proceedings No. 7, ICLARM, Manila, Philippines, pp. 83–113.

Lowe-McConnell, R.H. (2000) The role of tilapias in ecosystems. In: Beveridge, M.C.M. and McAndrew, B.J. (eds) *Tilapias: Biology and Exploitation.* Kluwer Academic Publishers, Dordrecht/Boston/London, pp. 129–162.

Lua, D.T., Te, B.Q. and Thanh, N.V. (1999) A study on parasites in different stages of three strains of cultured Nile tilapia (*Oreochromis niloticus*): Thai strain, GIFT strain and Viet strain in North Vietnam. Paper presented at the Fourth Symposium on Diseases in Asian Aquaculture: Aquatic Animal Health for Sustainability, 22–26 November, Cebu City, Philippines.

Luquet, P. (1989) Practical considerations on the protein nutrition and feeding of tilapia. *Aquatic Living Resources* 2, 99–104.

Lutz, C.G. (1996) Development of a tank-based tilapia industry in Louisiana. In: *Proceedings of the First International Conference on Recirculating Aquaculture*. US Department of Agriculture, Virginia Polytechnic Institute and State University, Blacksburg, Virginia, pp. 1–7.

McAndrew, B.J. (1993) Sex control in tilapiines. In: Muir, J.J.F. and Roberts, R.J. (eds) *Recent Advances in Aquaculture*, Vol. IV. Blackwell, London, UK, pp. 87–98.

McAndrew, B.J. (2000) Evolution, phylogenetic relationships and biogeography. In: Beveridge, M.C.M. and McAndrew, B.J. (eds) *Tilapias: Biology and Exploitation*. Kluwer Academic Publishers, Dordrecht/Boston/London, pp. 1–32.

McClain, W.R. and Gatlin, D.M. (1988) Dietary zinc requirement of *Oreochromis aureus* and effects of dietary calcium and phytate on zinc bioavailability. *Journal of the World Aquaculture Society* 19, 103–108.

McGeachin, R.B. and Stickney, R.R. (1982) Manuring rates for production of blue tilapia in simulated sewage lagoons receiving laying hen waste. *Progressive Fish Culturist* 44, 25–28.

McGeachin, R.B., Wicklund, R.I., Olla, B.L. and Winton, J.R. (1987) Growth of *Tilapia aurea* in seawater cages. *Journal of the World Aquaculture Society* 18, 31–34.

McGinty, A.S. (1986) Effects of size at stocking on competition and growth of all-male tilapia hybrids. *Journal of the World Mariculture Society* 16, 52–56.

McGinty, A.S. (1991) Tilapia production in cages: effects of cage size and number of noncaged fish. *Progressive Fish Culturist* 53, 246–249.

McGinty, A.S. and Alston, D.E. (1993) Multiple stocking and multiple harvesting of *Macrobrachium rosenbergii* and tilapia. In: Carrillo, M., Dahle, L., Morales, J., Sorgeloos, P., Svennevig, N. and Wyban, J. (eds) *From Discovery to Commercialization*. Special Publication No. 19, European Aquaculture Society, Ostend, Belgium, p. 148.

McGrogan, D.G., Ostland, V.E., Byrne, P.J. and Ferguson, H.W. (1998) Systemic disease involving an iridovirus-like agent in cultured tilapia, *Oreochromis niloticus* L. – a case report. *Journal of Fish Diseases* 21, 149–152.

Macintosh, D.J. and De Silva, S.S. (1984) The influence of stocking density and food ration on fry survival and growth in *Oreochromis mossambicus* and *O. niloticus* × *O. aureus* male hybrids reared in a closed circulated system. *Aquaculture* 41, 345–358.

Macintosh, D.J. and Little, D.C. (1995) Nile tilapia (*Oreochromis niloticus*). In: Bromage, N.R. and Roberts, R.J. (eds) *Broodstock Management and Egg and Larval Quality*. Blackwell Science, Oxford, UK, pp. 277–320.

Maclean, N., Rahman, M.A., Sohm, F., Hwang, G., Iyengar, A., Ayad, H., Smith, A. and Farahmand, H. (2002) Transgenic tilapia and the tilapia genome. *Gene* 295, 265–277.

McMahon, D.Z. and Baca, B. (1999) The effects of various salinities on the growth, pathology and reproduction of blue tilapia *Oreochromis aureus*. *World Aquaculture '99 – Book of Abstracts*. World Aquaculture Society, Baton Rouge, Louisiana, p. 508.

McNulty, S.T., Klesius, P.H. and Shoemaker, C.A. (2003) Hematological changes in Nile tilapia (*Oreochromis niloticus*) infected with *Streptococcus iniae* by nare inoculation. *Journal of the World Aquaculture Society* 34, 418–422.

Magid, A.M.A. and Babiker, M.M. (1975) Oxygen consumption and respiratory behaviour of three Nile fishes. *Hydrobiologia* 46, 359–367.

Maina, J.G., Beames, R.M., Higgs, D., Mbugua, P.N., Iwama, G. and Kisia, M. (2002) Digestibility and feeding value of some feed ingredients fed to tilapia *Oreochromis niloticus* (L.). *Aquaculture Research* 33, 853–862.

Mair, G.C. (2002) Tilapia genetics and breeding in Asia. In: Guerrero, R.D. III and Guerrero-del Castillo, M.R. (eds) *Tilapia Farming in the 21st Century. Proceedings of the International Forum on Tilapia Farming in the 21st Century (Tilapia Forum 2002)*. Philippines Fisheries Association, Los Baños, Laguna, Philippines, pp. 100–123.

Mair, G.C., Penman, D.J., Scott, A.G., Skibinski, D.O.F. and Beardmore, J.A. (1987) Hormonal sex-reversal and the mechanisms of sex determination in *Oreochromis*. In: Tiews, K. (ed.) *Proceedings of the World Symposium on Selection, Hybridization and Genetic Engineering in Aquaculture*, Vol. II. Heeneman, Berlin, pp. 301–312.

Mair, G.C., Estabillo, C.C., Sevilleja, R.C. and Recometa, R.D. (1993) Small-scale fry production systems for Nile tilapia, *Oreochromis niloticus* (L.). *Aquaculture and Fisheries Management* 24, 229–235.

Mair, G.C., Abucay, J.S., Beardmore, J.A. and Skibinski, D.O.F. (1995) Growth performance trials of genetically male tilapia (GMT) derived from `YY' males in *Oreochromis niloticus* L.: on-station comparisons with mixed sex and sex reversed male populations. *Aquaculture* 137, 313–322.

Mair, G.C., Abucay, J.S., Skibinski, D.O.F., Abella, T.A. and Beardmore, J.A. (1997) Genetic manipulation of sex ratio for the large-scale production of all-male tilapia, *Oreochromis niloticus*. *Canadian Journal of Fisheries and Aquatic Science* 54, 396–404.

Mair, G.C., Clarke, G.J.C., Morales, E.J. and Sevilleja, R.C. (2002) Genetic technologies focused on poverty? A case study of genetically improved tilapia (GMT) in the Philippines. In: Edwards, P., Little, D.C. and Demaine, H. (eds) *Rural Aquaculture*. CAB International, Wallingford, UK, pp. 197–225.

Majumdar, K.C., McAndrew, B.J., Fishelson, L. and Yaron, Z. (1983) Sex ratios from interspecific crosses within the tilapias. In: Fishelson, L. and Yaron, Z. (eds) *Proceedings of the International Symposium on Tilapia in Aquaculture*. Tel Aviv University, Tel Aviv, Israel, pp. 261–269.

Makhiessen, P. and Roberts, R.J. (1982) Histopathological change in the liver and brain of fish exposed to endosulfan insecticide during tsetse fly control operations in Botswana. *Journal of Fish Diseases* 5, 153–159.

Mang-Umphan, K. and Arce, R.G. (1988) Culture of Nile tilapia (*Oreochromis niloticus*) in a rice–fish culture system using chemical and commercial organic fertilizers. In: Pullin, R.S.V., Bhukaswan, T., Tonguthai, K. and Maclean, J.L. (eds) *Proceedings of the Second International Symposium on Tilapia in Aquaculture*. ICLARM Conference Proceedings No. 15, Department of Fisheries, Bangkok, Thailand, and ICLARM, Manila, Philippines, pp. 59–62.

Manna, G.K. and Sadhukhan, A. (1991) Genotoxic potential of the spores of three species of fungi experimented on cichlid fish, *Oreochromis mossambicus*. *Kromosomo* 2, 2129–2134.

Mansour, C.R. (1998) Nutrient requirements of red tilapia fingerlings. MSc thesis, Alexandria University, Alexandria, Egypt.

Mansour, C.R. (2001) Nutritional requirements of Nile tilapia broodstock reared at different water salinities. PhD thesis, Alexandria University, Alexandria, Egypt.

Marengoni, N.G. and Onoue, Y. (1998) Ultraviolet-induced androgenesis in Nile tilapia, *Oreochromis niloticus* (L.) and hybrid Nile × blue tilapia, *O. aureus* (Steindachner). *Aquaculture Research* 29, 359–366.

Marengoni, N.G., Onoue, Y. and Oyama, T. (1998) All-male tilapia hybrids of two strains of *Oreochromis niloticus*. *Journal of the World Aquaculture Society* 29, 108–113.

Marte, C.L., Cruz, P. and Flores, E.F.C. (2000) Recent developments in freshwater and marine cage aquaculture in the Philippines. In: Lin, C.K. and Liao, I.C. (eds) *Proceedings of the First International Symposium on Cage Aquaculture in Asia*. Asian Fisheries Society, Manila, Philippines, pp. 83–96.

Martin, J.M., Rakocy, J.E. and Cole, W.M. (2000) Greenwater tank culture of tilapia. In: Creswell, R.L. (ed.) *Proceedings of the Gulf and Caribbean Fisheries Institute*. No. 51, Gulf and Caribbean Fisheries Institute, c/o Harbor Branch, Oceanographic Institution, Florida, pp. 330–340.

Martin, J.M., Bailey, D.S., Shultz, R.C. and Rakocy, J.E. (2001) Effect of three stocking densities on the production of advanced-sized *Oreochromis niloticus* fingerlings in a greenwater tank system. *Aquaculture 2001 – Book of Abstracts*. World Aquaculture Society, Baton Rouge, Louisiana, p. 406.

Martinez, R., Juncal, J., Zaldivar, C., Arenal, A., Guillen, I., Morera, V., Carrillo, O., Estrada, M., Morales, A. and Estrada, M.P. (2000) Growth efficiency in transgenic tilapia (*Oreochromis* sp.) carrying a single copy of an homologous cDNA growth hormone. *Biochemical and Biophysical Research Communications* 267, 466–472.

Martinez Espinosa, M. (1994) *Between the Aquaculture of the 'More Poor' and That of the 'Less Poor'. A Methodologic Proposal for the Development of Type 2 Rural Aquaculture: Two Pilot Cases in Venezuela and Colombia*. FAO, Mexico City, Mexico.

Martinez Espinosa, M. (1995) Development of type 2 rural aquaculture in Latin America. *FAO Aquaculture Newsletter, Rome* 11, 6–10.

Marzouk, M.S. and Bakeer, A. (1991) Some investigations in Nile tilapia exposed to sublethal concentration of Bayluscide. *Journal of Egyptian Veterinary Medicine Association*, 51, 97–107.

Mather, P.B., Lal, S.N. and Wilson, J. (2001) Experimental evaluation of mass selection to improve red body colour in Fijian hybrid tilapia (*Oreochromis niloticus* × *Oreochromis mossambicus*). *Aquaculture Research* 32, 329–336.

Mazid, A.M., Tanaka, Y., Katayama, T., Rahman, A.M., Simpson, K.L. and Chichester, C.O. (1979) Growth response of *Tilapia zillii* fingerlings fed isocaloric diets with variable protein levels. *Aquaculture* 18, 115–122.

Mehrim, A.I.M. (2001) Effects of some chemical pollutants on growth performance, feed and nutrient utilization of tilapia. MSc thesis, Alexandria University, Egypt.

Mélard, C. (1995) Production of a high percentage of male offspring with 17α-ethynylestradiol sex-reversed *Oreochromis aureus*. 1. Estrogen sex-reversal and production of F2 pseudofemales. *Aquaculture* 130, 25–34.

Mélard, C. and Ducarme, C. (1993) The tilapia intensive rearing technology in Europe. In: Kestemont, P. and Billard, R. (eds) *Proceedings of the Workshop on Aquaculture of Freshwater Species (Except Salmonids)*. Special Publication No. 20, European Aquaculture Society, Ostend, Belgium, pp. 24–28.

Mélard, C., Desprez, D. and Philippart, J.C. (1995) Le comntrôle du sexe chez les tilapias: bilan et perspectives de recherche á la station de Tihange. *Cahiers d'ethologie appliquée* 13, 421–434.

Meyer, D.E. (2001) Small-scale tilapia culture in rural Central America. *Global Aquaculture Advocate* 4 (6), 47–49.

Meyer, D.E. (2002) Tilapia culture in Honduras. *Global Aquaculture Advocate* 5 (6), 36–37.

Meyer-Burgdorff, K., Osman, M.F. and Gunther, K.D. (1989) Energy metabolism in *Oreochromis niloticus*. *Aquaculture* 79, 283–291.

Micha, J.-C., Antoine, T., Wery, P. and Van Hove, C. (1988) Growth, ingestion capacity, comparative appetency and biochemical composition of *Oreochromis niloticus* and *Tilapia rendalli* fed with *Azolla*. In: Pullin, R.S.V., Bhukaswan, T., Tonguthai, K. and Maclean, J.L. (eds) *Proceedings of the Second International Symposium on Tilapia in Aquaculture*. ICLARM Proceedings No. 15, Department of Fisheries, Bangkok, Thailand, and ICLARM, Manila, Philippines, pp. 347–355.

Middendorp, A.J. (1988) Small-scale cage culture of tilapia (*Oreochromis niloticus*) in communal ponds in northeastern Thailand. In: Pullin, R.S.V., Bhukaswan, T., Tonguthai, K. and Maclean, J.L. (eds) *Proceedings of the Second International Symposium on Tilapia in Aquaculture*. ICLARM Conference Proceedings No. 15, Department of Fisheries, Bangkok, Thailand, and ICLARM, Manila, Philippines, p. 600.

Middendorp, A.J. (1995) Pond farming of Nile tilapia *Oreochromis niloticus* L. in northern Cameroon. Mixed culture of large tilapia (> 200 g) with cattle manure and cottonseed cake as pond inputs, and African catfish, *Clarias gariepinus* (Burchell), as police-fish. *Aquaculture Research* 26, 723–730.

Middendorp, A.J. and Verreth, J.A.J. (1992) The feasibility of small-scale hapa culture of tilapia *Oreochromis niloticus* as an additional income source for rice farmers in northeast Thailand. *Asian Fisheries Science* 5, 303–314.

Milstein, A. (1997) Do management procedures affect the ecology of warm water polyculture ponds? *World Aquaculture* 28 (3), 12–19.

Milstein, A. and Lev, O. (2004) Organic tilapia culture in Israel. In: Bolivar, R., Mair, G. and Fitzsimmons, K. (eds) *Proceedings of the Sixth International Symposium on Tilapia in Aquaculture*. Bureau of Fisheries and Aquatic Resources, Manila, Philippines, and American Tilapia Association, Charles Town, West Virginia, pp. 657–660.

Milstein, A., Avnimelech, Y., Zoran, M. and Joseph, D. (2001a) Growth performance of hybrid bass and hybrid tilapia in conventional and active suspension intensive ponds. *Israeli Journal of Aquaculture/Bamidgeh* 53, 147–157.

Milstein, A., Zoran, M., Kochba, M. and Avnimelech, Y. (2001b) Effect of different management practices on water quality of intensive tilapia culture in Israel. *Aquaculture International* 9, 133–152.

Ming, L.Q. and Yi, Y. (2004) Tilapia culture in China. In: Bolivar, R., Mair, G. and Fitzsimmons, K. (eds) *Proceedings of the Sixth International Symposium on Tilapia Aquaculture*. Bureau of Fisheries and Aquatic Resources, Manila, Philippines, and American Tilapia Association, Charles Town, West Virginia, pp. 18–27.

Miranova, N.V. (1977) Energy expenditure on egg production in young *T. mossambica* and the influence of maintenance conditions on their reproductive intensity. *Journal of Ichthyology* 17, 627–633.

Mires, D. (1982) Study of the problems of the mass production of hybrid tilapia fry. In: Pullin, R.S.V. and Lowe-McConnell, R.H. (eds) *The Biology and Culture of Tilapias*. ICLARM Conference Proceedings No. 7, ICLARM, Manila, Philippines, pp. 83–113.

Mires, D. and Amit, Y. (1992) Intensive culture of tilapia in quasi-closed water cycled flow-through ponds – the Dekel aquaculture system. *Israeli Journal of Aquaculture/Bamidgeh* 44, 82–86.

Mires, D. and Anjioni, C. (1997) Technical and economic comparative evaluation of two intensive closed water-cycled culture systems for tilapias in Israel. In: Fitzsimmons, K. (ed.) *Proceedings from the Fourth International Symposium on Tilapia in Aquaculture*. Northeast Regional Agriculture Engineering Service, Ithaca, New York, pp. 416–425.

Miyashita, T. (1984) *Pseudomonas fluorescens* and *Edwardsiella tarda* isolated from diseased tilapia. *Fish Pathology* 19, 45–50.

Miyazaki, T., Kubota, S.S. and Miyashita, T. (1984) A histopathological study of *Pseudomonas fluorescens* infection in tilapia. *Fish Pathology* 19, 161–166.

Moehl, J.F. (2002) Aquaculture development as a national strategy for poverty alleviation and improved food security in Africa. *World Aquaculture 2002 – Book of Abstracts*. World Aquaculture Society, Baton Rouge, Louisiana, p. 513.

Mohamed, G.A., Farag, M.E., Elghobashy, H.A. and Ali, M.A. (2004) Feminization of sexually undifferentiated progeny of *Oreochromis niloticus* and *O. aureus*. *Journal of Egyptian Academic Society for Environmental Development – C. Molecular Biology* 5, 31–44.

Molnar, J.J., Hanson, T.R. and Lovshin, L.L. (1996) *Social, Economic, and Institutional Impacts of Aquacultural Research on Tilapia: the PD/ACRSP in Rwanda, Honduras, the Philippines, and Thailand*. Research and Development Series

No. 40, International Center for Aquaculture and Aquatic Environments, Auburn University, Auburn, Alabama.

Moreau, J. (1983) Review of introductions of tilapia in open inland waters of Africa, their influence on ecology and fisheries. In: Fishelson, L. and Yaron, Z. (eds) *Proceedings of the International Symposium on Tilapia in Aquaculture*. Tel Aviv University, Tel Aviv, Israel, pp. 77–85.

Moreau, Y. (1996) Measurement of the apparent digestibility coefficients for *Oreochromis niloticus* of agro-industrial by-products available in Côte d'Ivoire. In: Pullin, R.S.V., Lazard, J., Legendre, M., Amon Kottias, J.B. and Pauly, D. (eds) *Proceedings of the Third International Symposium on Tilapia in Aquaculture*. ICLARM Conference Proceedings No. 41, ICLARM, Penang, Malaysia, pp. 204–210.

Morgan, J.D., Sakamoto, T., Grau, E.G. and Iwama, G.K. (1997) Physiological and respiratory responses of the Mozambique tilapia (*Oreochromis mossambicus*) to salinity acclimation. *Comparative Biochemistry and Physiology* 117A, 391–398.

Morgan, P.R. (1972) Causes of mortality in the endemic tilapia of Lake Chilwa (Malawi). *Hydrobiologia* 40, 101–119.

Moriarty, C.M. and Moriarty, D.J.W. (1973) Quantitative estimation of daily ingestion of phytoplankton by *Tilapia nilotica* and *Haplochromis nigripinnis* in Lake George, Uganda. *Journal of Zoology* 171, 15–24.

Moriarty, D.J.W., Darlington, P.E.C., Dunn, I.G., Moriarty, C.M. and Tevlin, M.P. (1973) Feeding and grazing in Lake George, Uganda. *Proceedings of Royal Society, Edinburgh (B)* 184, 299–319.

Morrison, J.R., Deavours, W.L., Jones, J.C. and Tabb, M.A. (1995) Maximizing first-year growth of mixed-sex blue tilapia *Oreochromis aureus* in polyculture with catfish *Ictalurus* spp. *Journal of the World Aquaculture Society* 26, 447–452.

Muir, J.F., Van Rijn, J. and Hargreaves, J. (2000) Production in intensive and recycle systems. In: Beveridge, M.C.M. and McAndrew, B.J. (eds) *Tilapias: Biology and Exploitation*. Kluwer Academic Publishers, Dordrecht/Boston/London, pp. 405–445.

Mukhi, S.K., Chandrika, V., Madhavi, B. and Nayak, B.B. (2001) Incidence of β-haemolytic streptococcal infection associated with mass mortalities of cultured tilapia, *Oreochromis mossambicus* in brackishwater ponds in India. *Journal of Aquaculture in the Tropics* 16, 373–383.

Müller-Belecke, A. and Hörstgen-Schwark, G. (1995) Sex determination in tilapia *Oreochromis niloticus*. Sex ratios in homozygous gynogenetic progeny and their offspring. *Aquaculture* 137, 57–65.

Müller-Belecke, A. and Hörstgen-Schwark, G. (2000) Performance testing of clonal *Oreochromis niloticus* lines. *Aquaculture* 184, 67–76.

Muratori, M.C.S., de Oliveira, A.L., Leite, R.C., Costa, A.P.R. and da Silva, M.C.C. (2000) *Edwardsiella tarda* isolated in integrated fish farming. *Aquaculture Research* 31, 481–483.

Myers, J.M. and Hershberger, W.K. (1991) Early growth and survival of heat-shocked and tetraploid-derived triploid rainbow trout (*Oncorhynchus mykiss*). *Aquaculture* 96, 97–107.

Myers, J.M., Penman, D.J., Basavaraju, Y., Powell, S.F., Baoprasertkul, P., Rana, K.J., Bromage, N. and McAndrew, B.J. (1995) Induction of diploid androgenetic and mitotic gynogenetic Nile tilapia (*Oreochromis niloticus* L.). *Theoretical and Applied Genetics* 90, 205–210.

Naegel, L.C.A. (1997) *Azolla* meal as a supplemental feed ingredient for tilapias. In: Fitzsimmons, K. (ed.) *Proceedings from the Fourth International Symposium on Tilapia in Aquaculture*. Northeast Regional Agriculture Engineering Service, Ithaca, New York, pp. 20–30.

Nafady, A., Shahata, A., Ibrahim, I.A. and Shabaan, A.A. (1986) Clinical signs, post-mortem findings and histopathological changes in *Tilapia nilotica* fish intoxicated with Bayluscide. *Assut Veterinary Medical Journal* 17, 126–132.

Nakayama, T., Liu, D.-J. and Ooi, A. (1992) Tension change of stressed and unstressed carp muscles in isometric rigor concentration and resolution. *Nippon Suisan Gakkaishi* 58, 1517–1522.

Nandeesha, M.C. (2002) Contribution of aquaculture to poverty alleviation and food security in India. *World Aquaculture 2002 – Book of Abstracts*. World Aquaculture Society, Baton Rouge, Louisiana, p. 535.

Natarajan, P. and Aravindan, C.M. (2002) Status of tilapia farming and its prospects in India. In: Guerrero, R.D. III and Guerrero-del Castillo, M.R. (eds) *Tilapia Farming in the 21st Century. Proceedings of the International Forum on Tilapia Farming in the 21st Century (Tilapia Forum 2002)*. Philippine Fisheries Association, Los Baños, Laguna, Philippines, pp. 78–86.

Natividad, J.M., Bondad-Reantaso, M.G. and Arthur, J.R. (1986) Parasites of Nile tilapia (*Oreochromis niloticus*) in the Philippines. In: Maclean, J.L., Dizon, L.B. and Hosillos, L.V. (eds) *Proceedings of the First Asian Fisheries Forum*. Asian Fisheries Society, Manila, Philippines, pp. 255–259.

Neira, I. and Engle, C. (2001) The Honduran market for tilapia: restaurant and supermarket surveys. In: *Aquaculture 2001 – Book of Abstracts*.

World Aquaculture Society, Baton Rouge, Louisiana, p. 469.

Newton, S.H., Mullins, J., Libey, G.S. and Kidd, M. (1998) A prototype tilapia/hydroponic greenhouse recirculating production system for institutional application. In: *Proceedings of the Second International Conference on Recirculating Aquaculture*. US Department of Agriculture, Virginia Polytechnic Institute and State University, Blacksburg, Virginia, pp. 405–406.

Ng, W.K. and Wee, K.L. (1989) The nutritive value of cassava leaf meal in pelleted feed for Nile tilapia. *Aquaculture* 83, 45–58.

Nguenga, D. (1988) A note on infestation of *Oreochromis niloticus* with *Trichodina* sp. and *Dactylogyrus* sp. In: Pullin, R.S.V., Bhukaswan, T., Tonguthai, K. and Maclean, J.L. (eds) *Proceedings of the Second International Symposium on Tilapia in Aquaculture*. ICLARM Proceedings No. 15, Department of Fisheries, Bangkok, Thailand, and ICLARM, Manila, Philippines, pp. 117–119.

Nitithamyong, C., Chiayvareesajja, J., Chiayvareesajja, S., Wongwit, C. and Tansakul, R. (1990a) Production of Nile tilapia (*Oreochromis niloticus*) in different culture and harvesting systems. In: De Silva, S. (ed.) *Fish Nutrition Research in Asia. Proceedings of the Fourth Asian Fish Nutrition Workshop*. Special Publication No. 5, Asian Fisheries Society, Manila, Philippines, pp. 169–174.

Nitithamyong, C., Chiayvareesajja, J., Chiayvareesajja, S., Tansakul, R. and Wongwit, C. (1990b) Growth and production of Nile tilapia (*Oreochromis niloticus*) in monoculture and polyculture with snakehead (*Channa striata*), and in integrated culture with pig and snakehead. In: De Silva, S.S. (ed.) *Fish Nutrition Research in Asia. Proceedings of the Fourth Asian Fish Nutrition Workshop*. Special Publication No. 5, Asian Fisheries Society, Manila, Philippines, pp. 175–180.

Njoku, D.C. and Ejiogu, C.O. (1999) On-farm trials of an integrated fish-cum-poultry farming system using indigenous chickens. *Aquaculture Research* 30, 399–408.

Noakes, D.L.G. and Balon, E.K. (1982) Life histories of tilapias: an evolutionary perspective. In: Pullin, R.S.V. and Lowe-McConnell, R.H. (eds) *The Biology and Culture of Tilapias*. ICLARM Conference Proceedings No. 7, ICLARM, Manila, Philippines, pp. 61–82.

Northcott, M.E., Beveridge, M.C.M. and Ross, L.G. (1991) A laboratory investigation of the filtration and ingestion rates of the tilapia, *Oreochromis niloticus*, feeding on two species of blue-green algae. *Environmental Biology of Fishes* 31, 75–85.

NRC (National Research Council) (1983) *Nutrient Requirements of Warmwater Fishes and Shellfishes*. National Academy of Science, Washington, DC, 102 pp.

NRC (National Research Council) (1993) *Nutrient Requirements of Fish*. National Academy of Science, Washington, DC, 141 pp.

Nwanna, L.C. and Daramola, J.A. (2000) Harnessing of shrimp head waste in Nigeria for low cost production of tilapia, *Oreochromis niloticus* (L.). In: Fitzsimmons, K. and Fhlho, J.C. (eds) *Tilapia Aquaculture in the 21st Century. Proceedings from the Fifth International Symposium on Tilapia in Aquaculture, Rio de Janeiro, Brazil*. American Tilapia Association, Charles Town, West Virginia, and ICLARM, Penang, Malaysia, pp. 174–178.

Oben, B.O., Ugwumba, O.A. and Oben, P.M. (1998) Food and feeding ecology of some tilapiine fishes from an inland tropical lake, Ibadan, Nigeria. Paper presented at the International Conference for the Paradi Association and the Fisheries Society of Africa, Grahamstown, South Africa, 13–18 September 1998.

Oduro-Boateng, F. (1998) Tilapia cage culture in Ghana: studies on economic and technical viability. In: Coetzee, L., Gon, J. and Kulongowski, C. (eds) *African Fishes and Fisheries Diversity and Utilisation*. International Conference for the Paradi Association and the Fisheries Society of Africa, Grahamstown, South Africa, p. 227.

Oduro-Boateng, F. and Bart-Plange, A. (1988) Pito brewery waste as an alternative protein source to fishmeal in feeds for *Tilapia busumana*. In: Pullin, R.S.V., Bhukaswan, T., Tonguthai, K. and Maclean, J.L. (eds) *The Second International Symposium on Tilapia in Aquaculture*. ICLARM Conference Proceedings No. 15, Department of Fisheries, Bangkok, Thailand, and ICLARM, Manila, Philippines, pp. 357–360.

Ofori, J.K. (1988) The effect of predation by *Lates niloticus* on overpopulation and stunting in mixed sex culture of tilapia species in ponds. In: Pullin, R.S.V., Bhukaswan, T., Tonguthai, K. and Maclean, J.L. (eds) *Proceedings of the Second International Symposium on Tilapia in Aquaculture*. ICLARM Conference Proceedings No. 15, Department of Fisheries, Bangkok, Thailand, and ICLARM, Manila, Philippines, pp. 69–73.

Ogbonna, C.I.C. and Alabi, R.O. (1991) Studies on species of fungi associated with mycotic infections of fish in a Nigerian freshwater fish pond. *Hydrobiologia* 220, 131–135.

Ogutu-Ohwayo, R. and Hecky, R.E. (1991) Fish introductions in Africa and some of their implications. *Canadian Journal of Fisheries and Aquatic Sciences* 48 (Suppl. 1), 8–12.

Okaeme, A.N. and Okojie, P.U. (1989) Parasites and diseases of feral *Oreochromis niloticus*,

Sarotherodon galilaeus and *Clarias* sp. of Lake Kainja. In: *Annual Report*. National Institute of Freshwater Fisheries Research, New Bussa, Nigeria, pp. 41–44.

Okaeme, A.N. and Olufemi, B.E. (1997) Fungi associated with tilapia (*Oreochromis niloticus*) culture ponds in Nigeria. *Journal of Aquaculture in the Tropics* 12, 267–274.

Okaeme, A.N., Obiekezie, A., Okojie, P.U. and Agbontale, J.J. (1989) Histopathology of normal and infected organs of tilapia by *Myxobolus ovariae*. In: *Annual Report*. National Institute of Freshwater Fisheries Research, New Bussa, Nigeria, pp. 37–40.

Okorie, O. (1973) Lunar periodicity and the breeding of *Tilapia nilotica* in the northern part of Lake Victoria. In: *Annual Report*, Appendix E. East African Freshwater Fish Research Organization, Nairobi, Kenya, pp. 50–58.

Oldorf, W., Kronert, U., Balarin, J., Haller, R., Hörstgen-Schwark, G. and Langholz, H.J. (1989) Prospects of selecting for late maturity in tilapia *Oreochromis niloticus*: II. Strain comparison under laboratory and field conditions. *Aquaculture* 77, 123–133.

Olvera-Novoa, M.A., Campos, G.S., Sabido, G.M. and Martínez-Palacios, C.A. (1990) The use of alfalfa leaf protein concentrate as a protein source in diets for tilapia (*Oreochromis mossambicus*). *Aquaculture* 90, 291–302.

Olvera-Novoa, M.A., Pereira-Pacheco, F., Olvira-castillo, L., Perez-Flores, V., Navarro, L. and Samano, J.C. (1997) Cowpea (*Vigna unguiculata*) protein concentrate as replacement for fish meal in diets for tilapia (*Oreochromis niloticus*) fry. *Aquaculture* 158, 107–116.

Olvera-Novoa, M.A., Domínguez-Cen, L.J., Olvira-castillo, L. and Martínez-Palacios, C.A. (1998) Effect of the use of the microalgae *Spirulina maxima* as fish meal replacement in diets for tilapia, *Oreochromis mossambicus* (Peters), fry. *Aquaculture Research* 29, 709–715.

Olvera-Novoa, M.A., Olvira-castillo, L. and Martínez-Palacios, C.A. (2002) Sunflower seed meal as a protein source in diets for *Tilapia rendalli* (Boulanger, 1896) fingerlings. *Aquaculture Research* 33, 223–229.

Omondi, J.G., Gichuri, W.M. and Veverica, K. (2001) A partial economic analysis for Nile tilapia *Oreochromis niloticus* L. and sharptoothed catfish *Clarias gariepinus* (Burchell 1822) polyculture in central Kenya. *Aquaculture Research* 32, 693–700.

Omoregie, E. and Ogbemudia, F.I. (1993) Effects of substituting fish meal with palm kernel meal on growth and food utilization of the Nile tilapia, *Oreochromis niloticus*. *Israeli Journal of Aquaculture/Bamidgeh* 45, 113–119.

Omoregie, E., Ufodike, E.B.C. and Onwuliri, C.O.E. (1995) A comparative survey of helminth parasites of the Nile tilapia, *Oreochromis niloticus* from a fish farm and petroleum polluted freshwater. *Journal of Aquatic Sciences* 10, 7–14.

Orachunwong, C., Thammasart, S. and Lohawantanakul, C. (2001) Recent developments in tilapia feeds. In: Subasinghe, S. and Singh, T. (eds) *Tilapia: Production, Marketing and Technical Developments. Proceedings of the Tilapia 2001 International Technical and Trade Conference on Tilapia. Infofish, Kuala Lumpur, Malaysia, pp. 113–122.*

Osman, M.F., Omar, E. and Nour, A.M. (1996) The use of leucaena leaf meal in feeding Nile tilapia. *Aquaculture International* 4, 9–18.

Oswald, M., Glasser, F., Sanchez, F. and Bamba, V. (1997) Reconsidering rural fish farming in Africa. In: Fitzsimmons, K. (ed.) *Proceedings from the Fourth International Symposium on Tilapia in Aquaculture*. Northeast Regional Agriculture Engineering Service, Ithaca, New York, pp. 499–511.

Palada, M.C., Cole, W.M. and Crossman, S.M.A. (1999) Influence of effluents from intensive aquaculture and sludge on growth and yield of bell peppers. *Journal of Sustainable Agriculture* 14, 85–103.

Pankhurst, N.W. and Van der Kraak, G. (1997) Effects of stress on reproduction and growth of fish. In: Iwama, J.K. and Pickering, A.D. (eds) *Fish Stress and Health in Aquaculture*. Cambridge University Press, Cambridge, UK, pp. 73–93.

Pant, J., Demaine, H. and Edwards, P. (2004) Assessment of the aquaculture subsystem in integrated agriculture–aquaculture systems in Northeast Thailand. *Aquaculture Research* 35, 289–298.

Paperna, I. (1974) Lymphocystis in fish from East African lakes. *Journal of Wildlife Diseases* 9, 331–335.

Paperna, I. and Smirnova, M. (1997) *Branchiomyces*-like infection in a cultured tilapia (*Oreochromis* hybrid, Cichlidae). *Diseases of Aquatic Organisms* 31, 233–238.

Paperna, I., Kim, S.H. and Hammerschlag, E. (1996) Liver lesions in cultured *Oreochromis* hybrids caused by amoeboid organisms similar to the aetiological agent of goldfish kidney granuloma. *Diseases of Aquatic Organisms* 25, 151–153.

Park, K.H. and Jeong, H.D. (1996) Enhanced resistance against *Edwardsiella tarda* infection in tilapia (*Oreochromis niloticus*) by administration of protein-bound polysaccharide. *Aquaculture* 143, 135–143.

Paull, G.C. and Matthews, R.A. (2001) *Spironucleus vortens*, a possible cause of hole-in-the-head disease in cichlids. *Diseases of Aquatic Organisms* 45, 197–202.

Pauly, D. (1976) The biology, fishery and potential for aquaculture of *Tilapia melanotheron* in a small West African lagoon. *Aquaculture* 7, 33–49.

Payne, A.I. (1983) Estuarine and salt tolerant tilapias. In: Fishelson, L. and Yaron, Z. (eds) *Proceedings of the International Symposium on Tilapia in Aquaculture*. Tel Aviv University, Tel Aviv, Israel, pp. 534–543.

Payne, A.I. and Collinson, R.I. (1983) A comparison of the biological characteristics of *Sarotherodon niloticus* (L.) with those of *S. aureus* (Steindachner) and other tilapia of the delta and lower Nile. *Aquaculture* 30, 335–351.

Payne, A.I., Ridgway, J. and Hamer, J.L. (1988) The influence of salt (NaCl) concentration and temperature on the growth of *Oreochromis spilurus*, *O. mossambicus* and a red tilapia hybrid. In: Pullin, R.S.V., Bhukaswan, T., Tonguthai, K. and Maclean, J.L. (eds) *Proceedings of the Second International Symposium on Tilapia in Aquaculture*. ICLARM Conference Proceedings No. 15, Department of Fisheries, Bangkok, Thailand, and ICLARM, Manila, Philippines, pp. 481–487.

Pekar, F., Be, N.V., Long, D.N., Cong, N.V., Dung, D.T. and Olah, J. (2002) Eco-technological analysis of fish farming households in the Mekong Delta of Vietnam. In: Edwards, P., Little, D. and Demaine, H. (eds) *Rural Aquaculture*. CAB International, Wallingford, UK, pp. 77–95.

Penman, D.J. and McAndrew, B.J. (2000) Genetics for the management and improvement of cultured tilapia. In: Beveridge, M.C.M. and McAndrew, B.J. (eds) *Tilapias: Biology and Exploitation*. Kluwer Academic Publishers, Dordrecht/Boston/London, pp. 227–266.

Perera, R.P., Johnson, S.K., Collins, M.D. and Lewis, D.H. (1994) *Streptococcus iniae* associated with mortality of *Tilapia nilotica* × *T. aurea* hybrids. *Journal of Aquatic Animal Health* 6, 335–340.

Perera, R.P., Johnson, S.K. and Lewis, D.H. (1997) Epizootiological aspects of *Streptococcus iniae* affecting tilapia in Texas. *Aquaculture* 152, 25–33.

Perry, W.G. and Avault, J.W. (1972) Comparisons of stripped mullet and tilapia for added production in caged catfish studies. *Progressive Fish Culturist* 34, 229–232.

Persand, S. and Bhikajee, M. (1997) Studies on an experimental fouling resistant mariculture cage for red tilapia in Mauritius. In: Fitzsimmons, K. (ed.) *Proceedings from the Fourth International Symposium on Tilapia in Aquaculture*. Northeast Regional Agriculture Engineering Service, Ithaca, New York, pp. 408–415.

Perschbacher, P.W. (1995) Algal management in intensive channel catfish production trials. *World Aquaculture* 26 (3), 65–68.

Perschbacher, P.W. and McGeachin, R.B. (1988) Salinity tolerance of red hybrid tilapia fry, juvenile and adults. In: Pullin, R.S.V., Bhukaswan, T., Tonguthai, K. and Maclean, J.L. (eds) *Proceedings of the Second International Symposium on Tilapia in Aquaculture*. ICLARM Conference Proceedings No. 15, Department of Fisheries, Bangkok, Thailand, and ICLARM, Manila, Philippines, pp. 415–420.

Peters, H.M. (1963) Untersuchungen zum problem des angeborenen verhaltens. *Naturwissenschaften* 50, 677–686.

Peters, H.M. (1983) *Fecundity, Egg Weight and Oocyte Development in Tilapia (Cichlidae, Teleostei)*. ICLARM Translations 2, ICLARM, Manila, Philippines, 28 pp.

Philippart, J.-C.L. and Ruwet, J.-C.L. (1982) Ecology and distribution of tilapias. In: Pullin, R.S.V. and Lowe-McConnell, R.H. (eds) *The Biology and Culture of Tilapias*. ICLARM Conference Proceedings No. 7, ICLARM, Manila, Philippines, pp. 15–59.

Phromkunthong, W., Musakopas, A., Supamattaya, K. and Chittiwan, V. (2004) Effects of phytase on enhancement of phosphorus utilization for feeds with plant materials in sex-reversed red tilapia (*Oreochromis niloticus* Linn.). In: *The 11th International Symposium on Nutrition and Feeding in Fish, Phuket Island, Thailand, 2–7 May*. Department of Fisheries, Bangkok, Thailand, p. 156.

Phuong, N.T., Yi, Y., Diana, J.S., Lin, C.K. and Bui, T.V. (2004) Integrated cage-cum-pond culture: stocking densities of caged climbing perch in Nile tilapia ponds. In: Bolivar, R., Mair, G. and Fitzsimmons, K. (eds) *Proceedings of the Sixth International Symposium on Tilapia in Aquaculture*. Bureau of Fisheries and Aquatic Resources, Manila, Philippines, and American Tilapia Association, Charles Town, West Virginia, pp. 597–598.

Pierce, B.A. (1980) Production of hybrid tilapia in indoor aquaria. *Progressive Fish Culturist* 42, 233–234.

Pillay, T.V.R. (1990) *Aquaculture Principles and Practices*. Fishing News Books, Blackwell Science, Oxford, UK, 575 pp.

Platt, S. and Hauser, W.J. (1978) Optimum temperature for feeding and growth of *Tilapia zillii*. *Progressive Fish Culturist* 40, 105–107.

Plumb, J.A. (1997) Infectious diseases of tilapia. In: Costa-Pierce, B.A. and Rakocy, J.E. (eds) *Tilapia Aquaculture in the Americas*, Vol. 1. World Aquaculture Society, Baton Rouge, Louisiana, pp. 212–228.

Plumb, J.A. (1999) Overview of warmwater fish diseases. *Journal of Applied Aquaculture* 9, 1–10.

Pollock, L.J. and Little, D. (2002) The nature of participation of the rural poor in poverty-focused aquaculture research: empirical finding from process monitoring and livelihoods studies with inland fishing communities in Sri Lanka. In: *World Aquaculture 2002 – Book of Abstracts*. World Aquaculture Society, Baton Rouge, Louisiana, p. 610.

Popma, T.J. and Lovshin, L.L. (1996) *Worldwide Prospects for Commercial Production of Tilapia*. Research and Development Series, No. 41, Department of Fisheries and Allied Aquaculture, Auburn University, Auburn, Alabama, 23 pp.

Popma, T.J. and Rodriquez, F.B. (2000) Tilapia aquaculture in Colombia. In: Costa-Pierce, B.A. and Rakocy, J.E. (eds) *Tilapia Aquaculture in the Americas*, Vol. 2. The World Aquaculture Society, Baton Rouge, Louisiana, pp. 141–150.

Popma, T.J., Phelps, R.P., Castillo, S., Hatch, L.U. and Hanson, T.R. (1995) Family-scale fish farming in Guatemala, Part 1: outreach strategies and production practices. *Journal of Aquaculture in the Tropics* 10, 43–56.

Popper, D. and Lichatowich, T. (1975) Preliminary success in predator control of *Tilapia mossambica*. *Aquaculture* 5, 213–214.

Potts, A.C. and Phelps, R.P. (1995) Use of diethylstilbestrol and ethynylestradiol to feminize Nile tilapia, *Oreochromis niloticus*. *Journal of Applied Ichthyology* 11, 111–117.

Pouomogne, V., Gabriel, T. and Pouemegne, J.-B. (1997) A preliminary evaluation of cacoa husks in practical diets for juvenile Nile tilapia (*Oreochromis niloticus*). *Aquaculture* 156, 215–223.

Pruginin, Y. (1967) *Report to the Government of Uganda on the Experimental Fish Culture Project in Uganda, 1965–1966*. Reports on Fisheries, TA Reports 2446, FAO/UNDP (Technical Assistance), FAO, Rome, 19 pp.

Pruginin, Y., Rothbard, S., Wohlfarth, G., Halevy, A., Moav, R. and Hulata, G. (1975) All-male broods of *Tilapia nilotica* and *T. aurea* hybrids. *Aquaculture* 6, 11–21.

Pruginin, Y., Fishelson, L. and Koren, A. (1988) Intensive tilapia farming in brackishwater from an Israeli desert. In: Pullin, R.S.V., Bhukaswan, T., Tonguthai, K. and Maclean, J.L. (eds) *Proceedings of the Second International Symposium on Tilapia in Aquaculture*. Conference Proceedings No. 15, Department of Fisheries, Bangkok, Thailand, and ICLARM, Manila, Philippines, pp. 75–81.

Puckhaber, B. and Hörstgen-Schwak, G. (1996) Growth and gonadal development of triploid tilapia (*Oreochromis niloticus*). In: Pullin, R.S.V., Lazard, J., Legendre, M., Amon Kottias, J.B. and Pauly, D. (eds) *Proceedings of the Third International Symposium on Tilapia in Aquaculture*. ICLARM Conference Proceedings No. 41, ICLARM, Penang, Malaysia, pp. 377–382.

Pudadera, B.J. Jr, Corre, K.C., Coniza, E. and Taleon, G.A. (1986) Integrated farming of broiler chickens with fish and shrimp in brackishwater ponds. In: Maclean, J.L., Dizon, L.B. and Hosillos, L.V. (eds) *Proceedings of the First Asian Fisheries Forum*. Asian Fisheries Society, Manila, Philippines, pp. 141–144.

Pulido, A.B., Iregui, C.C. and Figueroa, J. (1999) Report of *Streptococcus* in tilapia cultivated in Colombia. In: Cabrera, T., Jory, D. and Silva, M. (eds) *Aquaculture 99*. Congreso sur American de Acuicultura, World Aquaculture Society, Latin American Chapter, Puerto la Cruz, Venezuela, pp. 229–239.

Pullin, R.V.S. and Lowe-McConnell, R.H. (eds) (1982) *The Biology and Culture of Tilapias*. ICLARM Conference Proceedings No. 7, ICLARM, Manila, Philippines, 432 pp.

Pullin, R.S.V., Palomares, M.L., Casal, C.V., Dey, M.M. and Pauly, D. (1997) Environmental impacts of tilapias. In: Fitzsimmons, K. (ed.) *Proceedings from the Fourth International Symposium on Tilapia in Aquaculture*. Northeast Regional Agriculture Engineering Service, Ithaca, New York, pp. 554–570.

Purchase, C.F., Boyce, D.L. and Brown, J.A. (2000) Growth and survival of juvenile flounder *Pleuronectes ferrugineus* (Storer) under different photoperiods. *Aquaculture Research* 31, 547–552.

Purdom, C.E. (1995) *Genetics and Fish Breeding*. Chapman & Hall, London, 291 pp.

Qiuming, L. and Yi, Y. (2004) Tilapia culture in China. In: Bolivar, R., Mair, G. and Fitzsimmons, K. (eds) *Proceedings of the Sixth International Symposium on Tilapia in Aquaculture*. Bureau of Fisheries and Aquatic Resources, Manila, Philippines, and American Tilapia Association, Charles Town, West Virginia, pp. 18–27.

Rab, M.A. (1989) Intensive nursing of Nile tilapia (*Oreochromis niloticus*) fry. MSc thesis, Asian Institute of Technology, Bangkok, Thailand.

Rahman, M.A. and Maclean, N. (1999) Growth performance of transgenic tilapia containing an exogenous piscine growth hormone gene. *Aquaculture* 173, 333–346.

Rahman, M.A., Ronyai, A., Engidaw, B.Z., Jauncey, K., Hwang, G., Smith, A., Roderick, E., Penman, D., Varadi, L. and Maclean, N. (2001) Growth and nutritional trials on transgenic Nile tilapia containing an exogenous fish growth hormone gene. *Journal of Fish Biology* 59, 62–78.

Rajaratnam, S. and Balasundaram, C. (2001) Qualitative and quantitative analysis of fish and prawn maintained at different temperatures. In: *The Sixth Asian Fisheries Forum – Book of Abstracts*.

Asian Fisheries Society, Quezon City, Philippines, p. 19.

Rajavarthini, P.B. and Michael, R.D. (1996) The effect of monocrotophos (Nuvacron) on the antibody response to bovine serum albumin in *Oreochromis mossambicus* (Peters) *Journal of Aquaculture in the Tropics* 11, 27–32.

Rakocy, J.E. and Allison, R. (1981) Evaluation of a closed recirculating system for the culture of tilapia and aquatic macrophytes. In: Allen, L.J. and Kinney, E.C. (eds) *Proceedings of the Bio-engineering Symposium for Fish Culture*. American Fisheries Society, Bethesda, Maryland, pp. 296–307.

Rakocy, J.E. and Bailey, D.S. (2003) Initial economic analyses of aquaponic systems. Paper presented at 'Beyond Monoculture'. Trondheim, Norway, 6 June 2003.

Rakocy, J.E., Bailey, D.S., Martin, J.M. and Shultz, R.C. (2000a) Tilapia production systems for the Lesser Antilles and other resource-limited, tropical areas. In: Fitzsimmons, K. and Filho, J.C. (eds) *Tilapia Culture in the 21st Century. Proceedings from the Fifth International Symposium on Tilapia Aquaculture, Rio de Janeiro, Brazil.* American Tilapia Association, Charles Town, West Virginia, and ICLARM, Penang, Malaysia, pp. 651–662.

Rakocy, J.E., Shultz, R.C. and Bailey, D.S. (2000b) Commercial aquaponics for the Caribbean. In: Creswell, R.L. (ed.) *Proceedings of the Gulf and Caribbean Fisheries Institute*. No.51, Gulf and Caribbean Fisheries Institute, c/o Harbor Branch Oceanographic Institution, pp. 353–364.

Rakocy, J.E., Bailey, D.S., Shultz, R.C. and Martin, J.M. (2001) Improvements to a commercial-scale aquaponic system and preliminary evaluation of the production of red tilapia *Oreochromis niloticus* and 13 types of vegetables. *Aquaculture 2001 – Book of Abstracts*. World Aquaculture Society, Baton Rouge, Louisiana, p. 544.

Rakocy, J.E., Bailey, D.S., Shultz, R.C. and Thoman, E.S. (2004) Update on tilapia and vegetable production in the UVI aquaponic system. In: Bolivar, R., Mair, G. and Fitzsimmons, K. (eds) *Proceedings of the Sixth International Symposium on Tilapia in Aquaculture*. Bureau of Fisheries and Aquatic Resources, Manila, Philippines, and American Tilapia Association, Charles Town, West Virginia, pp. 676–690.

Ramadan, A., Afifi, N.A., Mostafa, M.M. and Samy, A.M. (1994) The effect of ascogen on the immune response of tilapia fish to *Aeromonas hydrophila* vaccine. *Fish and Shellfish Immunology* 4, 159–165.

Ramadan, H.H. (1991) Effect of host species, sex, length, diet and different seasons on the parasitic infection of *Tilapia* fish in Lake Manzala. *Journal of King Abdulaziz University for Marine Sciences* 2, 81–91.

Rana, K.J. (1981) Effects of rearing conditions, age and size on performance of normal hormone-treated *Sarotherodon mossambicus* fry, with observations on fry abnormalities. MSc thesis, University of Stirling, Stirling, UK.

Rana, K.J. (1986) *Tilapia Culture: Hatchery Methods for Oreochromis mossambicus and O. niloticus with Special Reference to All-Male Fry Production.* Institute of Aquaculture, University of Stirling, Stirling, UK, 154 pp.

Rana, K.J. (1990a) Reproductive biology and the hatchery rearing of tilapia eggs and fry. In: Muir, J.F. and Roberts, R.J. (eds) *Recent Advances in Aquaculture*, vol. 3. Croom Helm, London and Canberra, pp. 343–406.

Rana, K.J. (1990b) Influence of incubation temperature on *Oreochromis niloticus* (L.) eggs and fry. 1. Gross embryology, temperature tolerance and rates of embryonic development. *Aquaculture* 87, 165–181.

Rana, K.J. (1990c) Influence of incubation temperature on *Oreochromis niloticus* (L.) eggs and fry. 2. Survival, growth and feeding of fry developing solely on their yolk reserves. *Aquaculture* 87, 183–195.

Rana, K.J. and Macintosh, D.J. (1988) A comparison of the quality of hatchery-reared *Oreochromis niloticus* fry. In: Pullin, R.S.V., Bhukaswan, T., Tonguthai, K. and Maclean, J.L. (eds) *Proceedings of the Second International Symposium on Tilapia in Aquaculture*. ICLARM Conference Proceedings No. 15, Department of Fisheries, Bangkok, Thailand, and ICLARM, Manila, Philippines, pp. 497–502.

Rana, K.J. and Suliman, E.M. (1993) Artificial incubation of tilapia eggs and fry under semi-intensive and intensive aquaculture systems. In: *Proceedings of the First International Symposium on Aquaculture Technology and Investment Opportunities*. Ministry of Agriculture and Water, Riyadh, Saudi Arabia, pp. 208–218.

Randall, D.J., Wood, C.M., Perry, S.F., Bergman, H., Maloiy, G.M.O., Mommsen, T.P. and Wright, P.A. (1989) Urea excretion as a strategy for survival in a fish living in a very alkaline environment. *Nature* 337, 165–166.

Razak, S.A., Hwang, G.-L., Rahman, M.A. and Maclean, N. (1999) Growth performance and gonadal development of growth-enhanced transgenic tilapia *Oreochromis niloticus* (L.) following heat-shock-induced triploidy. *Marine Biotechnology* 1, 533–544.

Reddy, T.V., Ravindranath, K., Sreeraman, P.K. and Rao, M.V.S. (1994) *Aeromonas salmonicida* associated with mass mortality of *Cyprinus carpio* and *Oreochromis mossambicus* in a freshwater reservoir in Andhra Pradesh, India. *Journal of Aquaculture in the Tropics* 9, 259–268.

Redner, B.D. and Stickney, R.R. (1979) Acclimation to ammonia by *Tilapia aurea*. *Transactions of the American Fisheries Society* 108, 383–388.

Reigh, R.C., Robinson, E.D. and Brown, P.B. (1991) Effect of dietary magnesium on growth and tissue magnesium content of blue tilapia (*Oreochromis aureus*). *Journal of the World Aquaculture Society* 22, 192–200.

Reite, O.B., Maloiy, G.M.O. and Aasehaug, B. (1974) pH, salinity and temperature tolerance of Lake Magadi *Tilapia*. *Nature* 247, 315.

Riche, M., Trottier, N.L., Ku, P.K. and Garling, D.L. (2001) Apparent digestibility of crude protein and apparent availability of individual amino acids in tilapia (*Oreochromis niloticus*) fed phytase pretreated soybean meal diets. *Fish Physiology and Biochemistry* 25, 181–194.

Richter, H., Focken, U., Becker, K., Santiago, C.B. and Afuang, W.B. (1999) Analysing the diel feeding patterns and daily ration of Nile tilapia, *Oreochromis niloticus* (L.), in Laguna de Bay, Philippines. *Journal of Applied Ichthyology* 15, 165–170.

Ridha, M.T. and Cruz, E.M. (1989) Effect of age on the fecundity of tilapia *Oreochromis spilurus*. *Asian Fisheries Science* 2, 239–247.

Ridha, M.T. and Cruz, E.M. (1999) Effect of different broodstock densities on the reproductive performance of Nile tilapia, *Oreochromis niloticus* (L.), in a recycling system. *Aquaculture Research* 30, 203–210.

Ridha, M.T. and Cruz, E.M. (2000) Effect of light intensity and photoperiod on Nile tilapia *Oreochromis niloticus*. *Aquaculture Research* 31, 609–617.

Ridha, M.T. and Cruz, E.M. (2001) Effect of biofilter media on water quality and biological performance of the Nile tilapia *Oreochromis niloticus* L. reared in a simple recirculating system. *Aquacultural Engineering* 24, 157–166.

Ridha, M.T. and Lone, K.P. (1990) Effect of oral administration of different levels of 17α-methyltestosterone on the sex reversal, growth and food conversion efficiency of the tilapia *Oreochromis spilurus* (Guenther) in brackish water. *Aquaculture and Fisheries Management* 21, 391–397.

Ridha, M.T. and Lone, K.P. (1995) Preliminary studies on feminization and growth of *Oreochromis spilurus* (Gaunter) by oral administration of 17 β-ethynyloestradiol in sea water. *Aquaculture Research* 26, 479–482.

Ridha, M.T., Cruz, E.M., Al-Ameeri, A. and Al-Ahmad, A. (1998) Effect of controlling temperature and light duration on seed production in tilapia *Oreochromis spilurus* (Gunther). *Aquaculture Research* 29, 101–108.

Roberts, R.J. and Matthiessen, P. (1979) *Pathological Findings in Fishes of the Okavangu Swamp, Botswana, Following Endosulphan Spraying.* ODM Special Report, UK Ministry of Overseas Development, London, 9 pp.

Roberts, R.J. and Sommerville, C. (1982) Diseases of tilapias. In: Pullin, R.S.V. and Lowe-McConnell, R.H. (eds) *The Biology and Culture of Tilapias.* ICLARM Conference Proceedings No. 7, ICLARM, Manila, Philippines, pp. 247–262.

Robinson, E.H., LaBomascus, D., Brown, P.B. and Linton, T.L. (1987) Dietary calcium and phosphorus requirements of *Oreochromis aureus* reared in calcium-free water. *Aquaculture* 64, 267–276.

Rodriguez-Serna, M., Olvera-Novoa, M.A. and Carmona-Osalde, C. (1996) Nutritional value of animal by-product meal in practical diets for Nile tilapia *Oreochromis niloticus* (L.) fry. *Aquaculture Research* 27, 67–73.

Roem, A.J., Kohler, C.C. and Stickney, R.R. (1990) Vitamin E requirements of the blue tilapia *Oreochromis aureus* (Steindachner), in relation to dietary lipid levels. *Aquaculture* 87, 155–164.

Romana-Eguia, M.R.R. and Eguia, R.V. (1999) Growth of five Asian red tilapia strains in saline environments. *Aquaculture* 173, 161–170.

Ron, B., Shimoda, S.K., Iwama, G.K. and Gordon Grau, E. (1995) Relationships among ration, salinity, 17α-methyltestosterone and growth in the euryhaline tilapia, *Oreochromis mossambicus*. *Aquaculture* 135, 185–193.

Rosati, R., O'Rourke, P.D., Tudor, K. and Henry, R.D. (1993) Performance of a raceway and vertical screen filter while growing *Tilapia niloticus* under commercial conditions. In: Wang, J.-K. (ed.) *Techniques for Modern Aquaculture.* American Society of Agricultural Engineers, St Joseph, Michigan, pp. 303–314.

Rosati, R., O'Rourke, P.D., Foley, P. and Tudor, K. (1997) Operation of a prototype commercial-scale recirculating growout system for *Oreochromis niloticus*. In: Fitzsimmons, K. (ed.) *Proceedings from the Fourth International Symposium on Tilapia in Aquaculture.* Northeast Regional Agriculture Engineering Service, Ithaca, New York, pp. 330–347.

Rosenstein, S. and Hulata, G. (1992) Lack of sex inversion in tilapia immersed in estrogen solutions during embryonic development. In: *Aquaculture `92 – Book of Abstracts.* World Aquaculture Society, Baton Rouge, Louisiana, p. 197.

Rosenstein, S. and Hulata, G. (1994) Sex reversal in the genus *Oreochromis*: optimization of feminization protocol. *Aquaculture and Fisheries Management* 25, 329–339.

Ross, B. and Ross, L.G. (1983) The oxygen requirements of *Oreochromis niloticus* under adverse conditions. In: Fishelson, L. and Yaron, Z. (eds) *Proceedings of the International Symposium on Tilapia in Aquaculture*. Tel Aviv University, Tel Aviv, Israel, pp. 134–143.

Ross, L.G. (2000) Environmental physiology and energetics. In: Beveridge, M.C.M. and McAndrew, B.J. (eds) *Tilapias: Biology and Exploitation*. Kluwer Academic Publishers, Dordrecht/Boston/ London, pp. 89–128.

Rothbard, S.M. (1979) Observation on the reproductive behaviour of *Tilapia zillii* and several *Sarotherodon* spp. under aquarium conditions. *Bamidgeh* 31, 35–43.

Rothbard, S.M and Pruginin, Y. (1975) Induced spawning and artificial incubation of tilapia. *Aquaculture* 5, 315–321.

Rothbard, S., Solnik, E., Shabbath, S., Amado, R. and Grabie, I. (1983) The technology of mass production of hormonally sex-inversed all male tilapia. In: Fishilson, L. and Yaron, Z. (eds) *International Symposium on Tilapia in Aquaculture, Tel-Aviv University, Israel*, pp. 425–434.

Rothuis, A.J., Nam, C.Q., Richter, C.J.J. and Ollevier, F. (1998) Polyculture of silver barb, *Puntius gonionotus* (Bleeker), Nile tilapia, *Oreochromis niloticus* (L.), and common carp, *Cyprinus carpio* L., in Vietnamese ricefields: fish production parameters. *Aquaculture Research* 29, 661–668.

Rottmann, R.W., Shireman, J.V. and Chapman, F.A. (1991) *Induction and Verification of Triploidy in Fish*. Publication No. 427, Southern Regional Aquaculture Centre, Stoneville, Mississippi.

Roubach, R., Correia, E.S., Zaiden, S., Martino, R.C. and Cavalli, R.O. (2003) Aquaculture in Brazil. *World Aquaculture* 34 (1), 28–34.

Ruangpan, L., Kitao, T. and Yoshida, T. (1986) Protective efficacy of *Aeromonas hydrophila* vaccines in Nile tilapia. *Veterinary Immunology and Immunopathology* 12, 346–350.

Ruddle, K. (1996) The potential role of integrated management of natural resources in improving the nutritional and economic status of resource-poor farm households in Ghana. In: Prein, M., Ofori, J.K. and Lightfoot, C. (eds) *Research for the Future Development of Aquaculture in Ghana*. ICLARM Conference Proceedings No. 42, ICLARM, Manila, Philippines, pp. 57–85.

Sadek, S., Kallafalah, H. and Adell, F. (1992) Tilapia (*Oreochromis niloticus*) biomass yield in commercial farm using circular tanks. *Journal of Applied Ichthyology* 8, 193–202.

Sadiku, S.O.E. and Jauncey, K. (1995) Soybean flour–poultry meat meal blends as dietary protein source in practical diets of *Oreochromis niloticus* and *Clarias gariepinus*. *Asian Fisheries Science* 8, 159–168.

Saeed, M.O. (1993) Efforts to control outbreaks of diseases among cultured fish in Kuwait. In: *Proceedings of the First International Aquaculture Symposium: Technologies and Investment Opportunities*. Ministry of Agriculture and Waters, Riyadh, Saudi Arabia, pp. 400–409.

Saeed, M.O. and Al-Thobaiti, S.A. (1997) Gas bubble disease in farmed fish in Saudi Arabia. *Veterinary Record* 140, 682–684.

Saha, S.N. and Dewan, S. (1979) Food and feeding habits of *Tilapia nilotica* (Linnaeus) (Perciformes: Cichlidae). Types and amount of food taken by the fish and its size and patterns of feeding. *Bangladesh Journal of Zoology* 7, 53–60.

Sakata, T. and Hattori, M. (1988) Characteristics of *Vibrio vulnificus* isolated from diseased tilapia. *Fish Pathology* 23, 33–40.

Saleh, G., ElEraky, W. and Gropp, J.M. (1995) A short note on the effects of vitamin A hypervitaminosis and hypovitaminosis on health and growth of tilapia (*Oreochromis niloticus*). *Journal of Applied Ichthyology* 11, 382–385.

Samonte, G.P., Agbayani, R.F. and Tumaliuan, R.E. (1991) Economic feasibility of polyculture of tiger shrimp (*Penaeus monodon*) with Nile tilapia (*Oreochromis niloticus*) in brackishwater ponds. *Asian Fisheries Science* 4, 335–343.

Sanares, R.C., Katase, S.A., Fast, A.W. and Karpenter, K.E. (1986) Water dynamics in brackishwater shrimp ponds with artificial aeration and circulation. In: Maclean, J.L., Dizon, L.B. and Hosillos, L.V. (eds) *The First Asian Fisheries Forum*. Asian Fisheries Society, Manila, Philippines, pp. 83–86.

Santiago, C.B. and Laron, M.A. (2002) Growth and fry production of Nile tilapia, *Oreochromis niloticus* (L.), on different feeding schedules. *Aquaculture Research* 33, 129–136.

Santiago, C.B. and Lovell, R.T. (1988) Amino acid requirements for growth of Nile tilapia. *Journal of Nutrition* 118, 1540–1546.

Santiago, C.B. and Reyes, O.S. (1993) Effects of dietary lipid source on reproductive performance and tissue lipid levels of Nile tilapia *Oreochromis niloticus* (L.) broodstock. *Journal of Applied Ichthyology* 9, 33–40.

Santiago, C.B., Aldaba, M.B. and Laron, M.A. (1983) Effect of varying dietary crude protein levels on spawning frequency and growth of *Sarotherodon niloticus* breeders. *Fisheries Research Journal of Philippines* 8, 9–18.

Santiago, C.B., Aldaba, M.B., Abuan, E.F. and Laron, M.A. (1985) The effects of artificial diets on fry production and growth of *Oreochromis niloticus* breeders. *Aquaculture* 47, 193–203.

Santiago, C.B., Aldaba, M.B. and Reyes, O.S. (1987) Influence of feeding rate and diet form on growth and survival of Nile tilapia (*Oreochromis niloticus*) fry. *Aquaculture* 64, 277–282.

Santiago, C.B., Aldaba, M.B., Reyes, O.S. and Laron, M.A. (1988) Response of Nile tilapia (*Oreochromis niloticus*) fry to diets containing *Azolla* meal. In: Pullin, R.S.V., Bhukaswan, T., Tonguthai, K. and Maclean, J.L. (eds) *Proceedings of the Second International Symposium on Tilapia in Aquaculture*. ICLARM Proceedings No. 15, Department of Fisheries, Bangkok, Thailand, and ICLARM, Manila, Philippines, pp. 377–382.

Sarder, M.R.I., Thompson, K.D., Penman, D.J. and McAndrew, J. (2001) Immune response of Nile tilapia (*Oreochromis niloticus* L) clones: 1. Non-specific responses. *Developmental and Comparative Immunology* 25, 37–46.

Sarig, S. (1971) *Diseases of Warmwater Fishes*. T.F.H. Publications, Neptune City, New Jersey.

Sastry, B.N., DeLosReyes, A.A. Jr, Rusch, K.A. and Malone, R.F. (1999) Nitrification performance of a bubble-washed bead filter for combined solids removal and biological filtration in a recirculating aquaculture system. *Aquacultural Engineering* 19, 105–117.

Satoh, S.T., Takeuchi, T. and Watanabe, T. (1987) Requirement of tilapia for alpha tocopherol. *Nippon Suisan Gakkaishi* 53, 119–124.

Scheerer, P.D., Thorgaard, G.H., Allendorf, F.W. and Knudsen, K.L. (1986) Androgenetic rainbow trout production from inbred and outbred sperm sources show similar survival. *Aquaculture* 57, 289–298.

Schimittou, H.R. (1993) *High-density Fish Culture in Low-volume Cages*. MITA, Publication No. 518, AQ41, 1993/7, American Soybean Association, Singapore, 75 pp.

Schlechtriem, C., Bresler, V., Fishelson, L., Rosenfeld, M. and Becker, K. (2004) Protective effects of dietary l-carnitine on tilapia hybrids (*Oreochromis niloticus* × *Oreochromis aureus*) reared under intensive pond-culture conditions. *Aquaculture Nutrition* 10, 55–63.

Schramm, J.K. (2001) Generic marketing of tilapia. In: Subasinghe, S. and Singh, T. (eds) *Tilapia: Production, Marketing and Technical Developments. Proceedings of the Tilapia 2001 International Technical and Trade Conference on Tilapia*. Infofish, Kuala Lumpur, Malaysia, pp. 85–87.

Schwank, E. (1987) Lunar periodicity in the spawning of *Tilapia mariae* in the Ethiop River, Nigeria. *Journal of Fish Biology* 30, 533–537.

Scott, J.T., Scott, D.M. and Lawson, T.B. (1998) Rice hulls as solids removal media for aquaculture. In: *Aquaculture '98 – Book of Abstracts*. World Aquaculture Society, Baton Rouge, Louisiana, p. 480.

Scott, P.W. (1977) Preliminary studies on disease in intensively farmed tilapia in Kenya. MSc thesis, University of Stirling, Stirling, UK.

Scribner, K.T., Page, K.P. and Batron, M.L. (2001) Hybridization in freshwater fishes: a review of case studies and cytonuclear methods of biological inference. *Reviews in Fish Biology and Fisheries* 10, 293–323.

Seki, E. and Maly, R. (1994) *A Pilot Socio-economic Survey of Aquaculture in Ruvuma Region, Tanzania*. ALCOM Field Document No. 20, FAO, Harare, Zimbabwe, 107 pp.

Sen, S., van der Mheen, H. and van der Mheen-Sluijer, J. (1996) The place of aquaculture in rural development. In: *Expert Consultation on Small-Scale Rural Aquaculture*. FAO Fisheries Technical Report No. 548, FAO, Rome, pp. 91–118.

Sere, C. (1988) Pond culture as a branch of management in the small-farm agriculture of Colombia. PhD thesis, Hohenheim University, Stuttgart, Germany.

Sevilleja, R.C., Cagauan, A.G., Lopez, E.A., Dela Cruz, C.R. and Van Dam, A.A. (1992) Component technology research in rice–fish systems in the Philippines. In: Dela Cruz, C.R., Lightfoot, C., Costa-Pierce, B.A., Carangal, V.R. and Bimbao, M.A.P. (eds) *Rice–Fish Research and Development in Asia*. ICLARM Conference Proceedings No. 24, ICLARM, Manila, Philippines, pp. 373–384.

Shafland, P.L. and Pestrak, J.M. (1982) Lower lethal temperatures for fourteen non-native fishes in Florida. *Environmental Biology of Fishes* 7, 149–156.

Shelby, R.A., Shoemaker, C.A. and Klesius, P.H. (2002a) Detection of humoral response to *Streptococcus iniae* infection of Nile tilapia, *Oreochromis niloticus*, by a monoclonal antibody-based ELISA. *Journal of Applied Aquaculture* 12, 23–31.

Shelby, R.A., Klesius, P.H., Shoemaker, C.A. and Evans, J.J. (2002b) Passive immunization of tilapia, *Oreochromis niloticus* (L.), with anti-*Streptococcus iniae* whole sera. *Journal of Fish Diseases* 25, 1–6.

Shell, E.W. (1967) Relationship between rate of feeding, rate of growth and rate of conversion in feeding trials with two species of tilapia, *Tilapia mossambica* Peters and *Tilapia nilotica* Linnaeus. *FAO Fisheries Report* 44, 411–415.

Shevgoor, L., Knud-Hansen, C.F. and Edwards, P. (1994) An assessment of the role of buffalo manure for pond culture of tilapia. III. Limiting factors. *Aquaculture* 126, 107–118.

Shiau, S.-Y. (2002) Tilapia, *Oreochromis* spp. In: Webster, C.D. and Lim, C. (eds) *Nutrient Requirements and Feeding of Finfish for Aquaculture*. CAB International, Wallingford, UK, pp. 273–293.

Shiau, S.-Y. and Chen, M.J. (1993) Carbohydrate utilization by tilapia (*Oreochromis niloticus* × *O. aureus*) as influenced by different chromium sources. *Journal of Nutrition* 123, 1747–1753.

Shiau, S.-Y. and Chin, Y.H. (1999) Estimation of dietary biotin requirement of juvenile hybrid tilapia, *Oreochromis niloticus* × *O. aureus*. *Aquaculture* 170, 71–78.

Shiau, S.-Y. and Hsieh, H.L. (1997) Vitamin B_6 requirements of tilapia, *Oreochromis niloticus* × *O. aureus* fed two dietary protein concentrations. *Fisheries Science* 63, 1002–1007.

Shiau, S.-Y. and Hsieh, J.F. (2001) Quantifying the dietary potassium requirement of juvenile hybrid tilapia, *Oreochromis niloticus* × *O. aureus*. *British Journal of Nutrition* 85, 213–218.

Shiau, S.-Y. and Hsu, T.S. (1995) L-Ascorbyl- 2-sulfate has equal antiscorbutic activity as L-ascorbyl- 2-monophosphate for tilapia, *Oreochromis niloticus* × *O. aureus*. *Aquaculture* 133, 147–157.

Shiau, S.-Y. and Hwang, J.Y. (1993) Vitamin D requirement of juvenile hybrid tilapia, *Oreochromis niloticus* × *O. aureus*. *Nippon Suisan Gakkaishi* 59, 553–558.

Shiau, S.-Y. and Jan, F.L. (1992) Dietary ascorbic acid requirement of juvenile tilapia, *Oreochromis niloticus* × *O. aureus*. *Nippon Suisan Gakkaishi* 58, 671–675.

Shiau, S.-Y. and Lei, M.S. (1999) Feeding strategy does affect carbohydrate utilization by hybrid tilapia, *Oreochromis niloticus* × *O. aureus*. *Fisheries Science* 65, 553–557.

Shiau, S.-Y. and Liang, H.S. (1995) Carbohydrate utilization and digestibility by tilapia, *Oreochromis niloticus* × *O. aureus*, are affected by chromic oxide inclusion in the diet. *Journal of Nutrition* 125, 976–982.

Shiau, S.-Y. and Lin, S.F. (1993) Effects of supplemental dietary chromium and vanadium on the utilization of different carbohydrates in tilapia, *Oreochromis niloticus* × *O. aureus*. *Aquaculture* 110, 321–330.

Shiau, S.-Y. and Lo, P.S. (2000) Dietary choline requirements of juvenile hybrid tilapia, *Oreochromis niloticus* × *O. aureus*. *Journal of Nutrition* 130, 100–103.

Shiau, S.-Y. and Peng, C.Y. (1993) Protein sparing effect by carbohydrates in diets for tilapia, *Oreochromis niloticus* × *O. aureus*. *Aquaculture* 117, 327–334.

Shiau, S.-Y. and Shiau, L.F. (2001) Reevaluation of the vitamin E requirement of juvenile tilapia, *Oreochromis niloticus* × *O. aureus*. *Animal Science* 72, 529–534.

Shiau, S.-Y. and Shy, S.M. (1998) Dietary chromic oxide inclusion level required to maximize glucose utilization in hybrid tilapia *Oreochromis niloticus* × *O. aureus*. *Aquaculture* 161, 355–362.

Shiau, S.-Y. and Su, S.-L. (2005) Juvenile tilapia (*Oreochromis niloticus* × *O. aureus*) requires dietary myo-inositol for maximal growth. *Aquaculture* 243, 273–277.

Shiau, S.-Y. and Suen, G.S. (1992) Estimation of the niacin requirements for tilapia fed diets containing glucose or dextrin. *Journal of Nutrition* 122, 2030–2036.

Shiau, S.-Y. and Yu, Y.P. (1999) Dietary supplementation of chitin and chitosan depresses growth in tilapia, *Oreochromis niloticus* × *O. aureus*. *Aquaculture* 179, 439–446.

Shiau, S.-Y., Yu, H.L., Hwa, S., Chen, S.Y. and Hsu, S.I. (1988) The influence of carboxymethylcellulose on growth, digestion, gastric emptying time and body composition of tilapia. *Aquaculture* 70, 345–354.

Shiau, S.-Y., Kwok, C.C., Huang, J.Y., Chen, C.M. and Lee, S.L. (1989) Replacement of fish meal with soybean meal in male tilapia (*Oreochromis niloticus* × *O. aureus*) fingerling diet at a suboptimal protein level. *Journal of the World Aquaculture Society* 20, 230–235.

Shimura, S. and Asai, M. (1984) *Argulus americanus* (Crustacea: Branchiura) parasite on the bowfin, *Amia calva*, imported from North America. *Fish Pathology* 18, 199–213.

Shnel, N., Barak, Y., Ezer, T., Dafni, Z. and Rijn, J.V. (2002) Design and performance of a zero-discharge tilapia recirculating system. *Aquacultural Engineering* 26, 191–203.

Shoemaker, C.A. and Klesius, P.H. (1997) Streptococcal disease problems and control. In: Fitzsimmons, K. (ed.) *Proceedings from the Fourth International Symposium on Tilapia in Aquaculture*. Northeast Regional Engineering Service Coopertaive Extension, Ithaca, New York, pp. 671–680.

Shoemaker, C.A., Evans, J.J. and Klesius, P.H. (2000) Density and dose: factors affecting mortality to *Streptococcus iniae*-infected tilapia (*Oreochromis niloticus*). *Aquaculture* 188, 229–235.

Shrestha, M.K. and Lin, C.K. (1996a) Phosphorus fertilization strategy in fish ponds based on sediment phosphorus saturation level. *Aquaculture* 142, 207–219.

Shrestha, M.K. and Lin, C.K. (1996b) Determination of phosphorus saturation level in relation to clay content in formulated pond muds. *Aquacultural Engineering* 15, 441–459.

Siddiqui, A.Q. and Harbi, A.H. (1997) Effects of sex ratio, stocking density and age of hybrid tilapia on seed production in concrete tanks in Saudi Arabia. *Aquaculture International* 5, 207–216.

Siddiqui, A.Q., Howlander, M.S. and Adam, A.A. (1988) Effects of dietary protein levels on growth, diet conversion and protein utilization in fry and young Nile tilapia, *Oreochromis niloticus*. *Aquaculture* 70, 63–70.

Siddiqui, A.Q., Howlader, M.S. and Adam, A.A. (1991a) Management strategies for intensive culture of Nile tilapia (*Oreochromis niloticus* L.) in tanks using drainage water in Al Hassa region of Saudi Arabia. *Arab Gulf Journal of Scientific Research. B, Agricultural and Biological Sciences* 9, 149–163.

Siddiqui, A.Q., Howlader, M.S. and Adam, A.A. (1991b) Effects of water exchange on *Oreochromis niloticus* (L.) growth and water quality in outdoor concrete tanks. *Aquaculture* 95, 67–74.

Siddiqui, A.Q., Al-Harbi, A.H. and Hafedh, Y.S. (1997a) Effects of stocking density on patterns of reproduction and growth of hybrid tilapia in concrete tanks in Saudi Arabia. *Asian Fisheries Science* 10, 41–49.

Siddiqui, A.Q., Al-Harbi, A.H. and Al-Hafedh, Y.S. (1997b) Effects of food supply on size at first maturity, fecundity and growth of hybrid tilapia, *Oreochromis niloticus* (L.) × *Oreochromis aureus* (Steindachner), in outdoor concrete tanks in Saudi Arabia. *Aquaculture Research* 28, 341–349.

Siddiqui, A.Q., Al Hafedh, Y.S. and Ali, S.A. (1998) Effect of dietary protein level on the reproductive performance of Nile tilapia, *Oreochromis niloticus* (L.). *Aquaculture Research* 29, 349–358.

Sifa, L., Chenhong, L., Dey, M., Gagalac, F. and Dunham, R. (2002) Cold tolerance of three strains of Nile tilapia, *Oreochromis niloticus*, in China. *Aquaculture* 213, 123–129.

Silva, P.C., Souza, V.L., Padua, D.M., Dalacorte, P.C. and Goncalves, D.C. (2000) Effects of stocking density on growth and fillet composition of tetra hybrid red tilapia, Israeli strain. In: Fitzsimmons, K. and Filho, J.C. (eds) *Tilapia Aquaculture in the 21st Century. Proceedings from the Fifth International Symposium on Tilapia Aquaculture, Rio de Janeiro, Brazil.* American Tilapia Association, Charles Town, West Virginia, and ICLARM, Penang, Malaysia, pp. 341–345.

Silva-Garcia, A.J. (1996) Growth of juvenile gilthead seabraem (*Sparus auratus* L.) reared under different photoperiod regimes. *Israeli Journal of Aquaculture/Bamidgeh* 48, 84–93.

Sin, A.W. (1980) Integrated fish husbandry systems in Hong Kong, with case studies on duck–fish and goose–fish systems. In: Pullin, R.S.V. and Shehadeh, Z.H. (eds) *Integrated Agriculture – Aquaculture Farming Systems*. ICLARM Conference Proceedings No. 4, ICLARM, Manila, Philpppines, pp. 113–123.

Sin, A.W. and Chiu, M.T. (1983) The intensive monoculture of the tilapia hybrid, *Sarotherodon nilotica* (males) × *S. mossambica* (females) in Hong Kong. In: Fishelson, L. and Yaron, Z. (eds) *Proceedings of the International Symposium on Tilapia in Aquaculture*. Tel Aviv University, Tel Aviv, Israel, pp. 506–516.

Sin, Y.M., Ling, K.H. and Lam, T.J. (1994) Passive transfer of protective immunity against *Ichthyophthirius* from vaccinated mother to fry in tilapia, *Oreochromis aureus*. *Aquaculture* 120, 229–237.

Singh, T. and Daud, W.J.W. (2001) Live handling and marketing of tilapia. In: Subasinghe, S. and Singh, T. (eds) *Tilapia: Production, Marketing and Technological Developments: Proceedings of the Tilapia 2001 International Technical and Trade Conference on Tilapia*. Infofish, Kuala Lumpur, Malaysia, pp. 88–93.

Sinha, C.K. (1986) Occurrence of *Trypanosoma mukasai* Hoare, 1932 in *Tilapia mossambica* (Peters) from India. *Acta Protozoologica* 25, 449–452.

Sipaúba-Tavares, L.H., Yoshida, C.E. and de Souza Braga, F.M. (2000) Effects of continuous water exchange on limnology of tilapia (*Oreochromis niloticus*) culture tanks. In: Fitzsimmons, K. and Filho, J.C. (eds) *Tilapia Culture in the 21st Century. Proceedings from the Fifth International Symposium on Tilapia Aquaculture, Rio de Janeiro, Brazil.* American Tilapia Association, Charles Town, West Virginia, and ICLARM, Penang, Malaysia, pp. 279–287.

Siraj, S.S., Smitherman, R.O., Castillo-Galluser, S. and Dunham, R.A. (1983) Reproductive traits of three classes of *Tilapia nilotica* and maternal effects on their progeny. In: Fishelson, L. and Yaron, Z. (eds) *Proceedings of International Symposium on Tilapia in Aquaculture*. Tel Aviv University Press, Tel Aviv, Israel, pp. 210–218.

Sirol, R.N., Andrade, D.R. and Salaro, A.L. (2000) Growth and body composition of red tilapia fingerlings (*Oreochromis niloticus* × *O. aureus*) submitted to different feeding levels. In: Fitzsimmons, K. and Filho, J.C. (eds) *Tilapia Aquaculture in the 21st Century. Proceedings from Fifth International Symposium on Tilapia Aquaculture, Rio de Janeiro, Brazil.* American Tilapia Association, Charles Town, West Virginia, and ICLARM, Penang, Malaysia, pp. 382–389.

Skillicorn, P., Spira, W. and Journey, W. (1993) *Duckweed Aquaculture: A New Aquatic Farming System for Developing Countries*. World Bank, Washington, DC, 76 pp.

Sklan, D., Prag, T. and Lupatsch, I. (2004) Apparent digestibility coefficients of feed ingredients and their prediction in diets for tilapia *Oreochromis niloticus* × *Oreochromis aureus* (Teleostei, Cichlidae). *Aquaculture Research* 35, 358–364.

Skliris, G.P. and Richards, R.H. (1999) *Nodavirus* isolated from experimentally infected tilapia, *Oreochromis mossambicus* (Peters). *Journal of Fish Diseases* 22, 315–318.

Smith, A. (1840) *Illustration of the Zoology of South Africa*, vol.4, *Pisces*, Smith, Elder, London, p. 5.

Smith, C.E., Ramsey, D. and Speer, C.A. (1998) Histopathology associated with an irido-like virus infection of tilapia, *Tilapia nilotica*. In: *Aquaculture '98 – Book of Abstracts*. World Aquaculture Society, Baton Rouge, Louisiana, p. 504.

Smith, E.S. and Phelps, R.P. (1997) Reproductive efficiency, fry growth, and response to sex reversal of Nile and red tilapia. In: *Fourteenth Annual Technical Report*. Pond Dynamics/Aquaculture CRSP, Oregon State University, Corvallis, Oregon, pp. 112–119.

Smith, S.J., Watanabe, W.O., Chang, J.R., Ernst, D.H., Wicklund, R.I. and Olla, B.L. (1991) Hatchery production of Florida red tilapia seed in brackish-water tanks: the influence of broodstock age. *Aquaculture and Fisheries Management* 22, 141–147.

Snir, I.J. (2001) Value added tilapia products: deliberate policy or no other choice. In: Subasinghe, S. and Singh, T. (eds) *Tilapia: Production, Marketing and Technological Developments. Proceedings of the Tilapia 2001 International Technical and Trade Conference on Tilapia*. Infofish, Kuala Lumpur, Malaysia, p. 84.

Soliman, A.K. and Wilson, R.P. (1992a) Water-soluble vitamin requirements of tilapia. 1. Pantothenic acid requirement of blue tilapia, *Oreochromis aureus*. *Aquaculture* 104, 121–126.

Soliman, A.K. and Wilson, R.P. (1992b) Water-soluble vitamin requirements of tilapia. 2. Riboflavin requirement of blue tilapia, *Oreochromis aureus*. *Aquaculture* 104, 309–314.

Soliman, A.K., Jauncey, K. and Roberts, R.J. (1986) The effects of dietary ascorbic acid supplementation on hatchability, survival rate and fry performance in *Oreochromis mossambicus*. *Aquaculture* 59, 197–208.

Soliman, A.K., Jauncey, K. and Roberts, R.J. (1994) Water-soluble vitamin requirements of tilapia: ascorbic acid (vitamin C) requirement of Nile tilapia, *Oreochromis niloticus*. *Aquaculture and Fisheries Management* 25, 269–278.

Soliman, A.K., El-Horbeety, A.A.A., Essa, M.A.R., Kosba, M.A. and Kariony, I.A. (2000) Effects of introducing ducks into fish ponds on water quality, natural productivity and fish production together with the economic evaluation of the integrated and non-integrated systems. *Aquaculture International* 8, 315–326.

Souza, M.L.R. and Macedo-Viegas, E.M. (2000) Effects of filleting on processing yield of Nile tilapia (*Oreochromis niloticus*). In: Fitzsimmons, K. and Filho, J.C. (eds) *Tilapia Culture in the 21st Century. Proceedings from the Fifth International Symposium on Tilapia Aquaculture, Rio de Janeiro, Brazil*. American Tilapia Association, Charles Town, West Virginia, and ICLARM, Penang, Malaysia, pp. 451–457.

Souza, M.L.R., Marengoni, N.G., Pinto, A.A. and Cacador, W.C. (2000) Processing yield of Nile tilapia (*Oreochromis niloticus*): head cut types and two weight classes. *Acta Scientiarum* 22, 701–706.

Spataru, P. (1976) The feeding habits of *Tilapia galilaea* (Artedi) in Lake Kinneret (Israel). *Aquaculture* 9, 47–59.

Spataru, P. (1978) Food and feeding habits of *Tilapia zillii* (Gervais) (Cichlidae) in Lake Kinneret (Israel). *Aquaculture* 14, 327–338.

Spataru, P. and Zorn, M. (1978) Food and feeding habits of *Tilapia aurea* (Steindachner) (Cichlidae) in Lake Kinneret (Israel). *Aquaculture* 13, 67–79.

Stewart, J.A. (1993) *The Economic Viability of Aquaculture in Malawi: a Short-term Study for the Central and Northern Regions Fish Farming Project, Mzuzu, Malawi*. Institute of Aquaculture, University of Stirling, Stirling, UK.

Stewart, K.M. (1988) Changes in condition and maturity of the *Oreochromis niloticus* populations of Fergusons's Gulf, Lake Turkana, Kenya. *Journal of Fish Biology* 33, 181–188.

Stickney, R.R. and McGeachin, R.B. (1983) Effects of dietary lipid quality on growth and food conversion of tilapia. *Proceedings of the Annual Conference of Southeastern Association of Fish and Wildlife Agencies* 37, 352–357.

Stickney, R.R. and Wurts, W.A. (1986) Growth response of blue tilapia to selected levels of dietary menhaden and catfish oils. *Progressive Fish Culturist* 48, 107–109.

Stickney, R.R., McGeachin, R.B., Robinson, E.H., Arnold, G. and Suter, L. (1982) Growth of *Tilapia aurea* as a function of degree of dietary lipid saturation. *Proceedings of the Annual Conference of Southeastern Association of Fish and Wildlife Agencies* 36, 172–181.

Stickney, R.R., McGeachin, R.B., Lewis, D.H., Marks, J., Sis, R.F., Robinson, E.H. and Wurts, W. (1984) Response of *Tilapia aurea* to dietary vitamin C. *Journal of the World Mariculture Society* 15, 179–185.

Straus, D.L. and Griffin, B.R. (2001) Prevention of an initial infestation of *Ichthyophthirius multifiliis* in channel catfish and blue tilapia by potassium

permanganate treatment. *North American Journal of Aquaculture* 63, 11–16.
Subasinghe, R.P. and Sommerville, C. (1986) Acquired immunity of *Oreochromis mossambicus* to the ciliate ectoparasite *Ichthyophthirius multifiliis* (Fouquet). In: Maclean, J.L., Dizon, L.B. and Hosillos, L.V. (eds) *Proceedings of the First Asian Fisheries Forum*. Asian Fisheries Society, Manila, Philippines, pp. 279–283.
Subasinghe, R.P. and Sommerville, C. (1989) An experimental study into possible protection of fry from *Ichthyophthirius multifiliis* (Fouquet) infections during mouthbrooding of *Oreochromis mossambicus* (Peters). *Journal of Fish Diseases* 12, 143–149.
Subasinghe, R.P. and Sommerville, C. (1992) Susceptibility of *Oreochromis mossambicus* (Peters) fry to the ciliate ectoparasite *Ichthyophthirius multifiliis* (Fouquet). In: Shariff, M., Subasinghe, R.P. and Arthur, J.R. (eds) *Proceedings of the First Symposium on Diseases in Asian Aquaculture*. Asian Fisheries Society, Manila, Philippines, pp. 355–360.
Subasinghe, R.P., Curry, D., McGladdery, S.E. and Bartley, D. (2003) Recent technological innovations in aquaculture. In: *Review of the State of World Aquaculture*. FAO Fisheries Circular 886, Revision 2, FAO, Rome, pp. 59–74.
Sudharsan, R., Rani, A.S., Reddy, T.N., Reddy, P.U.M. and Raju, T.N. (2000) Effect of nitrite toxicity on the dehydrogenases in the fish *Tilapia mossambica*. *Journal of Environment and Pollution* 7, 127–130.
Sugita, H., Miyajima, C. and Deguchi, Y. (1990) The vitamin B_{12}-producing ability of intestinal bacteria isolated from tilapia and channel catfish. *Nippon Suisan Gakkaishi* 56, 701.
Suresh, A.V. and Lin, C.K. (1992a) Tilapia culture in saline waters: a review. *Aquaculture* 106, 201–226.
Suresh, A.V. and Lin, C.K. (1992b) Effect of stocking density on water quality and production of red tilapia in a recirculated water system. *Aquacultural Engineering* 11, 1–22.
Tacon, A.G.J. (1995) Feed formulation and on-farm feed management. In: New, M.B., Tacon, A.G.J. and Csavas, I. (eds) *Farm-made Aquafeeds*. FAO Fisheries Technical Paper 343, FAO, Rome, pp. 61–74.
Tacon, A.G.J. (2003) Aquaculture production trends analysis. In: *Review of the State of World Aquaculture*. FAO Fisheries Circular No. 886, Revision 2, FAO, Rome, pp. 5–29.
Tacon, A.G.J. and Jackson, A.J. (1985) Utilization of conventional and unconventional protein sources in practical fish feeds. In: Cowey, C.B., Mackie, A.M. and Bell, J.G. (eds) *Nutrition and Feeding in Fish*. Academic Press, London, pp. 119–145.
Tacon, A.G.J., Jauncey, K., Falaye, A., Pantah, M., MacGowen, I. and Stafford, E. (1983) The use of meat and bone meal, hydrolyzed feather meal and soybean meal in practical fry and fingerling diets for *Oreochromis niloticus*. In: Fishelson, L. and Yaron, Z. (eds) *Proceedings of the International Symposium on Tilapia in Aquaculture*. Tel Aviv University, Tel Aviv, Israel, pp. 356–365.
Tacon, P., Ndiaye, P., Cauty, C., Le Menn, F. and Jalabert, F. (1996) Relationships between the expression of maternal behavior and ovarian development in the mouthbrooding cichlid fish *Oreochromis niloticus*. *Aquaculture* 146, 261–275.
Takeuchi, T., Satoch, S. and Watanabe, T. (1983) Requirement of *Tilapia nilotica* for essential fatty acids. *Nippon Suisan Gakkaishi* 49, 1127–1134.
Tan, Y.-J. and Tong, H.-Y. (1989) The status of the exotic aquatic organisms in China. In: De Silva, S.S. (ed.) *Exotic Aquatic Organisms in Asia. Proceedings of a Workshop on Introduction of Exotic Aquatic Organisms in Asia*. Special Publication No. 3, Asian Fisheries Society, Manila, Philippines, pp. 35–43.
Tandler, A. and Helps, S. (1985) The effects of photoperiod and water exchange rate on growth and survival of gilthead seabream (*Sparus auratus*, Linnaeus; Sparidae) from hatching to metamorphosis in mass rearing systems. *Aquaculture* 48, 81–82.
Tavares-Dias, M., Moraes, F.R., Martins, M.L. and Kronka, S.N. (2001) Parasitic fauna of cultivated fishes in feefishing farm of Franca, São Paulo. II. Metazoans. *Revista Brasiliera de Zoologia* 18, 81–95.
Tayamen, M.M. (2004) Nationwide dissemination of GET EXCEL tilapia in the Philippines. In: Bolivar, R.B., Mair, G.C. and Fitzsimmons, K. (eds) *Proceedings of the Sixth International Symposium on Tilapia in Aquaculture*. Bureau of Fisheries and Aquatic Resources, Manila, Philippines, and American Tilapia Association, Charles Town, West Virginia, pp. 74–85.
Tayamen, M.M., Reyes, R.A., Danting, M.J., Mendoza, A.M., Marquez, E.B., Salguet, A.C., Gonzales, R.C., Abella, T.A. and Vera-Cruz, E.M. (2002) Tilapia broodstock development for saline waters in the Philippines. *Naga* 25, 32–36.
Te, B.Q., Lua, D.T., Viet, N.V. and Thanh, N.V. (1999) Parasitic fauna of cultured tilapia (*Oreochromis* spp) in Vietnam. Paper presented at the Fourth Symposium on Diseases in Asian Aquaculture: Aquatic Animal Health for Sustainability, Cebu City, Philippines, 22–26 November 1999.
Teichert-Coddington, D. and Green, B.W. (1993) Tilapia yield improvement through maintenance

of minimal oxygen concentrations in experimental grow-out ponds in Honduras. *Aquaculture* 118, 63–71.

Tendencia, E.A. and dela Peña, M. (2003) Investigations of some components of the green water system which makes it effective in the initial control of luminous bacteria. *Aquaculture* 218, 115–119.

Teshima, S. and Kanazawa, A. (1988) Nutritive value of methionine-enriched soy plastein for *Oreochromis niloticus* fry. In: Pullin, R.S.V., Bhukaswan, T., Tonguthai, K. and Maclean, J.L. (eds) *Proceedings of the Second International Symposium on Tilapia in Aquaculture*. ICLARM Conference Proceedings No. 15, Department of Fisheries, Bangkok, Thailand, and ICLARM, Manila, Philippines, pp. 393–399.

Teshima, S., Kanazawa. A. and Sakamoto, M. (1982) Essential fatty acids of *Tilapia nilotica*. *Memoirs of the Faculty of Fisheries, Kagoshima University* 31, 201–204.

Teshima, S., Kanazawa, A. and Uchiyama, Y. (1985a) Optimum protein levels in casein–gelatin diets for *Tilapia nilotica* fingerlings. *Memoirs of the Faculty of Fisheries, Kagoshima University* 34, 45–52.

Teshima, S., Kanazawa, A. and Uchiyama Y. (1985b) Effects of dietary protein, lipid and digestible carbohydrate levels on the weight gain, feed conversion efficiency and protein efficiency ratio of *Tilapia nilotica*. *Memoirs of the Kagoshima University, Research Centre for the South Pacific* 6, 56–71.

Thankur, D.P., Yi, Y., Diana, J.S. and Lin, C.K. (2004) Effects of fertilization and feeding strategy on water quality, growth performance, nutrient utilization and economic return in Nile tilapia (*Oreochromis niloticus*) ponds. In: Bolivar, R., Mair, G. and Fitzsimmons, K. (eds) *Proceedings of the Sixth International Symposium on Tilapia in Aquaculture*. Bureau of Fisheries and Aquatic Resources, Manila, Philippines, and American Tilapia Association, Charles Town, West Virginia, pp. 529–543.

Thomas, D.H.L. (1994) Socio-economic and cultural factors in aquaculture development: a case study from Nigeria. *Aquaculture* 119, 329–343.

Tian, X., Li, D., Dong, S., Yan, X., Qi, Z., Liu, G. and Lu, J. (2001) An experimental study on closed-polyculture of penaeid shrimp with tilapia and constricted tagelus. *Aquaculture* 202, 57–71.

Toguyeni, A., Fauconneau, B., Boujard, T., Fostier, A., Kuhn, E.R., Mol, K.A. and Baroiller, J.F. (1997) Feeding behaviour and food utilization in tilapia, *Oreochromis niloticus*: effects of sex ratio and relationship with the endocrine status. *Physiology and Behaviour* 62, 273–279.

Torrans, L., Meriwether, F., Lowell, F., Wyatt, B. and Gwinup, P.D. (1988) Sex-reversal of *Oreochromis aureus* by immersion in Mibolerone, a synthetic steroid. *Journal of World Aquaculture Society* 19, 97–102.

Torres, E.B. and Navera, E.R. (1985) Tilapia marketing in Central Luzon and Metro Manila, Philippines. In: Smith, I.R., Torres, E.B. and Tan, E.O. (eds) *Philippine Tilapia Economics*. ICLARM Conference Proceedings No. 15, ICLARM, Manila, Philippines, pp. 181–191.

Torres, N.J., Macabale, N.A. and Mercado, J.R. (1992) On-farm rice–fish farming systems research in Guimba, Nueva, Philippines. In: Dela Cruz, C.R., Lightfoot, C., Costa-Pierce, B.A., Carangal, V.R. and Bimbao, M.A.P. (eds) *Rice–Fish Research and Development in Asia*. ICLARM Conference Proceedings No. 24, ICLARM, Manila, Philippines, pp. 295–299.

Trewavas, E. (1982) Tilapias: taxonomy and speciation. In: Pullin, R.V.S. and Lowe-McConnell, R.H. (eds) *The Biology and Culture of Tilapias*. ICLARM Conference Proceedings No. 7, ICLARM, Manila, Philippines, pp. 3–13.

Trewavas, E. (1983) *Tilapiine Fishes of the Genera Sarotherodon, Oreochromis and Danakilia*. British Museum (Natural History), London, 340 pp.

Trippel, E.A. and Neil, S.R.E. (2002) Effect of photoperiod and light intensity on growth and activity of juvenile haddock (*Melanogrammus aeglefinus*). *Aquaculture* 217, 633–645.

Troell, M. and Berg, H. (1997) Cage fish farming in the tropical Lake Kariba, Zimbabwe: impact and biogeochemical changes in sediment. *Aquaculture Research* 28, 527–544.

Tsadik, G.G. and Kutty, M.N. (1987) *Influence of Ambient Oxygen on Feeding and Growth of the Tilapia,* Oreochromis niloticus *(Linnaeus)*. UNDP/FAO/NIOMR, Port Harcourt, Nigeria, 13 pp.

Tung, M.-C., Chen, S.-C. and Tsai, S.-S. (1987) General septicemia of streptococcal infection in cage-culture tilapia, *Tilapia mossambica*, in southern Taiwan. *COA Fisheries Series* 10, 187–197.

Tung, P.H. and Shiau, S.-Y. (1991) Effect of meal frequency on growth performance of hybrid tilapia, *Oreochromis niloticus* × *O. aureus*, fed different carbohydrate diets. *Aquaculture* 92, 343–350.

Tung, P.H. and Shiau, S.-Y. (1993) Carbohydrate utilization versus body size in tilapia, *Oreochromis niloticus* × *O. aureus*. *Comparative Biochemistry and Physiology* 104A, 585–588.

Turingan, J.E. and Kubaryk, J.M. (1992) The effect of high salt diet on survival and hatchability of Taiwanese red tilapia (*Oreochromis mossambicus* × *Oreochromis niloticus*) eggs upon direct transfer to seawater. In: *Aquaculture '92 – Book of*

Abstracts. World Aquaculture Society, Baton Rouge, Louisiana, p. 220.

Turner, G.F. and Robinson, R.L. (2000) Reproductive biology, mating systems and parental care. In: Beveridge, M.C.M. and McAndrew, B.J. (eds) *Tilapias: Biology and Exploitation*. Kluwer Academic Publishers, Dordrecht/Boston/London, pp. 33–58.

Twibell, R.G. and Brown, P.B. (1998) Optimal dietary protein concentration for hybrid tilapia (*Oreochromis niloticus* × *Oreochromis aureus*) fed all-plant diets. *Journal of the World Aquaculture Society* 29, 9–16.

Uchida, R.N. and King, J.E. (1962) Tank culture of tilapia. *US Fish and Wildlife Service Fisheries Bulletin* 199, 21–47.

Ulla Rojas, J.B. and Weerd, H.V. (1997) The growth and feed utilization of *Oreochromis aureus* fingerlings fed diets with various coffee pulp levels. In: Fitzsimmons, K. (ed.) *Proceedings from the Fourth International Symposium on Tilapia in Aquaculture*. Northeast Regional Agriculture Engineering Service, Ithaca, New York, pp. 40–49.

Ungsethaphan, T. (1995) An on-farm trial to investigate feeding strategies for Nile tilapia (*Oreochromis niloticus*) broodfish. MSc thesis, Asian Institute of Technology, Bangkok, Thailand.

United States Department of Agriculture (2004) *Aquaculture Outlook*. Economic Research Service, US Department of Agriculture, Washington, DC. Available at: www.ers.usda.gov

Uraiwan, S. (1988) Direct and indirect responses to selection for age at first maturation of *Oreochromis niloticus*. In: Pullin, R.S.V., Bhukaswan, T., Tonguthai, K. and Maclean, J.L. (eds) *Proceedings of the Second International Symposium on Tilapia in Aquaculture*. ICLARM Conference Proceedings No. 15, Department of Fisheries, Bangkok, Thailand, and ICLARM, Manila, Philippines, pp. 295–300.

van Dam, A.A., Beveridge, M.C.M., Azim, M.E. and Verdegem, M.C.J. (2002) The potential of fish production based on periphyton. *Reviews in Fish Biology and Fisheries* 12, 1–31.

Van der Mheen-Sluijer, J. (1991) *Adoption of Fish Farming: Promoting and Inhibiting Factors in Eastern Province, Zambia*. Field Document GCP/INT/436/SWE.13, FAO, Harare, Zimbabwe, 41 pp.

Van Dijk, P.L.M., Van den Thillart, G., Balm, P. and Wendelaar Bonga, S. (1993) The influence of gradual water acidification on the acid/base status and plasma hormone levels in carp. *Journal of Fish Biology* 42, 661–671.

Van Ginneken, V.J.T., Van Eersel, R., Balm, P., Nieveen, M. and Van den Thillart, G. (1997) Tilapia are able to withstand long-term exposure to low environmental pH, judged by their energy status, ionic balance and plasma cortisol. *Journal of Fish Biology* 51, 795–806.

Vannuccini, S. (2001) Global markets for tilapia. In: Subasinghe, S. and Singh, T. (eds) *Tilapia: Production, Marketing and Technical Developments. Proceedings of the Tilapia 2001 International Technical and Trade Conference on Tilapia*. Infofish, Kuala Lumpur, Malaysia, pp. 65–70.

Varadaraj, K. (1989) Feminization of *Oreochromis mossambicus* by the administration of diethylstilbestrol. *Aquaculture* 80, 337–341.

Varadaraj, K. and Pandian, T.J. (1987) Masculinization of *Oreochromis mossambicus* by administration of 17α-methyl- 5-androsten- 3β- 17β-diol through rearing water. *Current Science* 56, 412–413.

Varadaraj, K. and Pandian, T.J. (1989) First report on production of supermale tilapia by integrating endocrine sex reversal with gynogenetic technique. *Current Science* 58, 434–441.

Veerina, S.S., Nandeesha, M.C. and Gopal Roa, K. (1993) *Status and Technology of Indian Major Carp Farming in Andra Pradesh, India*. Special Publication No. 9, Asian Fishery Society. Indian Branch, Mangalore, India, 76 pp.

Vera-Cruz, E.M.V. and Mair, G.C. (1994) Conditions for effective androgen sex-reversal in *Oreochromis niloticus* (L.). *Aquaculture* 122, 237–248.

Verdegem, M.C. and McGinty, A.S. (1987) Effects of frequency of egg and fry removal on spawning of *Tilapia nilotica* in hapas. *Progressive Fish Culturist* 49, 129–131.

Veverica, K.L., Bowman, J. and Popma, T. (2001a) Global experiment: optimization of nitrogen fertilization rate in freshwater tilapia production ponds. In: Gupta, A., McElwee, K., Burke, D., Burright, J., Cummings, X. and Egna, H. (eds) *Eighteenth Annual Technical Report*. Pond Dynamics/Aquaculture CRSP, Oregon State University, Corvallis, Oregon, pp. 13–22.

Veverica, K.L., Liti, D., Were, E. and Bowman, J. (2001b) Fish yield and economic benefits of tilapia/*Clarias* polyculture in fertilized ponds receiving commercial feeds or pelleted agricultural by-products. In: Gupta, A., McElwee, K., Burke, D., Burright, J., Cummings, X. and Egna, H. (eds) *Eighteenth Annual Technical Report*. Pond Dynamics/Aquaculture CRSP, Oregon State University, Corvallis, Oregon, pp. 27–29.

Vincke, M.M.J. and Schmidt, U.W. (1991) *Report on the Mid-term Evaluation of the Central and Northern Regions Fish Farming Projects*. EEC/Government of Malawi, Zamba, Malawi, 67 pp.

Viola, S. and Zohar, G. (1984) Nutrition studies with market size hybrids of tilapia (*Oreochromis*) in intensive culture. *Bamidgeh* 36, 3–15.

Viola, S., Zohar, G. and Arieli, Y. (1986) Phosphorus requirements and its availability from different sources for intensive pond culture species in Israel. Part 1. Tilapia. *Bamidgeh* 38, 3–12.

Viola, S., Arieli, Y. and Zohar, G. (1988) Animal-protein-free feeds for hybrid tilapia (*Oreochromis niloticus* × *O. aureus*) in intensive culture. *Aquaculture* 75, 115–125.

Vromant, N., Rothuis, A.J., Cuc, N.T.T. and Ollevier, F. (1998) The effect of fish on the abundance of the rice caseworm *Nymphula depunctalis* (Guenée) (Lepidoptera: *Pyralidae*) in direct seeded, concurrent rice–fish fields. *Biocontrol Science and Technology* 8, 539–546.

Vromant, N., Nam, C.Q. and Ollevier, F. (2002) Growth performance and use of natural food by *Oreochromis niloticus* (L.) in polyculture systems with *Barbodes gonionotus* (Bleeker) and *Cyprinus carpio* (L.) in intensively cultivated rice fields. *Aquaculture Research* 33, 969–978.

Wang, J.-Q., Li, D., Dong, S., Wang, K. and Tian, X. (1998) Experimental studies on polyculture in closed shrimp ponds. 1. Intensive polyculture of Chinese shrimp (*Penaeus chinensis*) with tilapia hybrids. *Aquaculture* 163, 11–27.

Wang, J.-Q., Li, D., Dong, S., Wang, K., and Tian, X. (1999) Comparative studies on cultural efficiency and profits of different polycultural systems in Penaeid shrimp ponds. *Journal of Fisheries of China/Shuichan Xuebao* 23, 45–52.

Wang, K., Takeuchi, T. and Watanabe, T. (1985) Effect of dietary protein levels on growth of *Tilapia nilotica*. *Bulletin of the Japanese Society for Scientific Fisheries* 51, 133–140.

Wang, L.-H. and Tsai, C.-L. (2000) Effects of temperature on the deformity and sex differentiation of tilapia, *Oreochromis mossambicus*. *Journal of Experimental Zoology* 286, 534–537.

Wang, W.-S. and Wang, D.-H. (1997) Enhancement of the resistance of tilapia and grass carp to experimental *Aeromonas hydrophila* and *Edwardsiella tarda* infections by several polysaccharides. *Comparative Immunology and Microbiology of Infectious Diseases* 20, 261–270.

Wang, Z-G. and Xu, B-H. (1985) Studies on the pathogenic bacteria of the 'rotten-skin' diseases of the Nile tilapia (*Tilapia nilotica*). *Journal of Fisheries of China/Shuichan Xuebao* 9, 217–221.

Wangead, C., Greater, A. and Tansakul, R. (1988) Effects of acid water on survival and growth rate of Nile tilapia (*Oreochromis niloticus*). In: Pullin, R.S.V., Bhukaswan, T., Tonguthai, K. and Maclean, J.L. (eds) *Proceedings of the Second International Symposium on Tilapia in Aquaculture*. ICLARM Conference Proceedings No. 15, Department of Fisheries, Bangkok, Thailand, and ICLARM, Manila, Philippines, pp. 433–438.

Wan Johari, W.D., Ismail, A.A., Wan Ismail, W.I. and Tahir, S.M. (1999) *Transportation of Live Red Tilapia (Oreochromis niloticus) under Minimal Water Condition*. Occasional Paper No. 10, Malaysian Agricultural Research and Development Institute (MARDI), Kuala Lumpur, 3 pp.

Wardoyo, S.E. (1991) Effects of different salinity levels and acclimation regimes on survival, growth, and reproduction of three strains of *Tilapia nilotica* and a red *Tilapia nilotica* hybrid. PhD thesis, Auburn University, Auburn, Alabama.

Wassef, E.A., Plammer, G. and Poxton, M. (1988) Protease digestion of the meals of ungerminated and germinated soybeans. *Journal of Food Science and Agriculture* 44, 201–214.

Wassef, E.A., Sweilam, M.A. and Attalah, R.F. (2003) The use of fermented fish silage as a replacement for fish meal in Nile tilapia (*Oreochromis niloticus*) diets. *Egyptian Journal of Nutrition and Feeds* 6, 357–370.

Wassermann, G.J. and Afonso, L.O.B. (2003) Sex reversal in Nile tilapia (*Oreochromis niloticus* Linnaeus) by androgen immersion. *Aquaculture Research* 34, 65–71.

Watanabe, T. (1985) Importance of the study of broodstock nutrition for further development of aquaculture. In: Cowey, C.B., Mackie, A.M. and Bell, J.G. (eds) *Nutrition and Feeding in Fish*. Academic Press, London, pp. 395–414.

Watanabe, T. and Kiron, V. (1995) Red sea bream (*Pagrus major*). In: Bromage, N.R. and Roberts, R.J. (eds) *Broodstock Management and Egg and Larval Quality*. Blackwell Science, Oxford, UK, pp. 398–413.

Watanabe, T., Takeuchi, T., Murakami, A. and Ogino, C. (1980) The availability to *Tilapia nilotica* of phosphorus in white fish meal. *Bulletin of the Japanese Society for Scientific Fisheries* 46, 897–899.

Watanabe, T., Arakawa, T., Kitajima, C. and Fujita, S. (1984) Effect of nutritional quality of broodstock diets on reproduction of red sea bream. *Bulletin of Japanese Society for Scientific Fisheries* 50, 495–501.

Watanabe, T., Satoh, S. and Takeuchi, T. (1988) Availability of minerals in fish meal to fish. *Asian Fisheries Science* 1, 175–195.

Watanabe, T., Takeuchi, T., Satoh, S. and Kiron, V. (1996) Digestible crude protein contents in various feedstuffs determined with four freshwater fish species. *Fisheries Science* 62, 278–282.

Watanabe, W.O. and Kuo, C-M. (1985) Observations on the reproductive performance of Nile tilapia (*Oreochromis niloticus*) in laboratory aquaria at various salinities. *Aquaculture* 49, 315–323.

Watanabe, W.O., Kuo, C.-M. and Huang, M.-C. (1985) The ontogeny of salinity tolerance in the tilapias *Oreochromis aureus, O. niloticus* and *O. mossambicus* × *O. niloticus* hybrids, spawned and reared in freshwater. *Aquaculture* 47, 353–367.

Watanabe, W.O., Ellingson, L.J., Wicklund, R.I. and Olla, B.L. (1988a) The effects of salinity on growth, food consumption and conversion in juvenile, monosex male Florida red tilapia. In: Pullin, R.S.V., Bhukaswan, T., Tonguthai, K. and Maclean, J.L. (eds) *Proceedings of the Second International Symposium on Tilapia in Aquaculture*. ICLARM Conference Proceedings No. 15, Department of Fisheries, Bangkok, Thailand, and ICLARM, Manila, Philippines, pp. 515–523.

Watanabe, W.O., French, K.E., Ellingson, L.J. and Wicklund, R.I. (1988b) Further investigations on the effects of salinity on growth in Florida red tilapia: evidence for the influence of behaviour. In: Pullin, R.S.V., Bhukaswan, T., Tonguthai, K. and Maclean, J.L. (eds) *Proceedings of the Second International Symposium on Tilapia in Aquaculture*. ICLARM Conference Proceedings No. 15, Department of Fisheries, Bangkok, Thailand, and ICLARM, Manila, Philippines, pp. 525–530.

Watanabe, W.O., Clark, J.H., Dunhkam, J.B., Wicklund, R.I. and Olla, B.L. (1990) Culture of Florida red tilapia in marine cages: the effects of stocking and dietary protein on growth. *Aquaculture* 90, 123–134.

Watanabe, W.O., Ernst, D.H., Chasar, M.P., Wicklund, R.I. and Olla, B.L. (1993a) The effects of temperature and salinity on growth and feed utilization of juvenile, sex-reversed male Florida red tilapia cultured in a recirculating system. *Aquaculture* 112, 309–320.

Watanabe, W.O., Wicklund, R.I. and Olla, B.L. (1993b) Saltwater culture of Florida red tilapia: a summary of research at the Caribbean Marine Research Centre (1985–1991). In: *Proceedings of the First International Symposium on Aquaculture: Technology and Investment Opportunities*. Ministry of Agriculture and Water, Riyadh, Saudi Arabia, pp. 49–68.

Watanabe, W.O., Wicklund, R.I., Olla, B.L. and Head, W.D. (1997) Saltwater culture of the Florida red and other saline tolerant tilapias: a review. In: Costa-Pierce, B.A. and Rakocy, J.E. (eds) *Tilapia Aquaculture in the Americas*, Vol. 1. World Aquaculture Society, Baton Rouge, Louisiana, pp. 54–141.

Watanabe, W.O., Losordo, T.M., Fitzsimmons, K. and Hanley, F. (2002) Tilapia production systems in the Americas: technological advances, trends, and challenges. *Reviews in Fisheries Science* 10, 465–498.

Wee, K.L. and Tuan, N.A. (1988) Effects of dietary protein level on growth and reproduction of Nile tilapia (*Oreochromis niloticus*). In: Pullin, R.S.V., Bhukaswan, T., Tonguthai, K. and Maclean, J.L. (eds) *Proceedings of the Second International Symposium on Tilapia in Aquaculture*. ICLARM Conference Proceedings No. 15, Department of Fisheries, Bangkok, Thailand and ICLARM, Manila, Philippines, pp. 401–410.

Welcomme, R.L. (1988) *International Introductions of Inland Aquatic Species*. FAO Fisheries Technical Paper No. 294, FAO, Rome.

Wheaton, F.W. (1977) *Aquacultural Engineering*. John Wiley, New York, 708 pp.

Whitefield, A.K. and Blaber, S.J.M. (1976) The effects of temperature and salinity on *Tilapia rendalli* Boulenger 1896. *Journal of Fish Biology* 9, 99–104.

Whitefield, A.K. and Blaber, S.J.M. (1978) Resource segregation among iliophagous fish in Lake St Lucia, Zululand. *Environmental Biology of Fishes* 3, 293–296.

Whitefield, A.K. and Blaber, S.J.M. (1979) The distribution of the freshwater cichlid *Sarotherodon mossambicus* in estuarine systems. *Environmental Biology of Fishes* 4, 77–81.

Wijkstrom, U.N. and Larsson, R. (1992) *Fish Farmers in Rural Communities: Results of a Survey in North Western Province of Zambia*. FAO Field Document ALCOM No. 8, Harare, Zimbabwe. FAO, Rome, 40 pp.

Wilkie, M.P. and Wood, C.M. (1996) The adaptations of fish to extremely alkaline environments. *Comparative Biochemistry and Physiology* 113B, 665–673.

Wilson, R.P., Harding, D.E. and Garling, D.L. Jr (1977) The effect of dietary pH on amino acid utilization and the lysine requirement of fingerling channel catfish. *Journal of Nutrition* 107, 166–170.

Wilson, R.P., Freeman, D.W. and Poe, E.W. (1984) Three types of catfish offal meals for channel catfish fingerlings. *Progressive Fish Culturist* 46, 126–132.

Windmar, M., Jarding, S. and Paterson, R. (2000a) Current status of tilapia aquaculture and processing in Zimbabwe. In: Fitzsimmons, K. and Filho, J.C. (eds) *Tilapia Culture in the 21st Century. Proceedings from the Fifth International Symposium on Tilapia Aquaculture, Rio de Janeiro, Brazil.* American Tilapia Association, Charles Town, West Virginia, and ICLARM, Penang, Malaysia, pp. 595–597.

Windmar, M., Jarding, S. and Paterson, R. (2000b) Production of 480 tons of tilapia (*Oreochromis niloticus*) at Elanne farm, Zimbabwe. In: Fitzsimmons, K. and Filho, J.C. (eds) *Tilapia Culture in the 21st Century. Proceedings from the*

Fifth International Symposium on Tilapia Aquaculture, Rio de Janeiro, Brazil. American Tilapia Association, Charles Town, West Virginia, and ICLARM, Penang, Malaysia, pp. 602–604.

Winfree, R.A. and Stickney, R.R. (1981) Effects of dietary protein and energy on growth, feed conversion efficiency and body composition of *Tilapia aurea*. *Journal of Nutrition* 111, 1001–1012.

Wohlfarth, G.W. (1994) The unexploited potential of tilapia hybrids in aquaculture. *Aquaculture and Fisheries Management* 25, 781–788.

Wohlfarth, G.W. and Hulata, G. (1983) *Applied Genetics of Tilapia*, 2nd edn. ICLARM Studies and Reviews 6, ICLARM, Manila, Philippines, 26 pp.

Wohlfarth, G.W., Hulata, G. and Halevy, A. (1990) Growth, survival and sex ratio of some tilapia species and their interspecific hybrids. In: *European Aquaculture Society Special Publication* 11, European Aquaculture Society, Bredene, Belgium. pp. 87–101.

Wokoma, K. and Marioghae, I.E. (1996) Survival of *Tilapia guineensis* under conditions of low dissolved oxygen and low pH. In: Pullin, R.S.V., Lazard, J., Legendre, M., Amon Kottias, J.B. and Pauly, D. (eds) *Proceedings of the Third International Symposium on Tilapia in Aquaculture.* ICLARM Conference Proceedings No. 41, ICLARM, Penang, Malaysia, pp. 442–448.

Wolf, J.C. and Smith, S.A. (1999) Comparative severity of experimentally induced mycobacteriosis in striped bass *Morone saxatilis* and hybrid tilapia *Oreochromis* spp. *Diseases of Aquatic Organisms* 38, 191–200.

Wootton, R.J. (1979) Energy costs of egg production and environmental determinants of fecundity of teleost fishes. In: Miller, P.J. (ed.) *Fish Phynology: Anabolic Adaptiveness in Teleosts.* Symposium of the Zoological Society of London, No. 44, Academic Press, London, pp. 133–159.

Wright, J.R. Jr and Pohajdak, B. (2001) Cell therapy for diabetes using piscine islet tissue. *Cell Transplantation* 10, 125–143.

Wu, S.-M., Hwang, P.-P., Hew, C.-L. and Wu, J.-L. (1998) Effect of antifreeze protein on cold tolerance in juvenile tilapia (*Oreochromis mossambicus* Peters) and milkfish (*Chanos chanos* Forsskal). *Zoological Studies* 37, 39–44.

Wu, Y.V., Rosati, R.R., Sessa, D. and Brown, P.B. (1995) Utilization of corn gluten feed by Nile tilapia. *Progressive Fish Culturist* 57, 305–309.

Yada, T. and Ito, F. (1997) Difference in tolerance to acidic environments between two species of tilapia, *Oreochromis niloticus* and *O. mossambicus*. *Bulletin of the National Research Institute of Fisheries Science (Japan)* 9, 11–18.

Yakupitiyage, D. (1995) On-farm feed preparation and feeding strategies for carps and tilapias. In: New, M.B., Tacon, A.G.J. and Csavas, I. (eds) *Farm-made Aquafeeds.* FAO Fisheries Technical Paper 343, FAO, Rome, pp. 87–100.

Yambot, A.V. (1998) Isolation of *Aeromonas hydrophila* from *Oreochromis niloticus* during fish disease outbreaks in the Philippines. *Asian Fisheries Science* 10, 347–354.

Yashouv, A. (1960) Effect of low temperatures on *Tilapia nilotica* and *Tilapia galilaea*. *Bamidgeh* 12, 62–66.

Yavuzcan, Y.H. and Pulatsü, S. (1999) Evaluation of the secondary stress response in healthy Nile tilapia (*Oreochromis niloticus* L.) after treatment with a mixture of formalin, malachite green and methylene blue. *Aquaculture Research* 30, 379–383.

Yi, Y. and Lin, C.K. (1997) Finishing system for large tilapia. In: *Fourteenth Annual Technical Report.* Pond Dynamics/Aquaculture CRSP, Oregon State University, Corvallis, Oregon, pp. 146–156.

Yi, Y. and Lin, C.K. (2000) Integrated cage culture in ponds: concepts, practice and perspectives. In: Liao, C.I. and Lin, C.K. (eds) *Proceedings of the First International Symposium on Cage Aquaculture in Asia.* Asian Fisheries Society, Quezon City, Philippines, pp. 233–240.

Yi, Y. and Lin, C.K. (2001) Effects of biomass of caged Nile tilapia (*Oreochromis niloticus*) and aeration on the growth and yields in an integrated cage-cum-pond system. *Aquaculture* 195, 253–267.

Yi, Y., Lin, C.K. and Diana, J.S. (1996) Effects of stocking densities on growth of caged adult Nile tilapia (*Oreochromis niloticus*) and on yield of small Nile tilapia in open water in earthen ponds. *Aquaculture* 146, 205–215.

Yi, Y., Lin, C.K. and Diana, J.S. (2002) Recycling pond mud nutrients in integrated lotus–fish culture. *Aquaculture* 212, 213–226.

Yi, Y., Lin, C.K. and Diana, J.S. (2003a) Techniques to mitigate clay turbidity problems in fertilized earthen fish ponds. *Aquacultural Engineering* 27, 39–51.

Yi, Y., Lin, C.K. and Diana, J.S. (2003b) Hybrid catfish (*Clarias macrocephalus* × *C. gariepinus*) and Nile tilapia (*Oreochromis niloticus*) culture in an integrated pen-cum-pond system: growth performance and nutrient budget. *Aquaculture* 217, 395–408.

Yi, Y., Diana, J.S. and Lin, C.K. (2004a) Supplemental feeding for red tilapia culture in brackishwater. In: Bolivar, R., Mair, G. and Fitzsimmons, K. (eds) *Proceedings of the Sixth International Symposium on Tilapia in Aquaculture.* Bureau of Fisheries and Aquatic Resources, Manila, Philippines, and American Tilapia Association, Charles Town, West Virginia, pp. 451–462.

Yi, Y., Diana, J.S., Shrestha, M.K. and Lin, C.K. (2004b) Culture of mixed-sex Nile tilapia with predatory snakehead. In: Bolivar, R., Mair, G. and Fitzsimmons, K. (eds) *Proceedings of the Sixth International Symposium on Tilapia in Aquaculture*. Bureau of Fisheries and Aquatic Resources, Manila, Philippines, and American Tilapia Association, Charles Town, West Virginia, pp. 544–557.

Young, J.A. and Muir, J.F. (2000) Economics and marketing. In: Beveridge, M.C.M. and McAndrew, B.J. (eds) *Tilapias: Biology and Exploitation*. Kluwer Academic Publishers, Dordrecht, The Netherlands, pp. 447–487.

Zohar, G., Rappaport, U., Avnimelech, Y. and Sarig, S. (1984) Results of the experiments carried out in the Genosar Experimental Station in 1983. Cultivation of tilapia in high densities and with periodic flushing of the pond water. *Bamidgeh* 36, 63–69.

Zohar, Y., Harel, M., Hassin, S. and Tandler, A. (1995) Gilt-head sea bream (*Sparus aurata*). In: Bromage, N.R. and Roberts, R.J. (eds) *Broodstock Management and Egg and Larval Quality*. Blackwell Science, Oxford, UK, pp. 94–117.

General Index

Species Index